THE MAKING OF AMERICA'S CULTURE REGIONS

RICHARD L. NOSTRAND
University of Oklahoma

Cartography by David J. Morris

ROWMAN & LITTLEFIELD
Lanham • Boulder • New York • London

Executive Editor: Susan McEachern
Assistant Editor: Rebeccah Shumaker
Senior Marketing Manager: Kim Lyons
Interior Designer: Ilze Lemesis
Cover images, clockwise from upper left: Mormon Temple, Salt Lake City, Utah © iStock.com/AndreyKrav; Kirtland Temple, Ohio, photograph by the author; Mission Mountain farmhouse © iStock.com/MSMcCarthy_Photography; morning in mountains © iStock.com/ekolara; San Miguel Chapel, Santa Fe, New Mexico © iStock.com/gnagel; rural Vermont © iStock.com/SeanPavonePhoto; Greek revival mansion © iStock.com/miker8863
Back cover, left to right: US Capitol, Washington, DC, photograph by the author; I-style house, Utah, photograph by the author; Lewis and Clark's Fort Clatsop in old growth forest © iStock.com/LincolnRogers

Credits and acknowledgments for material borrowed from other sources, and reproduced with permission, appear on the appropriate page within the text.

Published by Rowman & Littlefield
A wholly owned subsidiary of The Rowman & Littlefield Publishing Group, Inc.
4501 Forbes Boulevard, Suite 200, Lanham, Maryland 20706
www.rowman.com

Unit A, Whitacre Mews, 26-34 Stannary Street, London SE11 4AB, United Kingdom

British Library Cataloguing in Publication Information Available

Library of Congress Cataloging-in-Publication Data Available
ISBN 978-1-5381-0396-8 (cloth : alk. paper)
ISBN 978-1-5381-0397-5 (electronic)

∞™ The paper used in this publication meets the minimum requirements of American National Standard for Information Sciences—Permanence of Paper for Printed Library Materials, ANSI/NISO Z39.48-1992.

Printed in the United States of America

Dedication

For John and Suzanne
Little Susan, Nicholas, Milana, and Natalie

Brief Contents

Detailed Contents

Figures and Regional Table

Figures

Regional Table

Maps

Geographer Spotlights

Preface

I am standing with my students on the crown of a hand-dug irrigation ditch in El Cerrito, New Mexico. This ditch is deeper than any of us is tall, and its age is suggested by the massive trunks of cottonwoods and box elders rooted along its banks. I want my field trippers to observe how water in this village lifeline flows by gravity for more than a mile from behind an upstream dam to agricultural fields below the village, all in perfect harmony with the natural environment. I want them to understand that, for protection from nomadic Indians, El Cerrito's first villagers built the houses we see in a tight rectangle that enclosed a church and small plaza. And I want them to appreciate that for these Spanish settlers, leaving Santa Fe and the Albuquerque area for this Pecos Valley frontier eight generations ago was a risky endeavor. Because I am passionate about history and geography, I want my students to know these things, and, with a little imagination, see this tiny, well-preserved village as a window on a past place and culture.

My passion for the historical geography of the United States runs deep and long. As a graduate student at UCLA and Syracuse University, I focused on this subdiscipline. In a long career at the University of Oklahoma, I taught a course called Historical Geography of the United States almost every semester, which meant fifty-two times. During those busy years I channeled my research time to my interest in the Hispanic or Latino peoples of the borderland, especially in New Mexico. But in retirement I have had the good fortune to be able to review materials, read recent publications, and write this new survey.

I have written this book with three convictions. The first is that the material should be organized regionally as opposed to topically or thematically. The reason for this is simple: I can comprehend and build on geographical concepts and information more readily if the material is organized region by region, and I think this is true for others. My second conviction is that maps should be new, simple, and closely tied to the text. However valuable they may be, old, reduced, and often dark and hard-to-read map reproductions are omitted. And my third conviction is that those historical geographers who have contributed substantially to our knowledge of given regions should be recognized and honored. Each of the twenty-one chapters shines a "spotlight" on a person and his or her work.

We will look at a total of sixteen regions, beginning with the Spanish Borderlands and ending with the Great Plains. Each region is developed around three themes: cultural ecology, or humans adjusting to their environments; cultural diffusion, or the spread of people and their cultural baggage; and cultural landscape, or the human imprint and its persistence in features and layouts. The story focuses on the formative periods for these regions—namely, the three centuries between 1600, when European colonization of the present-day United States effectively began, and 1900, when little good agricultural land remained and the settlement frontier was declared closed. A number of topics, like the evolution of plantations in the Lowland South or how the Ogallala Aquifer in the Great Plains came to be exploited, are carried beyond 1900. But our story of the making of America's culture regions essentially ends in 1900. By 1900 the Spanish and French regions have faded, and the fourteen regions that remain become the regional structure for the United States today. Since 1900 the regional boundaries of all fourteen have changed very little.

I am, of course, greatly indebted to those historical geographers who previously synthesized the literature. The first, Ralph H. Brown, produced a benchmark volume with his *Historical Geography of the United States* in 1948. Developed regionally, Brown's study drew upon primary materials and introduced the significance of perception in geographic decision-making. Brown died in 1948 and his book was never revised. In 1987 Robert D. Mitchell and Paul A. Groves coedited a second synthesis titled *North America: The Historical Geography of a Changing Continent.* Their approach was both regional and thematic, and the experts they assembled wrote some richly detailed essays. (In 2001 Thomas F. McIlwraith and Edward K. Muller would coedit a second edition of the same book; among the revisions would be six new chapters.) In a third synthesis, Michael P. Conzen paraded a largely new set of luminaries as editor of *The Making of the American Landscape*, in 1990. Developed for the most part thematically, its essays are often delightfully lively and innovative. Conzen brought forth an enlarged and in-color second edition of the book in 2010.

Chronologically, a fourth synthesis was provided by D. W. Meinig (see below), who completed his masterwork, which he organized both regionally and thematically, in 2004. In 2014, Craig E. Colten and Geoffrey L. Buckley brought together largely younger scholars in a fifth synthesis of the literature, their *North American Odyssey: Historical Geographies for the Twenty-First Century.* Organized thematically, the volume is insightful and up-to-date, and it emphasizes the twentieth century. Several chapters—for example, those by Robert Wilson on the changing use of animals in America, Joan M. Schwartz on photography in the nineteenth century, Derek H. Alderman and E. Arnold Modlin Jr. on racialized landscapes, and Mona Domosh on gendered landscapes—explore topics that broaden the range of interests typically considered by historical geographers. In several other chapters, topics are rather reinvented through innovative content reorganization. Examples are Karl Raitz on transportation, Buckley on forests and mines, Colten on waterways, and Michael P. Conzen on urbanization. Like most multiauthored works, this compendium has merit for having tapped the knowledge of experts, but it suffers from a lack of consistency in the development of themes. Its inattentiveness to maps, with several notable exceptions, is also worrisome.

Meinig stands tallest among those to whom I am indebted. Between 1986 and 2004 he published his truly comprehensive overview of the historical geography of the United States in a four-volume work titled *The Shaping of America: A Geographical Perspective on 500 Years of History.* Described by reviewers as "monumental" and "magisterial," Meinig's epic achievement is *the* recommended source for anyone seriously interested in America's changing geography. In Meinig's own words, "*The Shaping of America* offers the first detailed panoramic geographic view and assessment across the entire span of space and time relevant to the development of the United States."[1] But as Meinig also notes on the same page, "Few people will read a four-volume set of anything." I write this book with the thought that there is a need for a one-volume study—one that assumes less background about events in American history and less knowledge about sheer place location than Meinig's four volumes do.

The scores of scholars on whom I have drawn for specific information are cited in chapter endnotes and the comprehensive bibliography that follows chapter 21. Nearly all the maps and diagrams and a number of the photographs are my originals. For executing the maps, I thank David J. Morris, who quite expertly used ArcGIS 10.3.1 software and data drawn from the Environmental Systems Research Institute. I also thank Cody A. Taylor, who produced nearly all the diagrams and sketches, using Adobe Illustrator. Trisha L. Marlow made much appreciated last-minute revisions to both figures and maps. I am grateful to Lowell C. "Ben" Bennion for suggesting the book title I have used.

The following is an alphabetical list of the many individuals whom I thank for critically reading parts of the manuscript or helping in other ways: James P. Allen, Timothy G. Anderson, Alan R. Baker, Lowell C. Bennion, Martyn J. Bowden, Brock J. Brown, Craig E. Colten, Malcolm L. Comeaux, Michael P. Conzen, Linda Miles Coppens, Benjamin Y. Dixon, William E. Doolittle, Gary S. Dunbar, Richard Flint, Gregory Knapp, Alan P. Marcus, Kent Mathewson, Donald W. Meinig, Andrew Milson, Edward K. Muller, Antony R. Orme, Paul F. Starrs, Virginia Thompson, Richard W. Wilkie, David J. Wishart, Joseph S. Wood, William Wyckoff, and Wilbur Zelinsky. I appreciate the cheerful help of the professional staff at the University of Oklahoma, especially John Lovett and Jacquelyn (Jackie) Slater Reese in the Western History Collections, and Vicki Michener in Government Documents. I am grateful for the help and advice of the staff at Rowman and Littlefield, particularly Susan McEachern, Jehanne Schweitzer, and Rebeccah Shumaker. The attention to detail and editorial suggestions of the freelance copyeditor, Helen Subbio, both improved the manuscript and eliminated errors. And as always, my wife, Susan, has been at my side to help with my less-than-subtle requests, including typing every word.

Note

[1] Meinig, *The Shaping of America*, 4:xiv.

Overview 1

In content, Chapter 1 is quite unlike the chapters that follow; it is the chapter that provides important background. Each of the chapters in this book has four sections. Chapter 1's sections are titled (1) "Why Regions?" (2) "Historical Geography," (3) "The Natural Environment," and (4) "Native Americans." Under "Why Regions?" we try to make the case for the use of regions in our organizational framework. Under "Historical Geography," we focus on what historical geography is and how historical geographers have combined time and space. "The Natural Environment" of the United States—its geographic variety and how it struck European colonists—is found in the third section. And in "Native Americans," we generalize about five culture areas that existed circa 1500, as the Europeans began to arrive.

Why Regions?

A survey of America's historical geography can be organized thematically or regionally. If structured thematically, such topics as clearing the forests, surveying the public domain, or rural-to-urban migration leading to urbanization are considered individually. If structured regionally, these and other topics are synthesized within regional units such as New England, the Upland South, or the Great Plains. Using the regional approach is presently in disfavor within the discipline of geography. The argument is that the thematic approach allows for greater precision of analysis and has the potential for greater methodological rigor. But organizing information regionally also has its advantages, and we anticipate that the pendulum will swing back to a more balanced acceptance of both ways of organization. The sheer value of using regions in teaching geography would suggest as much.[1]

The rationale for structuring our information regionally is that regions enable us to understand *places* and appreciate their distinctive attributes. The United States may be one political nation-state, but it is not one homogeneous entity. Rather, it is composed of diverse units, each one different for its physical characteristics, such as landforms and climate, and for its cultural attributes, including economic ways of life, social mores and traditions, political values and behavior, and ethnic and national-origin population characteristics. Regional units allow us to focus on an array of topics in one place in order to analyze and explain the character of that place. And as noted in the preface, the regional approach has the additional benefit of providing a *context* for learning, retaining, and building on information. For many, sorting information within regional units simply has greater benefit than does sorting information by topic.

The term *region*, historian Fulmer Mood tells us, began to replace the somewhat broader term *section*, in use since colonial times, with the publication in 1895 of geologist John W. Powell's *Physiographic Regions of the United States*. The appearance of Powell's benchmark volume prompted generations of geographers, demographers, sociologists, and others to carve the United States into many sets of regions based on various kinds of criteria. The sixteen regions used here—whose names are identified in chapter titles—are rather large units areally, are relatively stable entities culturally, and, based on their ongoing recognition, would probably be considered by most geographers and historians to be traditional and mainstream (an issue we revisit in Chapter 21). Indeed, the use of regions by geographers has a parallel with the use of periods by historians. Whereas geographers use regions to sort out important differences across the land, historians use periods to sort out important changes over time. Historians identify the Jacksonian Period or the Great Depression or the New Deal era, while geographers see important differences between a Lowland South or an Old Northwest or a New Northwest.[2]

We employ three themes—cultural ecology, cultural diffusion, and cultural landscape—to bring focus to the analysis of our sixteen regions. *Cultural ecology* has to do with people adjusting to a new natural environment. Spaniards in the borderland, for example, were most challenged in the section between Florida and East Texas, where heat and humidity prompted them to switch from growing their favored wheat and raising sheep, both of which failed, to growing the better-suited Indian corn and raising hogs. *Cultural diffusion* concerns the movement or spread of ideas and material culture along some path between a source area and a destination. In geography circles, *migration* is used to describe the movement of people themselves, while *diffusion* captures the transfer of that people's cultural baggage. For the Spanish Borderlands, a long list of cultural items—crops, livestock, Spanish language, Roman Catholicism—jumped across the Atlantic from Spain (an example of relocation diffusion), and within the New World and the borderland, where these items were accepted, they either spread wavelike from people to people or they leapfrogged from community to community (two forms of expansion diffusion). Finally, *cultural landscape* has to do with the tangible features people build and the ways they lay out their farms and communities. In the borderlands, examples of the features and layouts that outlasted their Spanish builders include missions, presidios (garrisoned fortresses), and plaza-centered civil communities such as Santa Fe, San Antonio, and Los Angeles.

Our purpose in this book, then, is to use the three themes to help explain the development of sixteen traditional regions located in America's forty-eight contiguous states. The story begins with the significant presence of Europeans in America about 1600, and with exceptions it ends when European Americans have taken up the best of the available land about 1900. By then the fourteen lasting regions that form our present-day overarching regional structure had come to exist. (To develop what happened in the twentieth century—for example, urban-industrial changes and the formation of metropolitan regions—would take a second volume.) The contents are organized into three parts, as illustrated in figure 1.1: In Part 1, "Colonial America,"

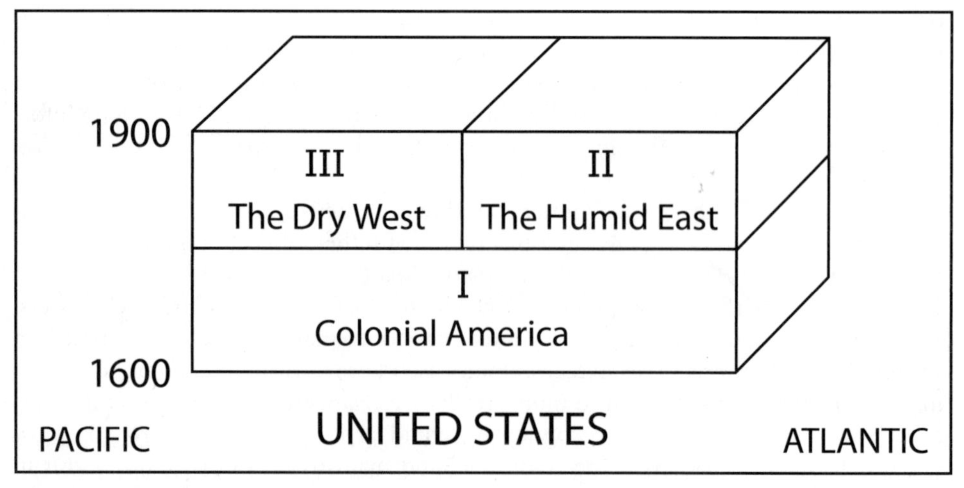

Figure 1.1 This Book.

we find notably Spaniards, French, and British and African peoples lightly colonizing large separate areas of the continent. In Part 2, "The Humid East," the dynamic largely British/African people spread west from the Atlantic Seaboard to the middle of the continent. And in Part 3, "The Dry West," the westward movement reaches the Pacific, leaving behind islands of settlement at places where critically important water was found. We will pause in our westward thrust to look back and compare, in chapters 7, 14, and 20. The closing of the frontier, as reported in the 1890 census and reasserted by the historian Frederick Jackson Turner in his famous 1893 essay, can be thought of as marking the end of the first epoch in America's historical geography. By that point America's regional cultural fabric had formed.[3]

Historical Geography

Geography, quite simply, is the study of differences from place to place. These differences can be about literally anything—physical phenomena like mountains or tornadoes, biological phenomena like forests or wildlife, or cultural phenomena like religions or house types. The qualifier, however, is that these variations exist within Earth's "shell" (figure 1.2). Most geographers study differences from place to place on the surface of Earth itself. This is, after all, where the action is. But the discipline embraces climatologists who study what happens in the atmosphere to about twenty-four miles (forty kilometers) above Earth's surface, and it also includes geomorphologists who study landforms and patterns of rock to depths of about two miles (3.3 kilometers) below Earth's surface. "Place" and "space" are the key words in defining what geography is.

Historical geography brings time to this spatial dimension. Understandably, most geographers study the present—that is, differences between current places. Historical geographers take this same spatial dimension and simply push it into the past. Geographer Harlan H. Barrows of the University of Chicago did more than anyone to launch the subdiscipline of historical geography. For thirty-eight years (1904–1942) he offered, normally twice a year, an extremely popular course called

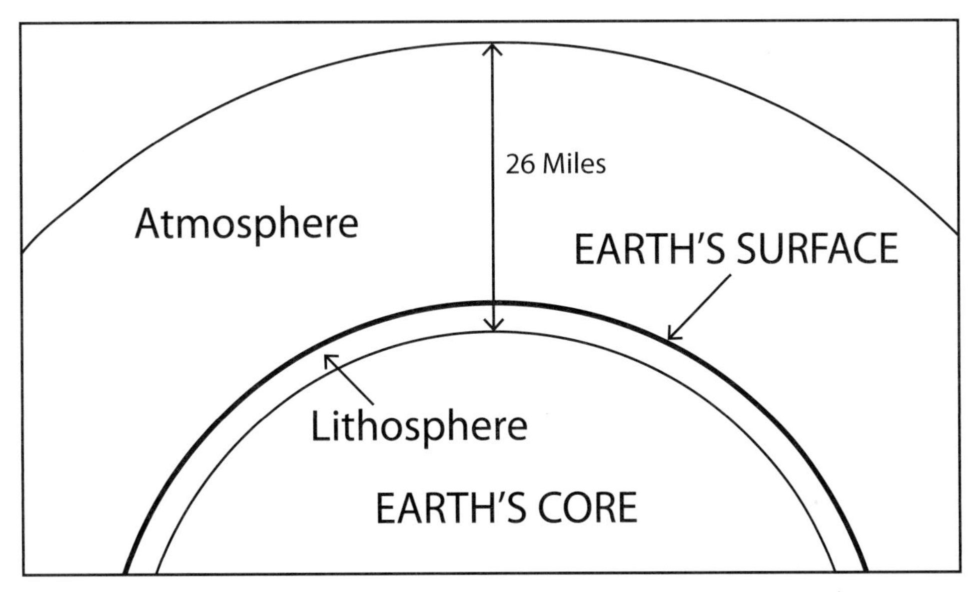

Figure 1.2 Earth's Shell. Geographers study patterns of all kinds in this area measuring twenty-six miles (forty-three kilometers) wide.

Historical Geography of the United States. In his class, Barrows used this definition: "Historical geography is the geography of the past." Since Barrows, some practitioners have placed more emphasis on geographical change through time while others have focused on only one cross section in time, but the definition used by Barrows captures both emphases and remains valid today. To avoid possible confusion, we can differentiate between historical geography, geographical history, and the history of geography. If "historical geography" is *geography* with an emphasis on the past, then "geographical history" is *history* with an emphasis on places. The borderland between the two is often blurred, and historical geographers consider geographical historians like Frederick Jackson Turner to be their scholarly brethren. The history of geography, on the other hand, is altogether different, for it deals only with the history of the discipline of geography.[4]

As alluded to when discussing "Why Regions?" geography and history are analogous fields. This is because they are both *methods* for structuring information. In geography, information is organized spatially, while in history, information is organized temporally. The analogy goes like this: *Time* is to history what *space* is to geography. Moreover, a *period* in time is to history what a *region* in space is to geography. And to carry the analogy even further, a *date* (1776) is to history what a specific *place* (New York City) is to geography. As methods for organizing information and thinking, geography and history cut across many systematically organized *subject matters* like economics or botany. For example, economic phenomena in history become economic history, and in geography they become economic geography. Botany approached by historians belongs in environmental history, and for geographers it is biogeography. As methods, history and geography are equally broad disciplines, and both are also fundamental dimensions for understanding humanity on Earth.[5]

Historical geography, then, combines time and space, and historical geographers have come up with four traditional, viable ways to do this. In the literature these are known as *approaches*, and they are illustrated diagrammatically in figure 1.3. In the *temporal cross section* approach, one reconstructs the total geography of one area at one time. Claimed by some to be historical geography in its most orthodox or pure form, those using this approach are obliged to justify the slice of time chosen lest their study lie static and "lifeless as a shut museum room," to quote geographer Robert M. Newcomb. In the more popular *vertical theme* approach, only one theme (for example, agriculture) is developed through time. Carl O. Sauer and his students at the University of California–Berkeley developed this dynamic way of combining time and space, which explains why some refer to it as the *Berkeley* approach. Also popular is the *successive cross section* approach, which actually combines the temporal cross section and the vertical theme. Practitioners must hone their organizational skills to be able to describe changes while pausing along the way to give summary reconstructions. Geographer David Ward once likened this approach to riding an up elevator in a building and pausing to exit at sample floors to make summary geographical assessments. Finally, with the little-used but always intriguing *relic geography* approach (also called the *retrogressive* approach), one begins with the present, mapping relic features in the field today in order to reconstruct the past. Suffice it here to mention but one example of each approach.[6]

In his classic *Mirror for Americans: Likeness of the Eastern Seaboard, 1810* (1943), Ralph H. Brown does a temporal cross section, reconstructing the total geography of the Atlantic coast at the time the United States was a new republic. Drawing on his lively imagination, Brown invents an author (Thomas Pownall Keystone) who in 1810 allegedly wrote the synopsis based on the several thousand books, pamphlets, and maps then published on the subject. A wealthy German, Christopher Daniel Ebeling (1741–1817), author of a seven-volume history and geography of

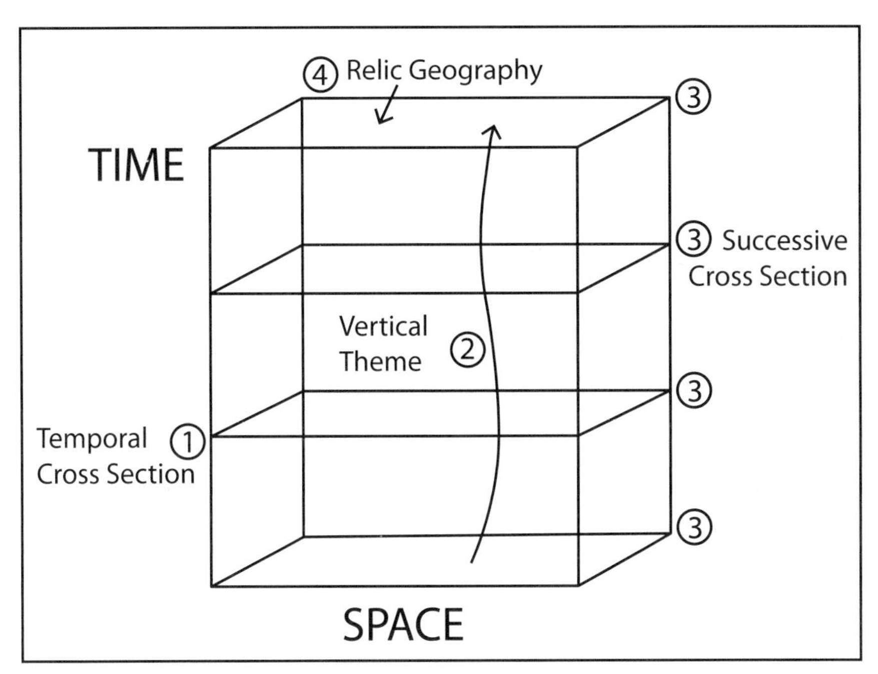

Figure 1.3 Approaches to Historical Geography. Four traditional approaches to combining time and space are illustrated.

the United States, had actually gathered the bulk of these sources, and Brown found them conveniently accessible because Harvard University had purchased Ebeling's library. Terry G. Jordan uses the vertical theme approach in his *North American Cattle-Ranching Frontiers* (1993) to develop five centuries in the changing geography of one theme: cattle ranching. Jordan drew on an astonishingly wide range of sources, only partly archival, to tell the complicated story of the diffusion of cattle ranching from four source areas in Europe and Africa to the West Indies, where these cattle-ranching styles converged and mixed, and from there to Mexico and the United States. D. W. Meinig (see spotlight) employs successive cross sections in each of the four volumes of his *The Shaping of America*. He pauses to map, using his core-domain-sphere morphology, the outcomes of change for circa 1800, the late 1850s, 1915, 1950, and 2000. Finally, in "A Town That Has Gone Downhill" (1927), James W. Goldthwait uses relic features in the field, like cellar holes, stone walls, graveyards, and abandoned roads to reconstruct the depopulation of the town of Lyme, New Hampshire (located one town north of Hanover, where Goldthwait taught at Dartmouth College). Settled in the 1760s, Lyme's population peaked in 1830. But the town's records before 1873 were completely destroyed by fire, so in 1925 Goldthwait and his students went into the field to determine where those Yankees had lived in that benchmark year of 1830. By 1925, Goldthwait learned, virtually everyone in Lyme above the 1,100-foot (310-meter) contour had moved "downhill" to join others in the environmentally less challenging Connecticut Valley.[7]

Relic geography relies on fieldwork, but the other three approaches depend on the use of archival materials. Most of the chapters that follow contain at least one example of why research in the archives is of value in historical geography. In "The Puritan Village Myth" in Chapter 4, for example, it is pointed out that those observing New England villages in the field got it wrong when they characterized New England colonial villages as nuclear. Those researching the villages in the archives got it right when characterizing the villages as (with exceptions) dispersed, which is what

Spotlight

D. W. Meinig (1924–)

D. W. Meinig. Photograph taken probably in the 1970s, courtesy of D. W. Meinig.

As a boy Donald Meinig worked on his father's Palouse Hills wheat farm in eastern Washington. Like his father, who used horses rather than tractors, Meinig (a self-described technophobe) wrote with a pen rather than use a typewriter or computer. But with that pen he "wrote like an angel," as UCLA geographer Henry J. Bruman once exclaimed. Earning his doctorate under mentor Graham Lawton at the University of Washington (1953), Meinig's two permanent faculty positions were in the University of Utah (1950–1959) and at Syracuse University (1959–1989). Meinig once told me that his most innovative piece, because it employed his core-domain-sphere morphology, was his interpretation of the Mormon Culture Region (1965). His most lasting work, informed by this same morphology, will certainly be his widely acclaimed, four-volume *The Shaping of America* (1986–2004). Meinig's astounding ability to master the details of American history and geography, and his brilliance in analyzing and interpreting this material geographically, make him a giant in the American geography profession. To our great benefit, D. W. Meinig has spent his entire life thinking as a geographer.

they were from the outset. My favorite examples of the value of archival research are in chapters 13 and 16. In Chapter 13, a recent local historian in Kansas followed the footsteps of the 1870 census enumerator to be able to locate the site of the original "Little House on the Prairie." And in Chapter 16, use of railroad archives allowed geographer James Vance Jr. to explain why the Union Pacific Railroad abandoned the route of the Oregon Trail through South Pass—in violation of the principle of locational succession. Drawing on archival sources is what makes teaching university classes in historical geography quite imperative, and geographer Cole Harris points out why: Without faculty whose historical proclivities find them steering their students to the archives, little archival research will get done; lacking historical researchers, geography as a discipline will find itself relinquishing the study of past places and environments to other disciplines.[8]

The Natural Environment

In the United States, there are differences from place to place in the natural environment. For example, Alabama has pine trees and summers that are oppressively hot while Arizona has desert shrubs and summers that are furnace hot. Disregarding landforms for now, these differences in the natural environment are governed primarily by climate—the long-term averages of temperatures and precipitation—and reflecting climate are vegetation and soils. During the last five hundred years the climate of the United States has changed. For example, temperatures rose just in the twentieth century by one degree Fahrenheit, and believers in the existence of global warming warn that temperatures will continue to rise at accelerating rates. Patterns of vegetation and soils have also changed, largely because of European American settlers. Forests have been cut down and soils have eroded away. It is estimated, for example, that since the grasslands of the Great Plains were plowed under in the 1870s and 1880s, an average of twenty-four inches of topsoil in the region has been lost to erosion, and it takes one thousand years for just one inch of new soil to regenerate under a grassland cover. Wind was partly responsible for this erosion, and John W. Morris, my predecessor in teaching historical geography at the University of Oklahoma, told me that in Norman, Oklahoma, during the Dust Bowl of the 1930s, on occasion the dust was so thick in his own tightly closed classroom that students sitting at the back of his room could not see the blackboard in the front of the room.[9]

Generalizing first about temperatures, three principles are at play. The first is that temperatures decrease with an increase in latitude (and with altitude). This explains why it is warmer in Texas and cooler in Minnesota (see figure 1.4). But what happens to upset nice, even latitudinal decreases in temperatures going poleward is a second principle: that land heats faster and cools faster than water (the specific gravity of land is greater than that of water). So, over the land surface of the continental United States, there are seasonal extremes in temperature—hot summers and cold winters—while temperatures over the adjacent oceans are milder. The sixty-degree-Fahrenheit isotherm passing from coast to coast across the United States in summer bends well to the north over land, and conversely in winter the thirty-two-degree-Fahrenheit isotherm bends well to the south. All this is simply the difference between continents and oceans. But then there is a third principle at work: The United States lies within the prevailing westerlies (winds are named for the direction from which they blow). This means that maritime temperatures over the Pacific are blown onto the West Coast. But the high Cascade Mountains–Sierra Nevada barrier that rises up just inland from the coast blocks the eastward penetration of mild ocean temperatures. Continentality begins abruptly on the leeward side of these mountains, and it continues all the way to the Atlantic coast; mild temperatures associated with the warm offshore Gulf Stream are usually kept off the Eastern Seaboard by the prevailing westerlies.

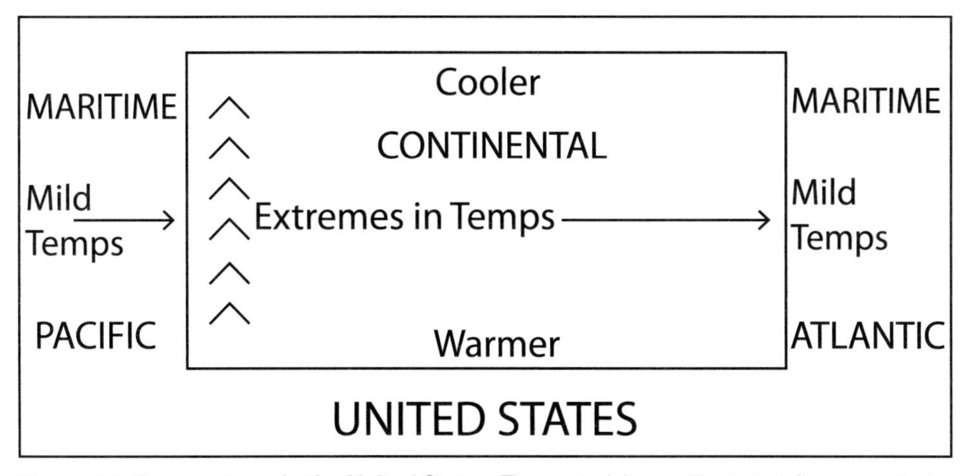

Figure 1.4 Temperatures in the United States. Three principles are illustrated diagrammatically. Arrows indicate westerlies.

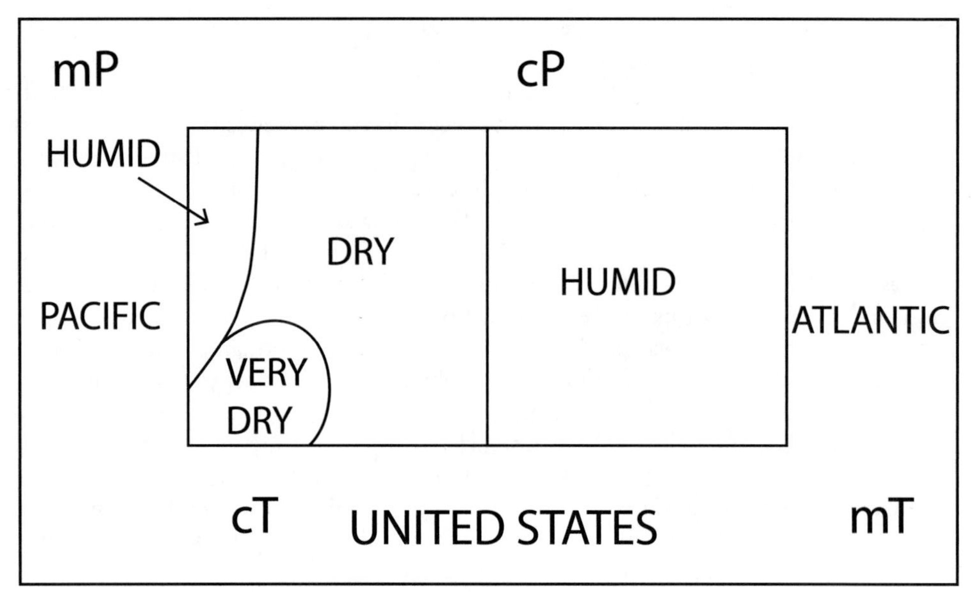

Figure 1.5 Precipitation in the United States. Four air masses play controlling roles.

Generalizing about precipitation, there are four air masses that play controlling roles. Two of these influence precipitation (and temperatures) in the eastern half of the United States (see figure 1.5). Maritime tropical air (mT), formed over the Atlantic and Gulf of Mexico, is warm and moist, and it brings rain and humid conditions to the East in the high-sun summer months. Continental polar air (cP), formed over northern North America, by contrast is cold and dry, and it dominates the East in the much drier low-sun winter months. Largely because of mT, the East is humid. By contrast, the West, except for along the coast, is dry. In the West two air masses are at play. Maritime polar air (mP), found over the north Pacific, is cold and moist (cold air potentially holds much less water vapor than does warm air), and in winter months it brings moisture to the West Coast as far south as northern California. Continental tropical air (cT), which is warm and dry, is only weakly developed because the land surface of Mexico is not large. This air mass in part explains the dry conditions in the interior West, but the Cascades–Sierra Nevada barrier is at least as important in walling off moisture. The southwestern corner of the United States (Arizona-Nevada-Utah) is an especially dry desert area. The result basically comprises three north-south bands that are humid-dry-humid and one corner that is very dry, as shown in figure 1.5.

We are talking here about *climate*, or the long-term averages of precipitation and temperature. *Weather* refers to day-to-day atmospheric conditions. An important yet little appreciated fact about the United States is that the humid eastern half of the country has the most violent weather on Earth. In the East cP and mT masses collide as they are being pulled from west to east beneath the stratospheric (thirty-two thousand feet plus aloft) jet stream. The two air masses could not be more different; cP is cold, heavy, and dry while mT is warm, light, and moist. When they collide in cyclonic fronts, within hours they can produce frightening thunderstorms, blizzards, and the most intense of all storms: tornadoes. Some 90 percent of all tornadoes in the world occur in North America, and roughly one-third of them strike Texas, Oklahoma, and Kansas, especially in April, May, and June. No wonder the National Severe Storms Laboratory is located in Norman, Oklahoma. The big point is that cP develops only over large land masses located at high latitudes, which means that cP develops only in northern North America and northern Eurasia; continents in the Southern Hemisphere taper off and end going poleward, precluding cP from forming. In Eurasia the high, east-west Himalaya Mountains block the interaction of cP with mT, which is formed over the Indian Ocean. But in the eastern United States there is no east-west mountain barrier to separate cP and mT. "Violent" is not too extreme a word to characterize the weather of the humid East.[10]

Whether an area is humid or dry has much effect on its vegetation and soils. Figure 1.6 shows how generalized patterns of vegetation and soils in the United States correlate quite directly with amounts of precipitation as shown in figure 1.5. Looking at native vegetation and soils before European American alteration, the East, because of its ample precipitation, had a forest cover and soils that were leached of their mineral content. The symbols for the trees in figure 1.6 represent, from south to north, southern pine; broadleaf, deciduous trees such as maples, oaks, and elms; and white pine. In the dry interior West, a grassland existed, yet pine trees covered higher and therefore more humid areas like the Rocky Mountains. Beneath the grasslands, soils were rich because decomposed organic matter called humus accumulated in the upper profiles. Along the Pacific coast, where conditions were again humid, a forest of fir trees existed—Douglas fir in western Washington and Oregon and redwoods in coastal northern California. Humid conditions also meant leached soils. And in the very dry Southwest, where low-lying desert shrubs prevailed, soils were rich in calcium and minerals.

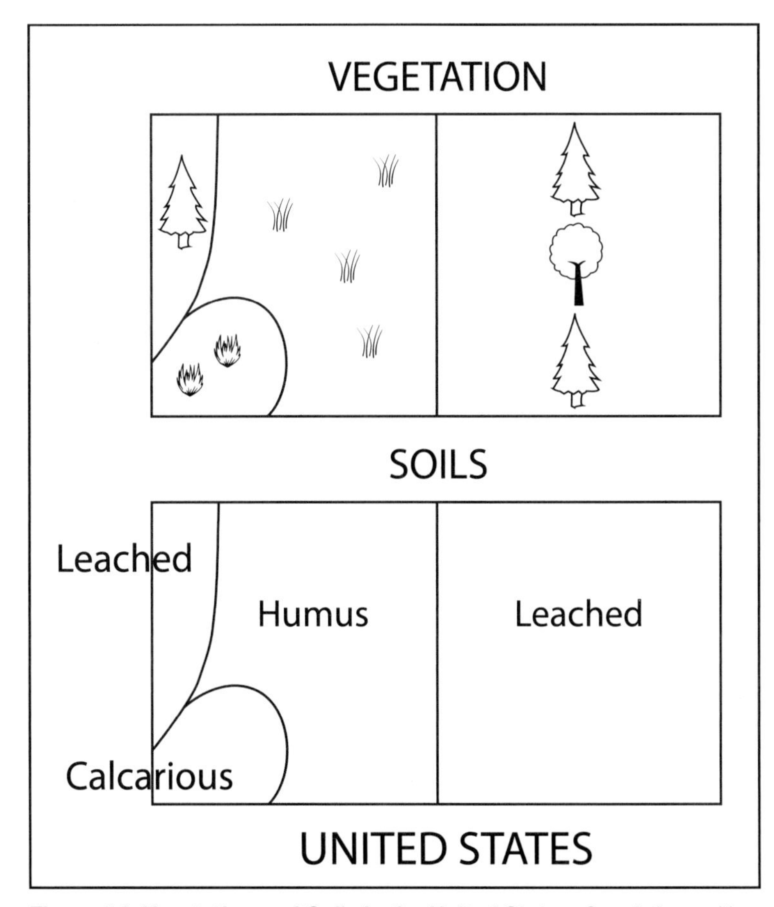

Figure 1.6 Vegetation and Soils in the United States. Correlations with precipitation are strong.

So this was the natural environment encountered by European Americans after 1500, and it is informative to ask how this natural environment struck them. First, how did it strike the English and French? As to *vegetation*, the similarities between the Old and New Worlds were striking. On both sides of the Atlantic could be found the same species of trees, but in the New World the variety was even greater. The grasses, however, were less hardy in America, for they could not withstand the trampling of livestock. As geographer Carl O. Sauer pointed out with reference to plants and animals, "It would be impossible, indeed, to cross an ocean anywhere else and find as little that is unfamiliar in nature on the opposite side." As to *climate*, what struck these Europeans were the differences. England and France had mild temperatures—warm (not hot) summers and cool (not cold) winters. Precipitation came mainly in winter, and, especially in England, it fell as a drizzle day after day. Along the Atlantic Seaboard, summers were long and hot and winters were long and bitterly cold, especially for the French in Canada. Rains came mainly in the summer and with greater intensity but shorter duration. The English described the rains as "tempestuous."[11]

The bitter cold winters and long, hot summers were what most surprised the English and French colonists. These Europeans knew that in crossing the Atlantic they had moved closer to the equator (see map 1.1). They knew that in latitude London (51°) and Paris (49°) were ten degrees poleward from Philadelphia (40°). They reasoned, correctly, that winters should be warmer, not colder, in the New World, and wondered why those tempestuous rains occurred in summer. Not until the end of the colonial period did they finally learn the explanation: West coasts of continents are influenced by maritime conditions while continentality dominates east coasts of continents. Georg Forster, who accompanied James Cook on Cook's second voyage to the Pacific Ocean (1772–1775), was the first to observe the similarity

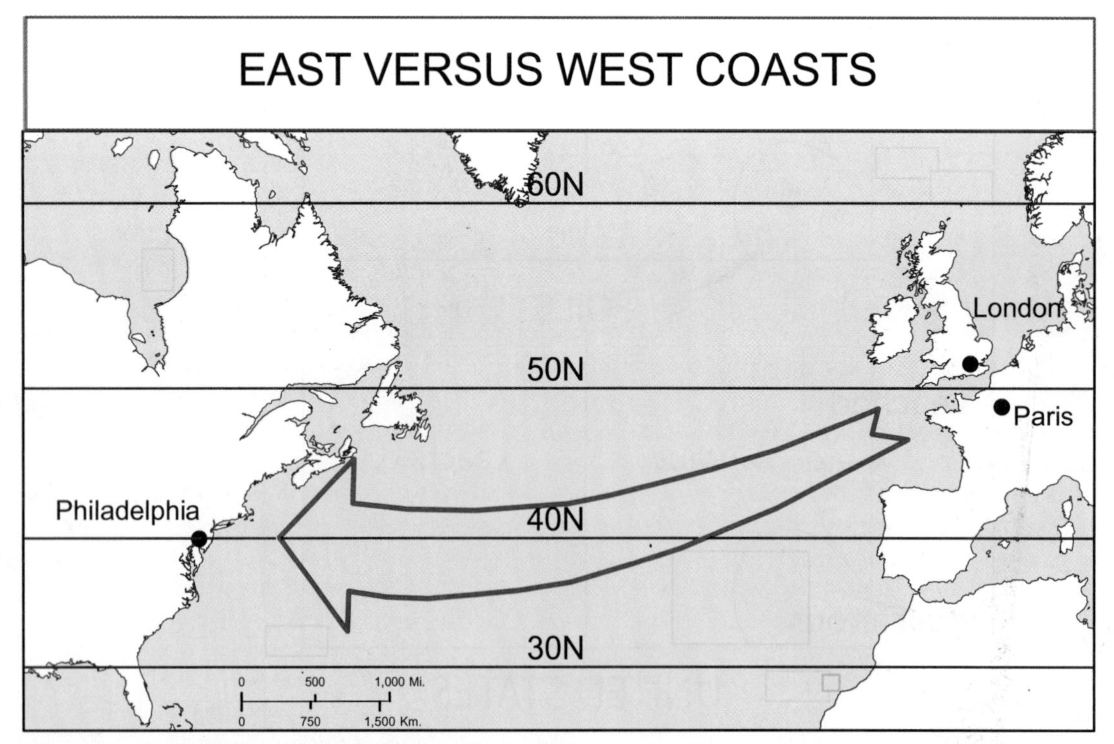

Map 1.1 East versus West Coasts. An entity's location on the east or the west side of a continent explained more about climatic differences than did latitude.

between temperatures and climate at the same latitudes on west coasts (e.g., London and Paris compared to Vancouver, BC, and Seattle) and likewise on east coasts (e.g., the American South at Charleston and China at Shanghai).[12]

How Spaniards were struck by their New World environments was more mixed. Coastal Spain had a Mediterranean climate, which meant only two seasons: a hot and dry summer and a mild and moist winter. In coastal Spain oak trees grew on grasslands, sometimes referred to as oak parklands. In the New World, coastal California had a Mediterranean climate and oak parkland vegetation. Spaniards from coastal Spain could not have been transplanted to a more familiar environment than to coastal California. High and dry New Mexico presented colder temperatures and less total precipitation than California, although Spaniards coming north from the central plateau of Mexico certainly knew the challenges of too little precipitation. What Spaniards found to be really different environmentally, as already noted, was the hot and wet Southeast between Florida and east central Texas. And efforts to colonize there met with limited success.

Native Americans

We can't be sure how the natural environment in what is now the United States struck Native Americans, but we have a clue: Their numbers were far greater in the tropical highlands of Middle America and South America where temperatures were temperate and precipitation modest. Many archaeologists believe that Native Americans arrived in this New World environment 20,000–30,000 years before present (BP). A few, including geographer George F. Carter, think they arrived about 100,000 BP. Archaeological evidence documents their presence in present-day Alaska and the Lower 48 to about 12,000–14,000 BP. Native Americans seem to have come from Asia, apparently over a land or ice bridge that is now the Bering Strait. They apparently spread slowly southward in a process that eventually found them occupying the entire Western Hemisphere to the tip of South America. About 1500, when Columbus encountered these real discoverers of the New World, these first Americans may very approximately be estimated to have numbered three million in the contiguous forty-eight states and Canada, twenty million in the highlands of Middle America, and ten million in the highlands of South America.[13]

About 1500, within the contiguous forty-eight states, Native Americans lived in five broad culture areas (see map 1.2). The Woodland Indians, such as the Algonquin, Iroquois, Creek, and Choctaw, inhabited the humid forested East. These peoples were hunters (most importantly of the white-tailed deer); gatherers of edible roots, nuts, and berries; and fishers, of whom a relatively dense population exploited shellfish around Chesapeake Bay. They were also farmers, growing corn, beans (kidney and navy), squash, and pumpkins. Corn, the major crop, seems to have been domesticated about 4,500 BP in the Mexican state of Puebla, from which it diffused north. According to geographers Myers and Doolittle, the revisionist "new narrative" account of Native American farming in the forested East is that, despite the absence of the plow and draft animals in Native American culture, *huge areas* were under *permanent cultivation* prior to the arrival of Europeans. The old narrative had Native Americans clearing small openings in the forest—some cultivated, some fallow, and some reverting to forest—in what was characterized as shifting cultivation. For the vestiges of agricultural clearings found throughout the forested East from Canada to Florida, Europeans used the term "Indian Old Fields" in their descriptions.[14]

Woodland Indian agriculture ended about where the Great Plains began—where tallgrass prairie gave way to shortgrass steppe. In the latter lived a second group, the Plains Indians, such as the Sioux, Kiowa, and Comanche (map 1.2). These Indians

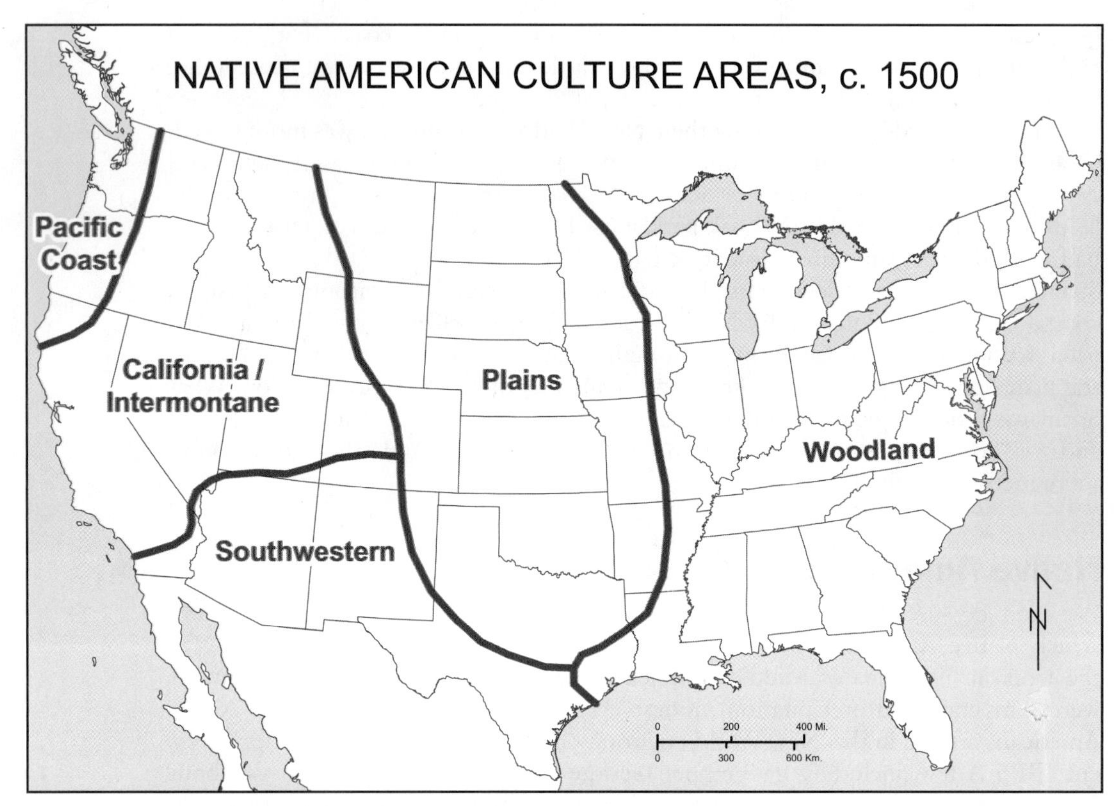

Map 1.2 Native American Culture Areas, c. 1500.

Sources: "Indians of North America," map supplement in *National Geographic* 142, no. 6 (December 1972): 739A; "American Indians," in *Hammond United States History Atlas*, map B, page U-4.

were nomadic. Their possessions, including their teepees, were portable and could be loaded on travois—two poles bearing a load pulled by dogs. Being mobile allowed Plains Indians to pursue buffalo (*Bison bison*), the focus of their livelihood. The introduction of the horse by Spaniards in the 1600s revolutionized their hunting. More effective hunters on horseback, the Plains Indians now played a role in what became a greatly expanded range of buffalo beyond the Great Plains. From a review of the historical, archaeological, and toponymic evidence, geographer Erhard Rostlund found that buffalo, apparently totally absent in the southeastern United States in 1500, began to migrate into the area after about 1550, and by 1700 reached all the way to the Gulf coast between Louisiana and central Florida and to the Atlantic coast of Georgia. After 1700, buffalo disappeared from the Southeast because settlers exterminated them in heavy hunting.[15]

The Southwestern Indians formed a third culture group (map 1.2). In the dry Southwest lived two contrasting peoples. They were the nomadic Indians, such as the Apache, Navajo (an offshoot of the Apache), and Ute, who hunted and were warlike; and the Pueblo, including the Zuñi, Hopi, and Pima, peaceful peoples who lived in multistory adobe and stone villages that were also called pueblos. Like the Woodland Indians, the sedentary Pueblo grew beans and squash and corn that they acquired from the same center of plant domestication in Mexico; but unlike their Woodland contemporaries, the Pueblo, in their dry environment, had to irrigate. They built irrigation canals and warped water onto fields from behind low rock dams, a practice called floodwater farming. These two peoples exchanged goods from the field for those from the hunt in a fragile relationship that was sometimes broken by nomadic Indian raids on Pueblo Indian storehouses.[16]

Farther west there lived the two final cultural groups: the California-Intermontane Indians and the Pacific Coast Indians (map 1.2). Along the California coast were the Chumash, who were hunter-gatherers and fishers. Indians such as the Shoshoni in the intermontane country between the Rocky Mountains and the Cascade–Sierra Nevada hunted and gathered in an especially harsh and forbidding environment. Along the marine-moderated Pacific coast going north from northern California lived the river-oriented, salmon-fishing Pacific Coast Indians. An abundance of salmon supported a relatively dense population—and one characterized by great linguistic fragmentation. People speaking completely different languages lived just a river basin apart, suggesting that for many centuries they had lived self-sufficiently and in isolation.

About 1500, then, fractionalization characterized the three million Native Americans who were spread thinly across the present-day forty-eight states and Canada. Language differences, especially great along the Pacific coast, differentiated Native Americans everywhere: Perhaps one hundred different linguistic families existed in North America. Socially, families lived in clans; politically, these clans belonged to many tribes or nations. Economically, two of the five broad cultural groups practiced agriculture while the rest were hunter-gatherers and fishers. The impact of Native Americans on the natural environment was quite substantial: Indian Old Fields were widespread in the forested East; in the Ohio and Mississippi valleys great mounds of earth marked the sites of cities that had flourished between 1100 and 1300 CE (common era); and in the Southwest there were vestiges of irrigation systems and multistory compact villages that continue to house the Pueblo Indians. What would eventually unite these fractionalized peoples would be their rejection of their mistreatment at the hands of European Americans.

For Native Americans, the arrival of European Americans proved disastrous and tragic. European Americans had attitudes of superiority and intolerance; they assigned Native Americans a subordinate status, and they refused to accommodate them. The English were especially contemptuous; the Spanish and French worked with Indians in their efforts to introduce Roman Catholicism, and Spanish and French males were more willing to form unions with Indian females. Nevertheless, the two civilizations collided, despite "the simple but long-neglected truth that Indians were critically involved in the creation of every European mainland colony," notably through their crops and methods of cultivation. The outcome: Indian numbers were drastically reduced by warfare and disease, Indians were dispossessed of their lands and were forcibly removed to reservations, and Indians were in all ways humiliated. Their treatment is an important example of European American imperialism, the theme of Meinig's four-volume *The Shaping of America*. European Americans in the nineteenth century did not admit to their imperialistic behavior, but the fact is that they used government authority to acquire Indian territory and to dominate these captive peoples. As Meinig points out, "the gross difference between . . . [government] promise and . . . performance . . . is of course one of the great scandals in American history."[17]

Notes

[1] For the "relative disfavor" of the regional approach because of its "alleged lack of precision," see McKnight, *Regional Geography of the United States and Canada*, xi.

[2] Mood, "The Origin, Evolution, and Application of the Sectional Concept, 1750–1900," 7–8, 83–85.

[3] Turner, "The Significance of the Frontier in American History," in *The Frontier in American History*, 1–38. Originally published in the *Proceedings of the State Historical Society of Wisconsin at Its Forty-First Annual Meeting Held December 14, 1893* (Madison: Democrat Printing Company, 1894), 79–112.

[4] Barrows never published a textbook for his popular course. The content of his lectures, however, was preserved when Kathlyn Purtell, a University of Chicago geography department secretary, audited the class in the winter quarter of 1933 and made a stenographic transcript of what Barrows said. Her transcript became the basis for the publication in 1962 of Barrows, *Lectures on the Historical Geography of the United States as Given in 1933*, edited by geographer William Koelsch. Barrows defined historical geography on page 6.

[5] For the analogy between geography and history, see Nostrand, "A Model for Geography," 13–17.

[6] Newcomb, "Twelve Working Approaches to Historical Geography," 30. For Ward's simile, see "Comment at Eastern Historical Geography Association."

[7] Brown, *Mirror for Americans*, xiii–xxxii; Jordan, *North American Cattle-Ranching Frontiers*; Meinig, *The Shaping of America*, 1:402, 2:424, 3:387, 4:192, 378–79; Goldthwait, "A Town That Has Gone Downhill," 527–52.

[8] Harris, "Archival Fieldwork," 334.

[9] For the one-degree-Fahrenheit rise in temperature in the twentieth century, Cowen, "Global Warming," 10–11. For the Great Plains example of soil loss, see the Spaeth interview.

[10] For 90 percent of the world's tornadoes occurring in North America, see Lineback, "Deadly Tornadoes," 11.

[11] Sauer, "The Settlement of the Humid East," 159.

[12] For breakthrough observations by Forster, see James, *All Possible Worlds*, 138.

[13] For Native Americans arriving 100,000 BP, see Carter, *Earlier Than You Think*, ix, xi, 3, 13ff. For archaeological evidence dating Native Americans in Alaska to 14,000 BP and in Oregon to 12,300 BP, see Butzer, "Retrieving American Indian Landscapes," 33. And for estimated numbers of Native Americans in the New World in 1492, see Denevan, *The Native Population of the Americas in 1492*, 291.

[14] For corn (maize) domestication in Puebla about 4,500 BP, see Wilford, "Corn in the New World," B5. For Myers and Doolittle's description of permanent cultivation, see their "The New Narrative on Native Landscape Transformations," 13 and 15.

[15] For the expansion of buffalo, see Rostlund, "The Geographic Range of the Historic Bison in the Southeast," 405–7.

[16] Meinig contrasts nomadic and Pueblo peoples quite eloquently in *Southwest*, 11.

[17] In *The Shaping of America*, Meinig discusses at length the tragedy of the Native American encounter with European Americans. See 1:213, for "the simple but long-neglected truth"; 2:99, for "the gross difference."

COLONIAL AMERICA I PART

The Spanish Borderlands 2

We begin our regional analysis with the Spanish Borderlands because, with the founding of St. Augustine in 1565, Spaniards initiated permanent European American settlement of the present-day United States. The Spanish Borderlands stretched between what are now Florida and California. The chapter's first section, "Borderlands Ecology," focuses on how Spaniards adjusted to the region's subtropical environment. In the second and third sections, "Spanish Florida" and "The Southwestern Borderlands," we examine reasons why Spaniards colonized and when and where they went. By the time Spanish rule ended in Florida in 1819 and in the Southwest in 1821, these people had stamped their cultural landscape with an indelible imprint, as related in the final section, "The Spanish Impress."

Borderlands Ecology

American historians use the term "Spanish Borderlands" to refer to those present-day American states between Florida and California where Spain once "held sway." The term became popular after well-known Berkeley historian Herbert Eugene Bolton (1870–1953) published his seminal *The Spanish Borderlands* in 1921. But the *Borderlands* term has a drawback: Its use is anachronistic, at least when applied to New Spain's vast northern frontier before any political border existed. *Spanish Frontier* is a more accurate label. Indeed, the titles of the three major books that have synthesized the region's history illustrate the decline of Spanish Borderlands: Bolton's *The Spanish Borderlands* (1921), John Francis Bannon's *The Spanish Borderlands Frontier* (1970), and David J. Weber's *The Spanish Frontier in North America* (1992). We use *Spanish Borderlands* here, however, because it usefully suggests the region's modern-day proximity to Mexico and the overlap of Hispanic and American cultures.[1]

The tier of Borderlands states between Florida and California lies in the subtropics (see map 2.1). This means that in the lowlands, summers are long and hot and winters are short and mild, as in St. Augustine, San Antonio, and Los Angeles. At higher elevations in the subtropics, as in Santa Fe, located at seven thousand feet (2,130 meters) above sea level, summers are merely warm and winters are long and cold. But a more fundamental environmental contrast concerns not temperature but precipitation. Between Florida and East Texas average annual precipitation everywhere exceeds thirty inches, and at St. Augustine it is more than fifty inches. Heat and high humidity in this humid subtropical half of the Borderlands made seven months of the year quite oppressive for Spaniards. Climate combined with sandy soils and a pine forest were handicaps to Spaniards colonizing the Southeastern Borderlands. By contrast, between central Texas and California, the average annual precipitation is generally less than twenty inches. Native vegetation is sparse and shrub-like, except at higher elevations. Crops have to be irrigated even in San Antonio, which, with 29.6 inches of precipitation, is classified as subhumid. Spaniards adapted more readily to these dry-land conditions in the dry, subtropical half of the Southwestern Borderlands.

In Florida and the Southeastern Borderlands the consequences of heat and humidity on agricultural adjustment were numerous. Attempts to grow wheat, the preferred grain of Spaniards; grapes for wine; and olives all failed. Spaniards turned instead to the corn, beans, and squash grown by the natives. Other Old World crops did well; they included oranges, peaches, and melons, particularly watermelons.

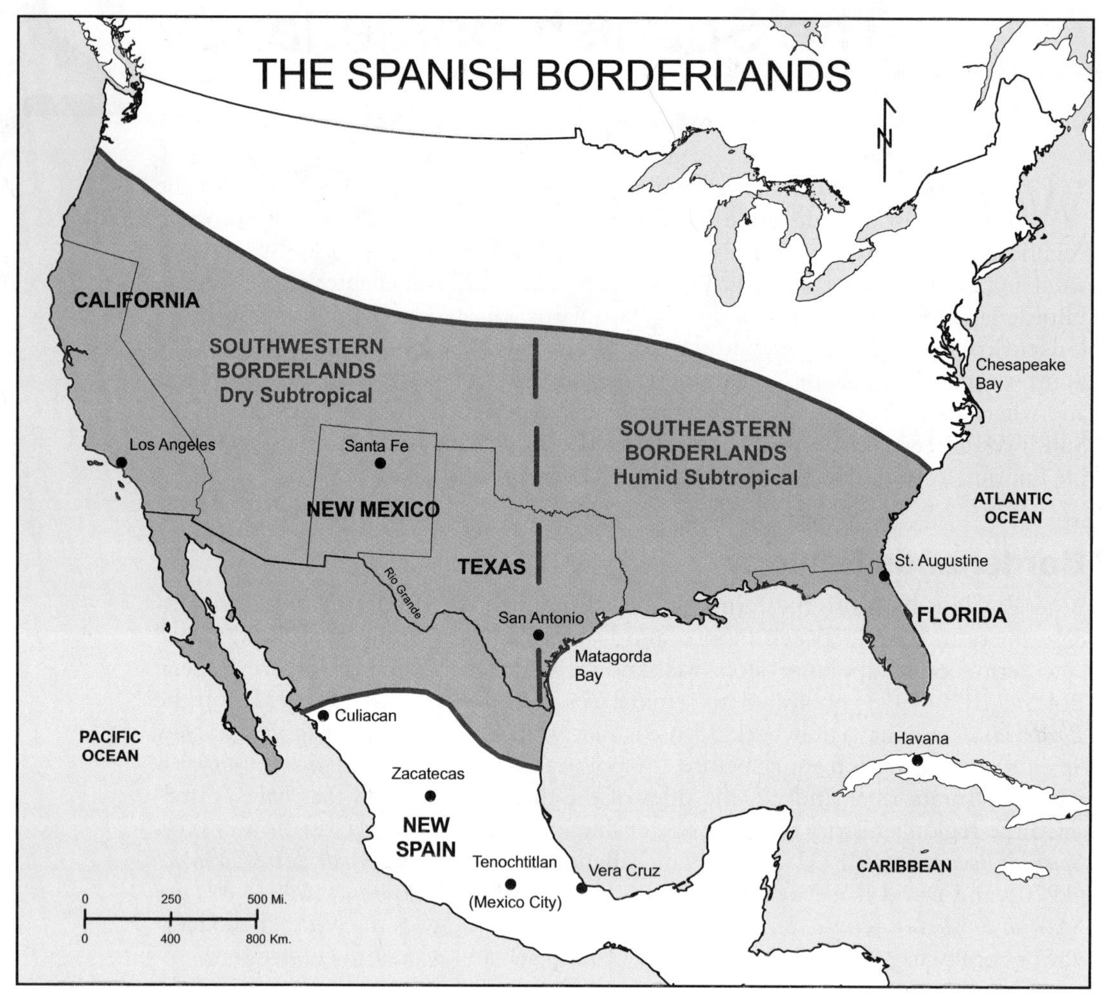

Map 2.1 The Spanish Borderlands. A humid, subtropical Southeast and a dry, subtropical Southwest separate at about San Antonio.

Spaniards in Florida also failed in the raising of sheep for lamb and mutton, their preferred meat. They did successfully raise hogs, cattle, and chickens and could supplement their diet with fish, venison, birds, and turtles.[2]

High and dry New Mexico in the Southwestern Borderlands had more familiar environmental conditions. With irrigation, Spaniards could grow their wheat, wine grapes, and deciduous fruits such as apricots and peaches. Up the Camino Real or Royal Highway that connected the frontier with central Mexico came chiles, tomatoes, and new varieties of corn and squash. Also diffusing from Mexico were food items with Nahuatl (the Aztec language) names including tortillas, tamales, and posole. Moreover, sheep did well in New Mexico's uplands, and sheep had the advantage of being slow and able to resist being stampeded in Indian raids. When low-elevation pastures dried up in summer, shepherds drove their sheep to higher pastures, a practice called *transhumance* that was brought from Spain. At elevations above six thousand feet (1,825 meters), ponderosa pine was available as a building material, but sun-dried adobe brick was preferred, even above six thousand feet. Geographer Charles F. Gritzner has cleverly correlated life zones and the building materials that were available to Spanish people in New Mexico (see figure 2.1).[3]

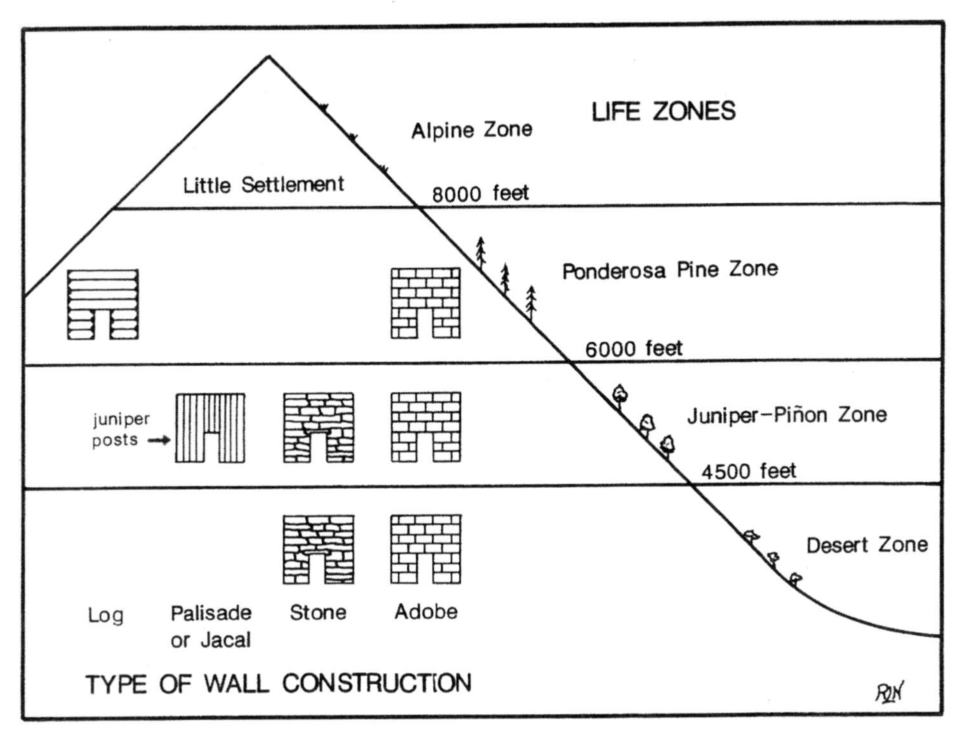

Figure 2.1 Life Zones and Building Materials in New Mexico.

Source: Charles Gritzner, "Construction Materials in a Folk Housing Tradition," 26. Courtesy of *Pioneer America* (now *Material Culture*, journal of the International Society for Landscape, Place and Material Culture) and the University of Oklahoma Press, publisher of the diagram in Nostrand, *El Cerrito, New Mexico*, 13.

In colonizing New Mexico, Spaniards laid out agricultural plots of land in what geographers call *long lots.* Long lots are an ingenious way to maximize the use of scarce water and limited arable valley floors. These long, ribbonlike fields stretch from an irrigation ditch at their uphill end to a stream or a river at their downhill end. Long lots are efficient in that they conduct water by gravity flow from the irrigation ditch through furrows containing their crops to a watercourse. Long lots are also equitable in that the fields of all farmers face the irrigation ditch, giving each farmer access to the precious water. And long lots accommodate population growth; in Spanish society, men and women inherited equally, and long lots could be subdivided into still narrower strips when being transferred to heirs. These advantages seem to explain why ribbonlike long lots came to cover virtually all valley floors in northern New Mexico. Long lots seem to have also existed in San Antonio, Texas, in the narrow interfluve between the upper San Antonio River and San Pedro Creek, and they definitely existed along the lower Rio Grande.

Significantly, long lots did not exist in Spain, and they are quite uncommon in Mexico. How, then, did they get to New Mexico and Texas? Something new in the culture of humans comes to exist in one of two ways: Either it is independently invented to fulfill a need or it diffuses from where it already exists. Geographers Alvar W. Carlson and Terry G. Jordan postulated that long lots in New Mexico (Carlson) and in San Antonio (Jordan) were independent inventions, and Jordan speculated that in the lower Rio Grande Valley of Texas, long lots diffused from the French, who employed them in Louisiana and elsewhere. At the University of Oklahoma, Jeffery E. Roth wrote a dissertation on this topic and concluded that in New Mexico long lots were not independent inventions but rather diffused by the French. In Roth's judgment diffusion likely explained long lots in San Antonio as well. Jean L'Archeveque and Jacques Grollet, two Frenchmen and survivors of LaSalle's exploits at Matagorda Bay in the mid-1680s, accompanied Spaniards in their reconquest of New Mexico in 1692. As residents of coastal France both must have had firsthand knowledge of long lots used there, and circumstantial evidence persuaded Roth that these two Frenchmen suggested the use of long lots, probably first at Santa Cruz in 1695. The possible diffusion of long lots to San Antonio awaits a future researcher.[4]

In California Spaniards had the fewest environmental challenges because California, like Spain, has a Mediterranean climate. As indicated in Chapter 1, Mediterranean climates have two seasons: hot, dry summers and mild, moist winters. To grow wheat and other crops in the dry summer growing season, Spaniards in California had to irrigate, as they did in Spain. Curiously *californios* called their irrigation ditches *zanjas*, while in New Mexico and Texas *nuevomexicanos* and *tejanos* called them *acequias*—a difference in need of explanation. Lacking timber for building houses in their oak-parkland coastal hills, californios built with adobe, as they did

Spotlight

Carl O. Sauer (1889–1975)

Carl O. Sauer in 1952. Photograph courtesy of Daniel Plumlee, Collections Manager, Department of Geography, University of California–Berkeley.

Most American historical geographers would probably acknowledge that Carl Sauer was their subdiscipline's most influential practitioner in the twentieth century. Because "Mr. Sauer," as he was known, engaged in fieldwork in the Borderland of the American Southwest and northwestern Mexico, and because of his keen interest in cultural ecology—including the origin and dispersal of domesticated plants such as maize or corn coming from Mexico to the Southwest—it is appropriate to profile him here. Born in Missouri of German parentage, Sauer earned his doctorate at the University of Chicago (1915) with a dissertation about Missouri's Ozark Highland. He taught first at the University of Michigan (1915–1923) and then at the University of California—Berkeley (1923–1957). As Berkeley's department chair (1923–1954), Sauer ran geography with a heavy hand. One of his thirty-seven doctoral advisees, Joseph E. Spencer, told several of us at UCLA that at Berkeley, graduate students would eventually learn what degree they would be earning when one day Mr. Sauer would ask what topic they planned to work on—for their master's thesis—or for their doctoral dissertation. Those students whom Sauer never approached eventually disappeared with no degree. Sauer's selectivity, however undemocratic, paid off in quality: The list is long of those among his advisees who perpetuated the cultural-historical, or Berkeley, approach, or who otherwise became leaders in the discipline. Sauer's own publications contributed mightily to scholarship on "man-land" themes and to a more humane use of Earth's resources.

in Spain. And in California, unlike in New Mexico, where temperatures dipped too low for frost-sensitive crops, seedling oranges, introduced in 1769, did well, as did lemons, grapefruit, and avocados when these fruits were introduced there. Longhorn cattle and horses multiplied quickly in California's grass-covered coastal hills, and cattle would become the basis for the hide and tallow trade in California's Mexican period. Geographer Carl O. Sauer (see spotlight) and his University of California field-trippers knew well these coastal hills near Berkeley.[5]

Spanish Florida

When Christopher Columbus "encountered" the New World in 1492—Native Americans had "discovered" it much earlier—he initiated sustained contact between the Old and the New Worlds and, of course, the two-way diffusion of countless examples of material culture. The Caribbean soon became a "Spanish lake," with Havana as the political seat of a governing captaincy general after 1515 (see map 2.1). In 1519 Hernán Cortés led a major expedition to the mainland of New Spain where, with superiority gained from horses and firearms, he toppled the Aztec ruler Moctezuma II, whose empire was centered in the Valley of Mexico. Out of the ruins of the Aztec capital, Tenochtitlán, rose Mexico City, seat of the Viceroyalty of New Spain after 1535. In their search for the mineral wealth witnessed at Tenochtitlán, Spaniards advanced north, and in 1546 they discovered at Zacatecas perhaps the most fabulous silver deposits in all the Americas.[6]

As Spanish galleons began to carry bars of silver to Spain, several interesting developments set the stage for Borderlands colonization. Alvar Núñez Cabeza de Vaca and three other survivors of a shipwrecked expedition to Florida, between 1528 and 1536, literally walked from the Texas coast to newly founded Culiacán on the Pacific. Estéban, a Moor who Indians believed possessed magical powers due to his dark skin, saved the four men many times. Cabeza de Vaca reported to an excited Viceroy Antonio de Mendoza in Mexico City that great treasure was to be found to the north in what natives had reported to be the Seven Cities of Cíbola (buffalo). So Mendoza sent Francisco Vásquez de Coronado on a major expedition (1539–1542) to find Cíbola, but Coronado returned to acknowledge that he had found only the villages of New Mexico's Pueblo Indians, specifically Zuñi and Pecos, and no mineral wealth. As a result, interest in the Southwestern Borderlands temporarily ended, and permanent settlement of Spain's northern frontier began instead in Florida.

In the early 1500s Spanish Florida was much larger than the present-day state of Florida. It stretched along the Gulf of Mexico and Atlantic coast from at least the Rio Grande to the Chesapeake Bay. However, much of what happened under the Spanish took place in what is now Florida (see map 2.2). The peninsula was flat and not much above sea level. Off its coast and parallel to it were barrier islands, long, narrow, sandy strips that stretched discontinuously from Maine to Texas. The Florida coast west of these islands was often covered with a dense subtropical shrub called mangrove, which made it difficult to reach the pine forests and sandy soils farther inland.

South of the peninsula lay the Straits of Florida, also called the Bahama Channel, a passage only ninety miles (150 kilometers) wide between the Florida Keys and Cuba. This channel was the route taken by Spain's silver-laden galleons as they sailed in biannual convoys from Vera Cruz, on the coast of Mexico, to Cádiz, the deep-water port for Seville, Spain. Ships had to sail north far enough along the Florida and Carolina coasts to reach the prevailing westerlies and the powerful Gulf Stream, thus to be carried across the Atlantic. To protect their galleons from pirates and storms, including the occasional hurricane, Spain needed to establish coastal forts

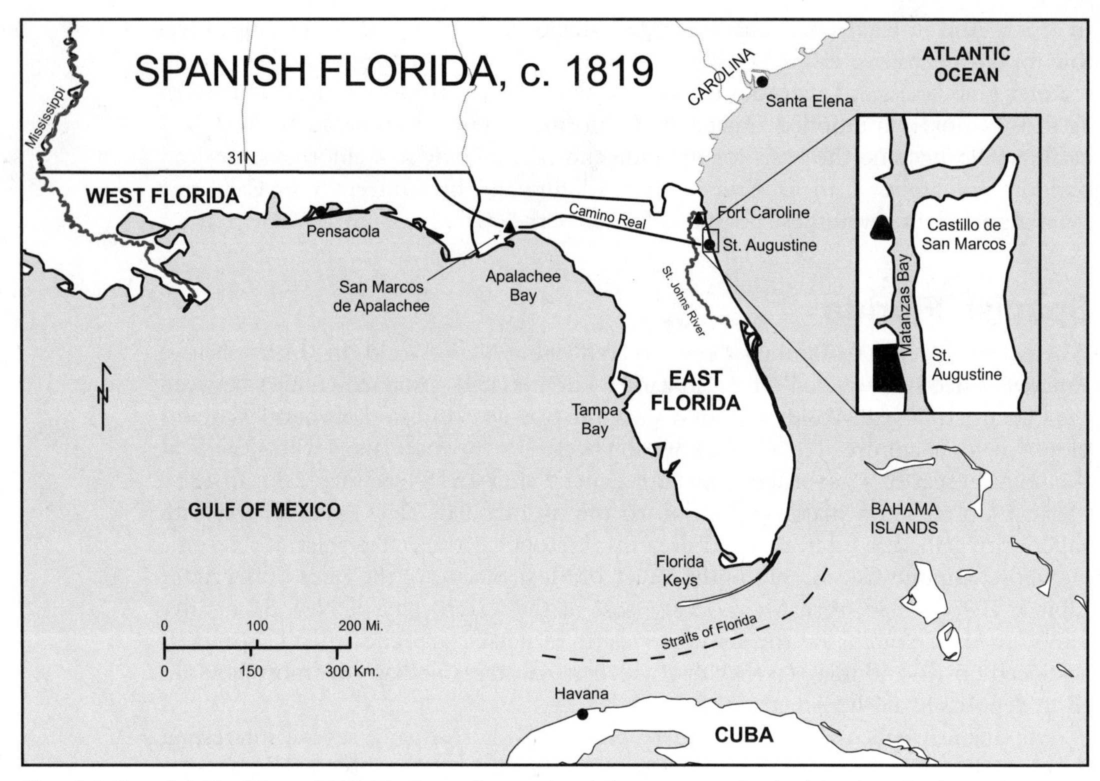

Map 2.2 Spanish Florida, c. 1819. The inset shows a break between two barrier islands at St. Augustine.

and settlements. Spaniards made several attempts to do this near Pensacola and along the Carolina coast, but their efforts were short-lived and nothing permanent happened until St. Augustine.

A threat posed by France, whose New World claims also included Florida, was what prompted the founding of St. Augustine. In 1564 Spaniards learned that France had established Fort Caroline at the mouth of the St. Johns River near present-day Jacksonville (map 2.2). The following year Pedro Menéndez de Avilés led a large expedition from Spain to establish St. Augustine some thirty miles south of Fort Caroline, and Menéndez then cruelly executed most of the French. Within several years St. Augustine's permanent site had been chosen on the mainland behind a break in two barrier islands, and the Castillo de San Marcos, a fort (or presidio) was built to protect the entrance to Matanzas Bay (map 2.2 inset). St. Augustine was laid out as a walled gridiron community with a plaza facing the bay, a Roman Catholic church in the plaza, a governor's house facing the plaza, and eventually some 120 houses and shops within the grid. Menéndez actually governed La Florida temporarily from a second community, Santa Elena, which he founded in 1566. After Santa Elena's abandonment in 1587, the location of its forts and some sixty houses was forgotten—until 1979, when the lumps that kept reappearing under the golf course at the Marine Corps base on Parris Island in South Carolina were excavated.

In St. Augustine, Franciscan missionaries built an *iglesia* (church) and a *convento* (quarters for the friars), and from this headquarters complex they expanded, building missions that were accompanied for protection by presidios. Florida's Indians, including the Timucua who lived near St. Augustine, inhabited villages. At the villages themselves, Franciscans founded iglesias and conventos south along the coast to Tampa Bay on the Gulf coast, behind St. Augustine along the St. Johns River, and in the 1600s across Florida's peninsula along a Camino Real to Apalachee Bay and into present-day Alabama. Ranchos and haciendas awarded by authorities to colonists for cattle raising accompanied the expanding mission-presidio frontier. Franciscan activity reached a high point in 1655, when seventy friars staffed the mission compounds. Between 1680 and 1706, however, almost the entire Florida mission system collapsed, owing to the ceaseless raids of the English and their Indian allies in Carolina. Only St. Augustine and its immediate nearby villages survived; before 1763 St. Augustine's Hispanic population probably never exceeded three thousand.[7]

By 1763 Florida was little more than a military outpost anchored at St. Augustine and dependent on subsidies from the Spanish treasury. In the same year, Spain and France lost the Seven Years War, and Florida became British. The British created two administrative units, East Florida and West Florida (map 2.2). Twenty years later (1783), the Americans defeated Great Britain in the Revolutionary War, and Florida was restored to Spain. Retaining the two British divisions of East and West Florida, Spain reoccupied St. Augustine in East Florida, and in West Florida it reoccupied the earlier-founded settlements of San Marcos de Apalachee and Pensacola, West Florida's capital. Unable to defend Florida against Americans who were bent on controlling, indeed "destroying," runaway slaves, Seminole Indian refugees, and white renegades in northern Florida, Spain ceded Florida to the United States in the Adams-Onís Treaty, signed in 1819 by the ministers of state John Quincy Adams and Luis de Onís. Compared to its Southwestern Borderlands, Spain's impress on Florida after two and one-half centuries was quite modest.[8]

The Southwestern Borderlands

Settlement of the present-day Southwest was not the result of one grand march to the north but of thrusts in four directions over more than a century and a half (see map 2.3). The first thrust went to New Mexico and those sedentary Pueblo Indians reported on by Coronado. Persuaded that there must be mineral riches to be found, wealthy Juan de Oñate, son of an original Zacatecas silver magnate, petitioned the viceroy for permission to lead a private colonizing venture to New Mexico. The majority of the 210 documented colonists recruited in the Mexico City and Zacatecas areas had been born in southern and central Spain, especially in Andalusia and Castile, although Oñate, of Basque parentage, had been born in New Spain. Reaching New Mexico in 1598, Oñate set up headquarters at pueblo villages located at the confluence of the Chama River and the Rio Grande. Oñate explored between the Gulf of California and central Kansas, but like Coronado, he found no mineral wealth. Accused of mistreating the Pueblos, he was recalled in 1606 (he returned briefly in early 1610)—but not before selecting a more defensible site for a new capital at Santa Fe. Formally founded in 1610 by Oñate's successor, Pedro de Peralta, the Villa Real de Santa Fe had a central plaza fronted by a Roman Catholic church and the Palace of the Governors. Like St. Augustine in Florida, Santa Fe would become New Mexico's largest community and its hub of activity.[9]

Following Oñate's recall, New Mexico might have been abandoned if Spanish officials had not been persuaded that the sizable Pueblo population—estimated to be forty-five thousand living in eighty villages as of 1540—needed Christianizing. So Franciscans, some having come with Oñate, in the 1600s established iglesias and conventos right at the villages of the Pueblo, much as they had done in Florida. By 1629 forty-six friars staffed some twenty-five missions, which probably was the high point in Franciscan activity in New Mexico before the revolt of 1680. Meanwhile, officials awarded soldier-settlers land grants, called *estancias* and haciendas, for stock raising; however, unlike in Florida, they also awarded *encomiendas*, which were not

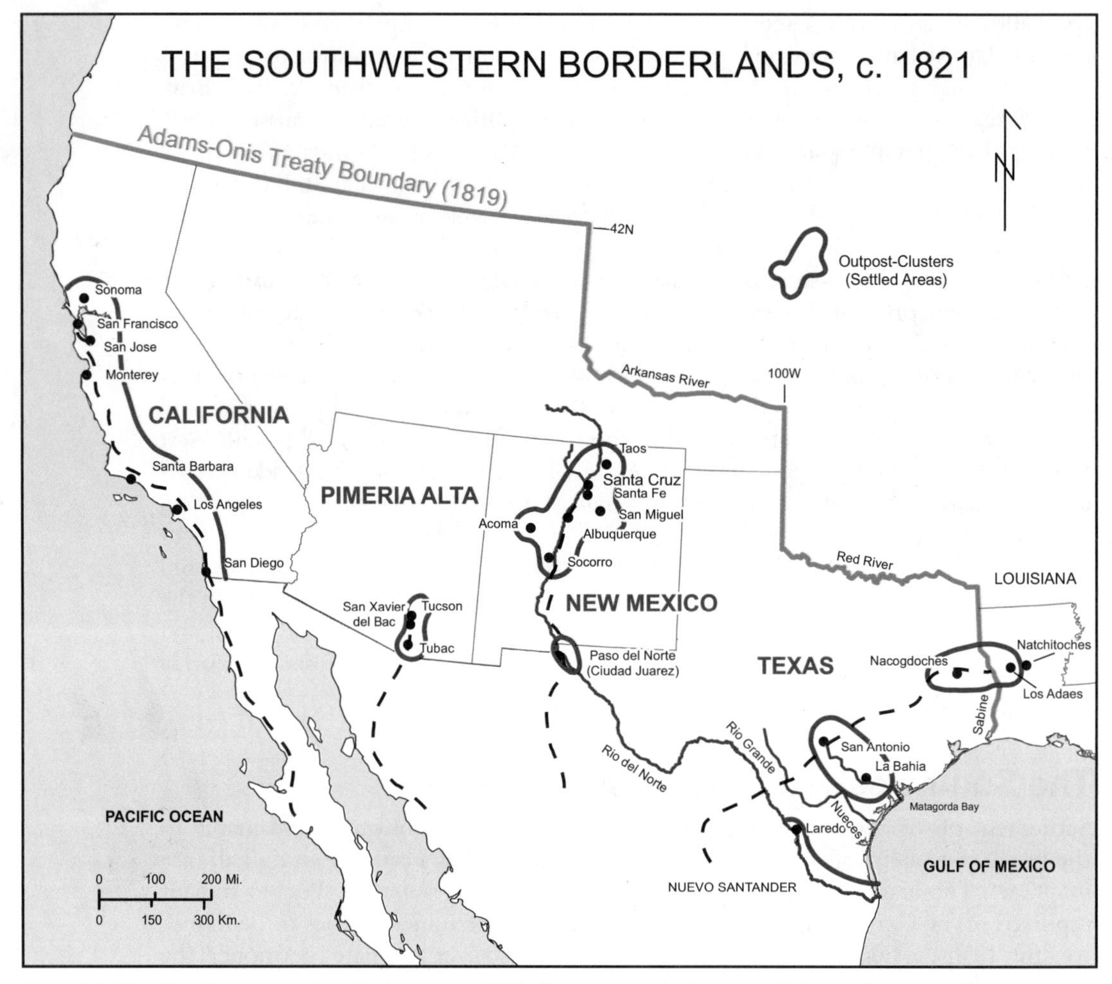

Map 2.3 The Southwestern Borderlands, c. 1821. Shown are missions, presidios, civil communities, or a combination.

land grants but rather "entrustments" of Pueblos living in a given village. The Indians living on encomiendas were often forced by the *encomendero* to serve as laborers, to grow corn, and to weave textile squares called *mantas.* Because of Indian abuse, encomiendas were no longer awarded in New Spain after about 1700, and in the Borderlands they existed only in New Mexico. In 1680, in an unprecedented show of unity, the Pueblo Indians revolted, killing several hundred Spaniards, including friars, and forcing some two thousand survivors to retreat to the Paso del Norte (present-day Ciudad Juárez) district.[10]

Twelve years later (1692), Diego de Vargas led the reconquest of the upper Rio Grande basin, starting at Santa Fe, which had to be taken by force. In the 1700s, Franciscans supervised the rebuilding (with Pueblo labor) of many missions, but now the friars were less intent on proselytizing and stamping out the religious practices of a much-reduced Pueblo people, and they spent more time ministering to a growing Spanish population. The Pueblos and the Spaniards increasingly went their separate ways. To make New Mexico more secure, officials in 1693 authorized the building of a presidio behind the Palace of the Governors (a fort in Santa Fe before 1680 was manned by volunteers, not soldiers, and technically was not a presidio). Officials founded new villas at Santa Cruz (1695) and at Albuquerque (1706), and long lots came to characterize the pattern of agricultural plots as linear villages began to stretch along irrigation ditches on the flat valley floors. During the 1700s raids increased between Spaniards and the nomadic Apache, Comanche, and Navajo Indians, as did the number of captives each side took. Indian captives among the Spanish, known as *genízaros*, became household servants, and their occasional intermarriage with Spanish people produced a mestizo element unique to New Mexico. When the Spanish period ended in 1821, some thirty thousand people of Spanish descent and mestizos lived in an area delimited by Taos, San Miguel, Socorro, and the Pueblo village of Acoma (map 2.3). But at around the same time, Franciscans staffed only five missions at the twenty remaining Pueblo villages.[11]

The second thrust into the Borderlands was to Pimería Alta, the high country of the Pima Indians in present-day southern Arizona (map 2.3). Jesuit missionaries had pushed north along Mexico's Pacific-facing slopes, and in 1700 they founded Mission San Xavier del Bac, south of Tucson. Energetic Father Eusebio Francisco Kino founded additional missions in the area, but following his death in 1711, Jesuit activity languished, owing to the raids of Apache Indians, enemies of the Pima and Spanish alike. Like the Pueblo, the Pima had their own revolt against Spanish subjugation, and in response Spaniards constructed two presidios, Tubac in 1751 and Tucson in 1776. Meanwhile, in 1768, Franciscans replaced the Jesuits; because of overly aggressive Jesuit policies, King Carlos III had expelled all Jesuits from Spain and Spain's New World colonies in 1767. Despite the new presidios, Spanish presence in Pimería Alta advanced and retreated at the pleasure of the Apaches, and by 1821 fewer than one thousand soldier-settlers clustered precariously at the two presidios.

The third thrust was to Texas. In the mid-1680s the French once again triggered the Spanish into action. René-Robert Cavelier, Sieur de La Salle for a short time had a fort on Matagorda Bay (map 2.3). In the early 1690s the Spanish responded by sending Franciscans and soldiers to establish several missions and presidios among the Caddo Indians in East Texas, efforts that were ephemeral. The French appeared again in 1714, when they advanced up the Red River to establish a trading post at Natchitoches, in present-day Louisiana. To counter them, in 1716 Spaniards founded a mission and later a presidio fifteen miles west of Natchitoches at Los Adaes, now called Robeline, Louisiana. Officials made Los Adaes the capital of Texas, and the small Arroyo Hondo running between these French and Spanish outposts became the international boundary. Thereafter, missions and presidios came to exist at Nacogdoches (1716), San Antonio (1718), and La Bahia (1722), the last of which Spaniards relocated from Matagorda Bay to the lower San Antonio River in 1749. At San Antonio, besides its mission-presidio pair (surprisingly, the Alamo was a mission, not a presidio), Spaniards also founded a plaza-centered villa called Béjar in 1718.

San Antonio's site at the base of the Balcones Escarpment had perennial springs that fed the San Antonio River and San Pedro Creek. In 1731 Canary Islanders (*isleños*) arrived to revive Béjar, which they renamed the villa of San Fernando. On the gentle interfluve between the two incipient streams, the isleños set about to irrigate crops, vineyards, and orchards in fields that apparently were laid out as long lots. Within ten miles down the San Antonio River, four additional missions were built (one in 1720 and three in 1731) among the local Coahuiltecan Indians, and ranchos were awarded for stock raising. In 1768 San Antonio replaced Los Adaes as the capital of Texas (a move officially recognized in 1773), and like St. Augustine and Santa Fe, San Antonio became the provincial focus, the seat of political power, and the largest community. Facing one of San Antonio's two nearly adjacent plazas was a Palace of the Governor. During Spanish times the Nueces River, not the Río del Norte (Rio Grande), marked the southern Texas border, and so the founding of Laredo in 1755 (elevated to a villa in 1767) took place in the province of Nuevo Santander, not in Texas. Political upheaval in the decade prior to 1821 saw the tejano population fall from four thousand to two thousand, with the largest concentration at San Antonio.[12]

The fourth and final thrust went to Alta California in 1769 (map 2.3). A relatively dense population of Chumash and other Indians lived as gatherers and fishers along California's coast. Neither Spanish horses nor guns had diffused to this isolated corner of the Borderlands. Under Fray Junípero Serra, the able Franciscan leader in California from 1769 until his death in 1784, Franciscans made California into a largely missionary province. In well-chosen valleys along California's coast, Franciscans built their mission compounds, which included *reducciones*, as the communities for the Indian recruits were known. With ample Indian labor at their disposal, they produced surpluses of grain and cattle hides that they shipped to Mexico, and with the profits they bought clothing, tools, luxury goods (chocolate and snuff), and trade goods that enabled them to recruit still more Indians. Father Serra and his successors were successful in stifling civilian population growth and military buildup; for example, officials awarded only some thirty stock-raising ranchos by 1821. Thus, late in the Spanish colonial era, Franciscans controlled an isolated and prosperous mission frontier made possible by Indian labor that, unlike other parts of the Borderlands, could be renewed when diseases carried off many in the reducciones.[13]

In their final Borderlands thrust, then, Spaniards converged by land and sea at San Diego, where in 1769 they built a mission and a presidio. Thereafter, they founded three more such pairs at Monterey (1770), designated the capital; San Francisco (1776), whose bay was first "discovered" by an overland party and not through the Golden Gate; and Santa Barbara (presidio in 1782, mission in 1786) along California's only east-west stretch of coast. Altogether, twenty missions were founded before 1821 in the Spanish period, and Sonoma (1823), established early in the Mexican period as the most northerly mission, added the final link in the concatenation. Felipe de Neve, California's able governor between 1775 and 1782, in responding to the need for a civil population to supply foodstuffs for soldiers at the four presidios, took charge of the founding of San Jose (1777) and Los Angeles (1781), civil communities strategically placed between presidio pairs. Los Angeles grew to be California's largest community, even larger than the capital at Monterey, which was uncharacteristic, but still only a reported 3,200 people of Spanish descent inhabited California when Mexico took over in 1821. California's Indian population declined from an estimated three hundred thousand in 1769 to an estimated two hundred thousand in 1821.[14]

By 1821, then, on a distant northern frontier—the Southwestern Borderlands—Spanish settlement took the form of four outpost-clusters. New Mexico was by far the oldest (1598) and largest (thirty thousand), Pimería Alta (1700) was but a tiny foothold (less than one thousand), Texas (1716, at Los Adaes) apparently had recently shrunk (only two thousand), and a quite recent (1769) California (three thousand). Each of these outpost-clusters was remote from Mexico City and only tenuously connected to it along royal highways (and by ship to California), and on that frontier, each was totally isolated from the others. Because Spain maintained a closed frontier, at least until late in its rule when it initiated an *empresario* policy, any contact with non-Spaniards took place illegally, which only reinforced isolation while also stifling private enterprise. Just before 1821, in the Adams-Onís Treaty signed in 1819, a new border was drawn well north of these four outpost-clusters (map 2.3). It followed a line between the Pacific Ocean and the Gulf of Mexico along latitude 42° and then south to the headwaters of the Arkansas River, along the Arkansas to longitude 100°, and then south to the Red River, which it followed until striking south to the Sabine River and the gulf. Important because it defined what the United States had purchased from France in 1803, the Adams-Onís boundary would increasingly be breached by Anglos in their aggressive westward movement.

The Spanish Impress

Wherever Spaniards colonized in the New World, they employed settlement institutions that were prescribed in the Laws of the Indies, first codified in 1573. The prescribed settlement institutions gave a rigidity and a uniformity to Spanish facilities and layouts not found where French and British peoples settled. In the vanguard on a new frontier were the mission and the presidio, institutions of Christianization and conquest that the Spanish crown would eventually phase out. Then came the civil community, the mainstay of settlement. Designated as *pueblos* (places), *villas* (villages), and *ciudades* (cities), according to their projected size and importance, civil communities were to contain the farmers and artisans. Also to be permanent were various kinds of land grants awarded mainly for stock raising. Down to the present day, vestiges of these institutions persist to shape the cultural landscape of the Borderlands and to make the region distinctive.

In historical geography, the idea that a cultural landscape is handed down from earlier generations is embodied in the *doctrine of first effective settlement.* Geographer Wilbur Zelinsky is often credited with the doctrine, but Zelinsky acknowledged that others before him articulated the same concept but in different words. The doctrine states that the first people to effectively colonize an area will be of singular importance in shaping the cultural landscape of that area, no matter how small that group may have been or if it was later dislodged or submerged by other groups. A corollary to this doctrine, called *geographical persistence*, underscores that those tangible things that people build, and the ways in which people organize their space, will outlast the builders. Examples of the role Spaniards played in shaping today's cultural landscape of the Borderlands are numerous.[15]

Take, for example, the mission. In the Borderlands, Franciscans took charge of founding missions, although Jesuits preceded them briefly in Florida (1566–1572) and in Pimería Alta (1700–1767). In Florida and New Mexico, as we have seen, Franciscans established their churches and friaries at the Indian villages themselves, while in Pimería Alta, Texas, and in California, they built compounds to which the Indians were attracted. Figures 2.2 and 2.3 illustrate both situations: the Zuñi village and church, in New Mexico, and Mission Santa Inés, in California. That California had twenty-one missions is well known; in New Mexico there must have been fifty missions, and in Florida at least thirty. These missions, especially those with a presidio nearby, became the nuclei around which many communities grew, like San Diego and San Francisco. Many Borderlands missions crumbled away and their sites have been forgotten, but where they survived they inspired a "Mission Revival" architectural style, especially in California. All twenty-one of California's missions have been restored and draw many thousands of tourists annually. Significantly, in San Antonio, the Alamo, with three million visitors a year, has become the number-one tourist destination in Texas.[16]

Although outnumbered by missions, presidios also left their impress. These garrisoned fortresses were massive structures built as rectangles, containing ramparts (thick walls) surmounted by parapets (short outer walls) flanked in opposite corners by bastions (towerlike fortifications). Like many missions, some presidios have disappeared. In Santa Fe apparently nothing remains of the walled presidio that once

Figure 2.2 Roman Catholic Church and Zuñi Village, New Mexico, 1873. The church was in ruins when it was photographed by Timothy H. O'Sullivan. Franciscans worked among the Zuñi between 1629 and 1680 and again from the early 1700s to 1821.

"Old Mission at Zuñi, 1873," cat. no. 61.13.802, from the Joseph Imhof Photograph Collection. Courtesy of the Maxwell Museum of Anthropology, University of New Mexico.

Figure 2.3 Mission Santa Inés, California, Latter Nineteenth Century. Built in 1804 in present-day Santa Barbara County. Behind the colonnade was a compound to which Indians had been attracted by Franciscans.

Photograph by William Henry Jackson, courtesy of the Denver Public Library, Western History Collection, WHJ-10004.

extended north behind the Palace of the Governors. In contrast, the Castillo de San Marcos near St. Augustine is a restored presidio gem made all the more important because it displays the local use of coquina or tabby, a whitish lime mixture made from shells and sand that was employed as a mortar. Also carefully restored is the rural Presidio La Bahia at Goliad on the lower San Antonio River in Texas. Peering over a parapet while standing on a rampart at this presidio, one can see Mission La Bahia, also restored, on the opposite bank of the river some two miles away—a distance, one is told, that would hopefully keep soldiers separated from the mission's Indian women.[17]

Civil communities were to be lasting institutions. As prescribed in the Laws of the Indies, they were grants of land of four square leagues, or 27.56 square miles. Near the centers of the grants, a village was to be platted in a gridiron. The corners of the blocks of the grid were to point in the four cardinal directions, as explained in the Laws of the Indies, to deflect the four principal winds—most notably, the cold wind blowing out of the north. This orientation would also have found houses casting longer shadows into the streets, an important temperature consideration in the tropics and subtropics. A central plaza was to take up four blocks of the grid. Facing the plaza were to be the Roman Catholic church, elevated so as to be more venerated; the palace of the governor (if the seat of government); perhaps a customs house, an arsenal, and a granary; and the *solares* (house lots) of citizens. And beyond the village, the four-square-league grant was to be apportioned to *ejido* (common land), *suertes* (agricultural fields), *dehesa* (pasture), *bosque* (woodland), and *realenga* (royal land) reserved by the crown for future assignment.[18]

The lasting impress of civil communities is remarkable. Best known among the civil communities in the Borderlands are St. Augustine (founded in 1565, even before the Laws of the Indies were codified), Santa Fe, Albuquerque, San Antonio, and Los Angeles. Each has a central plaza (San Antonio actually has two plazas, one civil and one military). Located on or facing each plaza is a Roman Catholic church (St. Francis Cathedral in Santa Fe faces a half of the plaza that in the Spanish era was encroached upon by buildings). Seats of government, such as St. Augustine, Santa Fe, and San Antonio, all have a palace of the governor (see figure 2.4). And beyond

Figure 2.4 Palace of the Governors, Santa Fe, New Mexico, c. 1885. This one-story adobe structure faces the plaza.

Photograph by Dana B. Chase, courtesy of the Denver Public Library, Western History Collection, Z-4144.

Map 2.4 Los Angeles Street Pattern, c. 2000. Locate the Pueblo of Los Angeles plaza about one-half mile west of the Los Angeles River, and from the plaza measure 2.625 miles—the length of one Spanish league—in each cardinal direction. At those distances, Henry Hancock in 1858 surveyed the pueblo's four-square-league (27.56-square-mile) boundaries in lines that followed cardinal directions. These boundaries are preserved in the street pattern, most notably along Hoover and Indiana streets. Notice how streets in the Spanish grid, which are oriented at a bias to the cardinal directions, "straighten out" at the American-era pueblo grant boundaries. (The shift is not perfect—see Olympic and Venice boulevards where they intersect with Hoover Street.) The Pueblo of Los Angeles street pattern as confined by the Hancock survey boundaries illustrates nicely the doctrine of first effective settlement.

each plaza, streets have at least a semblance of a grid pattern. In the case of Los Angeles, the shift in the grid pattern from the Spanish orientation (corners of blocks pointing in cardinal directions) to the American layout (where the streets themselves point in cardinal directions) is especially clear along Hoover Street (El Pueblo de Los Angeles west boundary) and Indiana Street (east boundary; see map 2.4).

Finally, Coronado's Trail has become a timely example of the Spanish impress. In 1986, in anticipation of the Columbus Quincentennial celebration in 1992,

Congress asked the National Park Service (NPS) to identify for tourists the route taken by Coronado on his major expedition through the Southwest from 1539 to 1542. After several years of study, the NPS responded that, except for Coronado's known encounters with the Zuñi and Pecos Indians at their villages and their known winter stay in the Bernalillo-Albuquerque area, so little had been documented about where Coronado traveled that it would be impossible to map his route. Since well before 1992, Richard and Shirley Flint, Coronado scholars and research historians at the University of New Mexico and experts in analyzing sixteenth-century Spanish documents, have galvanized efforts to learn more about just where Coronado did go. They and their colleagues have put on conferences, given lectures, and hosted "traveling road shows" to seek possible public input. And their efforts have paid off: Artifacts have been found that include strange-looking "caret [^] head" nails used perhaps in shoeing horses; distinctive copper crossbow dart tips; and lead balls shot from firearms—all of which point specifically to Coronado's travels half a century before Spanish colonists arrived. Now, a number of new campsites can be mapped that specifically identify Coronado's Trail. The day is nearing when people will have a map showing Coronado's Trail between eastern Arizona and central Kansas.[19]

Notes

[1] That Spain "held sway" in the Borderlands is from Bolton, *The Spanish Borderlands*, vii. That Bolton did not coin the term "Spanish Borderlands" is noted by Bannon, *Herbert Eugene Bolton*, 120.

[2] The examples of crops and animals in Florida are from Weber, *The Spanish Frontier in North America*, 315–16.

[3] Ibid., 303 and 316, for crop and animal adaptations in New Mexico; 310 (citing Simmons, *Albuquerque: A Narrative History*, 114), for sheep being slow and not easily stampeded.

[4] Carlson speculated that long lots were developed by Spanish officials in New Mexico's Rio Arriba in the mid-1700s, in "Long-Lots in the Rio Arriba," 53. Jordan thought that long lots in San Antonio in 1731 were likely an independent invention, and he attributed those laid out in 1767 on the lower Rio Grande to diffusion from Louisiana, in "Antecedents of the Long-Lot in Texas," 82. For Roth on diffusion to New Mexico, see "Long Lots in New Mexico and Texas," 55–77.

[5] For use of *acequia* in South Texas and the El Paso District, see Jordan-Bychkov and Domosh, *The Human Mosaic*, 118. For the years in which plants were introduced to Southern California, see Raup, "Transformation of Southern California," 71.

[6] Much here on Florida and the Southwest I draw from Nostrand, "The Spanish Borderlands," 47–63.

[7] See Weber, *The Spanish Frontier in North America*, 95 and 100, for Jesuits in Florida; 100, for Franciscan numbers (seventy in 1655); 101, for Timucua villages; 107, for Franciscan iglesias and conventos; 122 and 144, for the English in Carolina; 183 and 199, for St. Augustine's population.

[8] Meinig develops Spanish Florida's evolution in *The Shaping of America*, 1:191–93, 311, 332–38; and 2:24–32. Ibid., 2:31, for some Americans' desire to "even utterly destroy some specific peoples of Florida."

[9] I analyze the places of birth of Oñate's 210 documented colonists (130 in 1598 and 80 in 1600) in Nostrand, *The Hispano Homeland*, 27–29. Weber notes Oñate's choice of "a more defensible site in an unpopulated valley" for Santa Fe in *The Spanish Frontier in North America*, 78 and 90 (quote). Simmons concludes that Oñate deserves at least partial credit for the founding of Santa Fe in *The Last Conquistador*, 182–83, 195.

[10] For forty-five thousand Pueblos living in eighty villages about 1540, see Spicer, *Cycles of Conquest*, 155. For the presence of forty-six Franciscans in 1629, see Nostrand, *The Hispano Homeland*, 53.

[11] See Nostrand, *The Hispano Homeland*, 4 and 24, for New Mexico's nomadic Indian admixture; 44, 63–64, for genízaros; 58, for Franciscans staffing only five of twenty missions in 1832.

[12] See Weber, *The Spanish Frontier in North America*, 163, for San Antonio's site; 193, for Canary Islanders; 194, for Laredo; 211, for moving the capital to San Antonio; 299, for the population drop from four thousand in 1803 to two thousand in 1820. Weber notes (ibid., 306) that all five of San Antonio's missions were secularized between 1793 and the 1820s; Mission San Antonio de Velaro came to be called the Alamo after secularization.

[13] Ibid. 258–65; Weber's reinterpretation of Alta California under Franciscans is superb.

[14] Ibid., 263, for Indian numbers (three hundred thousand in 1769 and two hundred thousand in 1821); 265, for 3,200 Hispanics in 1821. Geographer R. Louis Gentilcore analyzes the

microenvironments and economies of California's twenty-one missions in "Missions and Mission Lands of Alta California," 46–72.

[15] Zelinsky articulated the doctrine of first effective settlement in his *The Cultural Geography of the United States*, 13–14, 20, 34. Both Meinig, *The Shaping of America*, 1:221, and Conzen, ed., *The Making of the American Landscape* (1990), 245, credit Zelinsky with the concept. Meinig, however, notes that John Porter had earlier enunciated the same concept with reference to Canada. Fred Kniffen ("Folk Housing," 551) spoke of "initial occupance," for which he credited W. G. McIntire, who stated, "The first . . . permanent settlement . . . [was] long lasting." I asked Zelinsky in correspondence (March 21, 2000) whether Kniffen had somehow inspired his own doctrine. His response (March 30, 2000) was that Kniffen "could claim priority" because the concept was "self-evident" and "axiomatic." Zelinsky added that he had been "rather embarrassed" for the amount of attention he had received for the concept, and that Kniffen had been "a bit annoyed" that Zelinsky had "stolen his thunder."

[16] See Weber, *The Spanish Frontier in North America*, 95, for dates Jesuits were active in the Borderlands; 343, for the origination of Mission Revival style in California in the 1880s. Geographer David Hornbeck's diagram of a California mission circa 1820 is insightful; see "Spanish Legacy in the Borderlands," 58. See Bryce, "Critics Remember Alamo Differently," 1, for the Alamo as the leading tourist attraction in Texas.

[17] Historian Max L. Moorhead, *The Presidio*, v, defines the presidio as a "garrisoned fortification." Inside the heavily gated rectangle were typically a chapel, guardhouse, barracks, corrals, and the House of the Captain.

[18] Zelia Nuttall, "Royal Ordinances," 249–54, translates the twenty-eight laws (ordinances 110–37) from the Laws of the Indies that relate to founding civil communities. Ordinance 114 states that the *corners* of the plaza and of other blocks in the grid were to point in the cardinal directions "because thus the streets diverging from the plaza will not be directly exposed to the four principal winds, which would cause much inconvenience" (250). Geographer Dan Stanislawski discusses grid-pattern towns in two articles, "The Origin and Spread of the Grid-Pattern Town," 105–20; and "Early Spanish Town Planning in the New World," 94–105.

[19] For a review of the major advances in our knowledge of the Coronado Expedition since the mid-1980s, see Flint and Flint, eds., *The Latest Word from 1540*, 1–9.

New France 3

New France embraced a large part of North America, but its focus was a crescent-shaped area that stretched from the mouth of the St. Lawrence to the mouth of the Mississippi (see map 3.1). This well-watered and largely forested land reached from the subarctics to the subtropics. Continental glaciers had reworked landforms south to about the Missouri and Ohio rivers, leaving behind troughs filled by the Great Lakes and other numerous waterways of which the French made heavy use. We begin with "Acadia and Canada," two of four distinct areas of New France that, although not within the present-day United States, were central to French activities to the south. We then turn to "Louisiana" and "Illinois Country" in the present-day United States. A discussion of "Long Lots," the hallmark of the French landscape impress, ends the chapter.

Acadia and Canada

French people went to Acadia and Canada mainly from the northwest part of France. Normandy supplied the largest numbers before about 1663, and its port of Rouen on the Seine River led as the point of embarkation (see map 3.1 inset). La Rochelle, located on the central coast of France, then surpassed Rouen as the leading port through which emigrants departed, and it and its hinterland became the main French source area after about 1663. Paris also contributed a sizable number of emigrants, especially single women from orphanages and poorhouses. An estimated five hundred French people went to Acadia during its French regime, and some ten thousand went to Canada before 1763. Most who did so were young men; French women were always in short supply. Moreover, many came as *engagés* (committed to sponsors) or as released soldiers. On the whole, these emigrants represented the lower socioeconomic strata of French society, and many were landless and nearly destitute. Perhaps only one-quarter had had farming experience, the occupation most would follow in Acadia and Canada.[1]

Acadia was the French name for the present-day Maritime Provinces. The focus of French Acadian activity was the Bay of Fundy, particularly the bay's southern margins (map 3.2). Attracting French people were tidal marshlands with their potentially productive soils and an absence of forest that would need to be cleared. The problem to overcome, however, was the Bay of Fundy's extraordinarily high tides. The bay is a cul-de-sac that is exposed to the north-flowing Gulf Stream. Twice a day in the bay's upper reaches in the Minas Basin, normal tides rise and fall over forty feet (fourteen to fifteen meters), and in the Annapolis Valley they rise and fall fifteen feet (8.5 meters). The tidal range at Halifax, by contrast, is only a generous meter (four feet). It took major efforts on the part of relatively few French Acadians working as extended families to reclaim in increments small units of grass-covered marshlands from the surging and ebbing salt water. Acadians constructed log and sod dikes at tidal openings, dug surface drainage canals in the floodplains, and built one-way sluice gates to regulate the in- and outflow of salt and fresh water. Their efforts paid off once the soils had freshened up, as early as the second year.[2]

French activity in the Bay of Fundy begins in 1605 with the French building a fort some eight kilometers (five miles) downstream from Port Royal in the Annapolis Valley (map 3.2). Geographer Karl W. Butzer reports on how Isaac de Razilly (1587–1635) introduced engagés to La Hève in 1632, how nearly all these French people returned

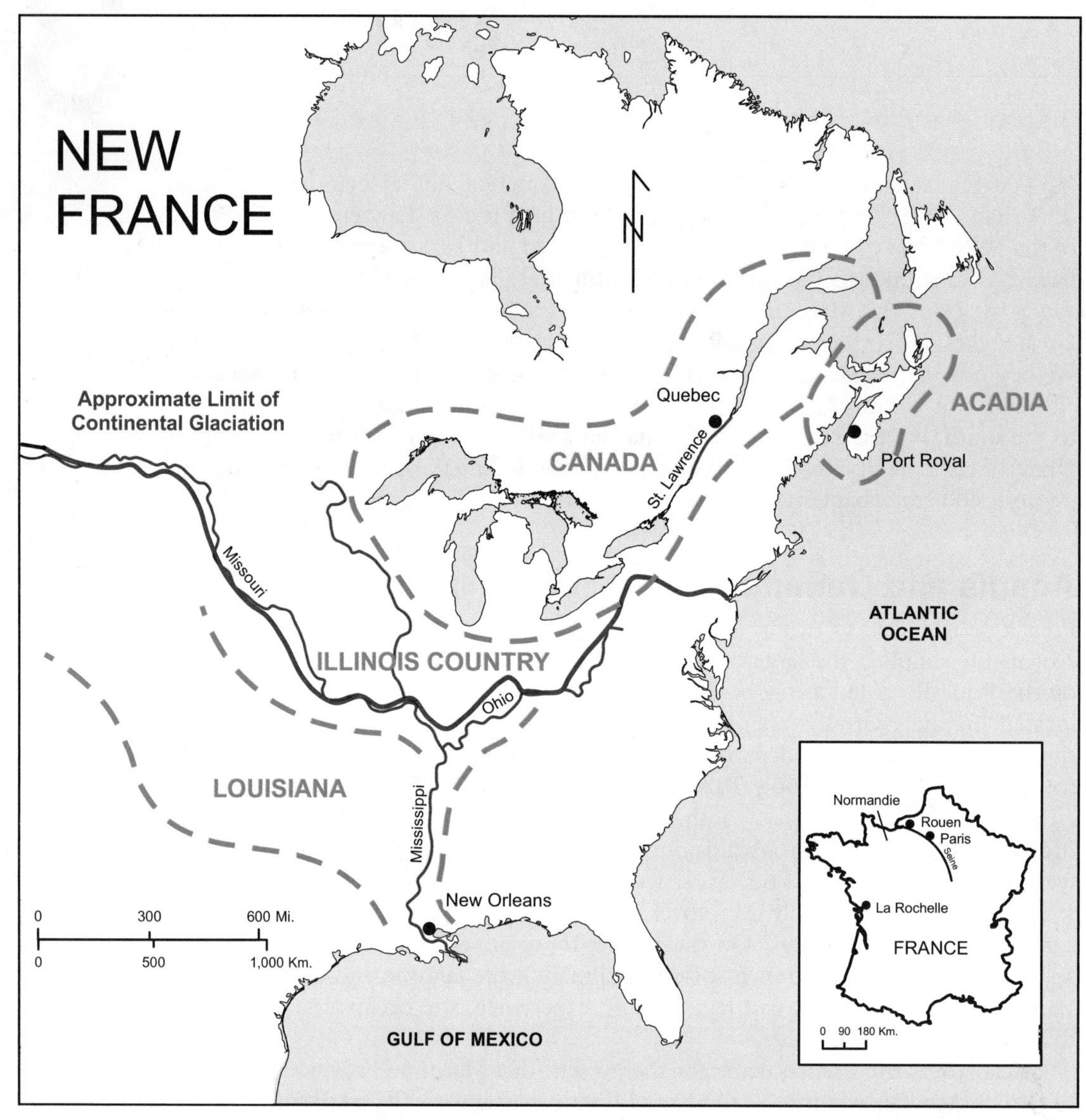

Map 3.1 New France. Boundaries for Acadia, Canada, Louisiana, and Illinois Country are dashed to be suggestive.

to France, and how a small number led by Charles de Menou d'Aulnay (c. 1604–1650) moved to Port Royal between 1635 and 1640. A very few Acadians expanded their reclamation efforts beyond the Annapolis Valley in the Beaubassin in 1671, and in the Minas Basin a decade later—both located at the upper reaches of the Bay of Fundy. In reclaimed fields Acadians grew wheat, their preferred grain, peas, and hay, and they raised cattle and sheep. Most Acadians were farmers, and a few engaged in fur trade with the local Mi'kmaq Indians. There was some lumbering, but there was no commercial fishing in the ice-free yet tide-plagued Bay of Fundy.[3]

From the beginning, the English also laid claim to French-settled Acadia. In a coastal trade based in Boston, they sailed into the Bay of Fundy to exchange their sugar, molasses, and brandy for Acadian wheat and livestock. And on three occasions Great Britain seized control of Acadia—in 1613, again between 1654 and 1670, and finally and permanently in 1710. France recognized British rule in the

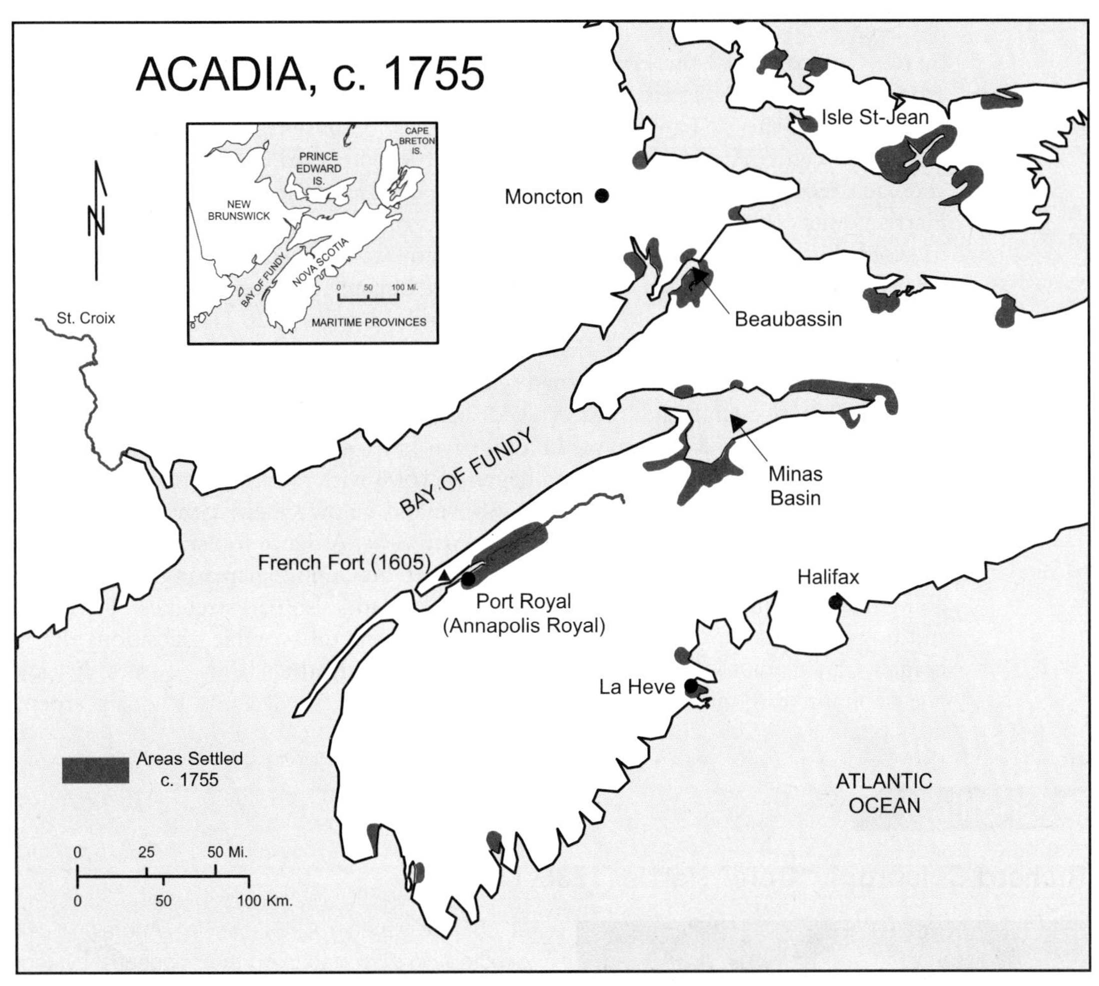

Map 3.2 Acadia, c. 1755. The Bay of Fundy was the heart of Acadia, and Port Royal was its capital.

Source: Meinig, *The Shaping of America*, 1:272, for areas settled c. 1755.

Treaty of Utrecht, signed in 1713 at the end of the War of Austrian Succession. Peninsular Acadia then became Nova Scotia, and Port Royal was renamed Annapolis Royal. Beginning in 1755, by which time the French Acadian population had grown through high birth rates to ten thousand, the British forcibly and tragically rounded up and deported the vast majority of the Acadians in a movement known as *le gran dérangement*. The diaspora, D. W. Meinig writes, "touched in some degree nearly every part of the English and French Atlantic worlds." Acadians were shipped to New England and Virginia and from there to England, France, and elsewhere; others fled to the wilderness north of the Bay of Fundy, to Isle St-Jean (Prince Edward Island), and to Quebec. After 1764, some Acadians were allowed to return, but they had to settle new districts because the English now occupied their farms. During this long time of disruption some four thousand Acadians reached Louisiana where, to this day, they retain their Acadian cultural identity as "Cajuns."[4]

During Acadia's French regime officials introduced the *seigneurial* system, a carry-over from feudal times for awarding land in return for annual rents and charges. But in Acadia the system was weak and left few traces on the landscape. By contrast, in the rest of Canada the seigneurial system helped shape a distinctive landscape, although geographer Cole Harris (see spotlight) cautions that its importance has been overstated. *Seigneurs* were individuals who were awarded land in Canada by the Company of New France between 1627 and 1663 and by the French government after 1663. Their land grants, called *seigneuries*, fronted on and were at right angles to the St. Lawrence and its tributaries, because travel was by river (water and ice). Seigneurs introduced *censitaires* (those who paid a *cens*) to their grants, and each received without charge a farm, called a *roture*, in return for certain fees. Harris points out that in France a censitaire was broadly equated with a peasant, and to avoid this pejorative association, *habitant* came to be used early in Canada as the term for the owner of a small farm. Altogether during the French period some 250 seigneuries were awarded along some two hundred miles (320 kilometers) of river. Because each habitant wanted river frontage, seigneurs awarded farms in narrow, ribbonlike long lots that were aligned with the seigneuries to stretch back from the river. A distinctive landscape thus evolved: Farmsteads that were located at the river-ends of farms arranged themselves in long rows known as *côtes* (lines ["coasts"]).[5]

Permanent settlement of Canada began in 1608 with the founding of Quebec by Samuel de Champlain (c. 1570–1635). Above and below Quebec, the St. Lawrence lowland stretched along both river shores (map 3.3). A dense forest, composed primarily of broadleaf trees around Montreal but becoming increasingly coniferous going downstream, covered this lowland. The lowland also had stretches of marshes and bogs and poorly drained soils, the result of some unfavorable glaciations. Early French Canadians took on the arduous task of clearing their lands of trees; it took one habitant roughly one year to clear about two arpents of land (1 square arpent

Spotlight

Richard Colebrook "Cole" Harris (1936–)

Cole Harris leading an International Conference of Historical Geography field trip, West Vancouver, British Columbia, August 19, 1992. Photograph by the author.

Born in Vancouver, British Columbia, Cole Harris earned his BA at the University of British Columbia (UBC) in 1958, and his MS (1961) and PhD (1964) at the University of Wisconsin. Andrew Clark directed his dissertation on the seigneurial system in Canada. Harris taught historical geography at the University of Toronto (1964–1971) and at UBC (1971–2001). Two of his many academic honors include a Guggenheim Fellowship (1968–1969) and election as president of the Canadian Association of Geographers (1985–1986). His books and articles have greatly enhanced our knowledge about the varied aspects of the lives of ordinary French people, especially in Canada before 1763. In more recent years Harris has focused on the encounter between Native Americans and Europeans in his home province of British Columbia. He notes in his *The Resettlement of British Columbia* that his boyhood memories of his grandfather's ranch in remote southeastern British Columbia are for him the "western Canadian counterpart of an ancient French village." It is fair to say that Cole Harris, for all his scholarship, is historical geography's Canadian counterpart to D. W. Meinig in the United States.

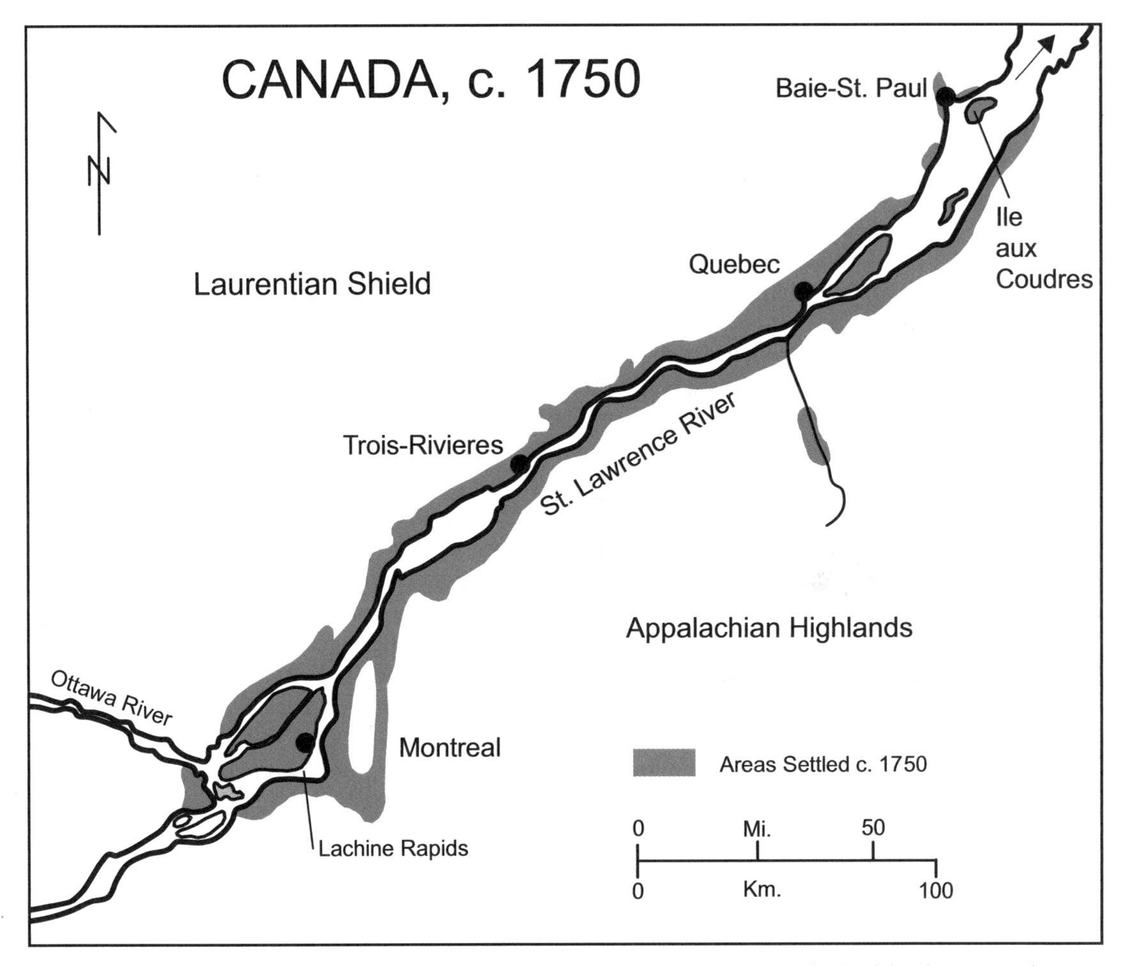

Map 3.3 Canada, c. 1750. The St. Lawrence Lowland—wider and less isolated south of the river—was the focus of Canada.

Source: Meinig, *The Shaping of America*, 1:112, for areas settled c. 1750.

equals about 0.85 acres). A typical roture in Canada's French regime had only some twenty-five acres of cleared land, about half of which was in crops and half in pasture. Progress in clearing land was slow because in Canada, unlike in France, land was cheap but labor was expensive. In Canada, moreover, markets for any agricultural surpluses were local and limited. Farming thus developed slowly along the St. Lawrence lowland, and visitors remarked that farms appeared "ragged and unkempt."[6]

Habitants in Canada, as in Acadia, grew their preferred wheat, and during the French period wheat comprised approximately three-quarters of all Canadian land in crops. However, the St. Lawrence lowland is located on the extreme northern climatic margin for wheat cultivation. Around Montreal, which has a growing season of 150 frost-free days (and some of the lowland's best soils), wheat did fairly well. But at Quebec, with only 130 frost-free days, and below Baie-St. Paul, where the frost-free days numbered barely 100, wheat was often frost-damaged. As a result, hardy crops such as barley, oats, and rye replaced wheat downstream from Quebec. Farmers in Canada sold their surplus wheat, beef, and butter in Quebec and Montreal. Winters in Canada were even more severe than in Acadia: In Quebec and Montreal, the January mean temperature was only ten degrees Fahrenheit (minus-eight degrees Celsius), fifteen degrees Fahrenheit lower than around the Bay of Fundy (and twenty-five degrees Fahrenheit below that in Paris). Pear, peach, and walnut trees did not survive Canada's severe winters. Ironically, because firewood was abundant in Canada, Canadian habitants may have stayed warmer in their sturdy, steep-roofed, Norman farmhouses than did their peasant counterparts in rural France.[7]

Along the St. Lawrence and its tributaries, then, a distinctive rural landscape evolved. Rows of whitewashed Norman farmhouses lined the rivers and their parallel paths or roads. Behind the houses, as one looked inland from the river, were wooden barns, stables, and sheds. Barns in Canada were larger than in France because of the need to store hay over the long winter. Beyond the farmsteads were the cleared, albeit rough, fields and pasture. And beyond them was forest that stretched onto the Laurentian Shield to the north or the Appalachian highland to the south. The string of farmhouses was occasionally broken by a "knot" where a hamlet contained a Roman Catholic church, conspicuous for its steeple, and perhaps a gristmill and a blacksmith. Only three urban centers evolved in the French regime: Quebec (1608), the capital, located at the "head of navigation" for deepwater ships, with its constricted lower town of commercial warehouses and its more expansive upper town containing churches, government buildings, and high society; Montreal (1642), the fur trade entrepôt, located at the head of navigation for smaller ships and boats, also with its lower and upper towns; and between them Trois-Rivières (1634), which remained but a village, bypassed by boats carrying major trade. In 1763, when France lost Canada to Great Britain at the end of the Seven Years' War, an estimated seventy thousand Canadians lived along the St. Lawrence, and at least 90 percent of them lived within one mile of its shore.[8]

As they worked their individual rotures in the St. Lawrence lowland, French Canadians before 1763 had minimal contact with the outside world. In isolation, certain cultural forms came to be preserved while French culture evolved in France and elsewhere. Interestingly, in much the same way, now-archaic cultural attributes also persisted among Spanish Americans in New Mexico after 1848. A comparative study between these two peoples would reveal some striking similarities. For example, Canada and New Mexico, the core areas of their Latin-derived societies, when forcibly combined into English-speaking-majority societies (in 1763 and 1848), each had some sixty thousand to seventy thousand people. Each had spoken languages characterized by archaic vocabularies, expressions, and pronunciations. New Mexico and presumably also Canada had folklore examples—tales, riddles, proverbs, superstitions—no longer current beyond those areas. In New Mexico, archaic given names such as Esquipula, Secundino, Onofre, and Belarmino were still in use in 1900. On the isolated L'Ile-aux-Coudres near Baie-St. Paul, geographer Roderick Peattie observed two-wheeled carts that were introduced from Normandy, where they had been preferred by some to the more heavily taxed four-wheeled carts still in use in the summer of 1916. On the other hand, one difference between French Canadians and Spanish New Mexicans is that Canadian influence spread widely through the North American continent, especially through fur traders, while New Mexican influence spread in minor ways only to California, reaching it over the Spanish Trail in the Mexican period (Chapter 15).[9]

Louisiana

Unlike the St. Lawrence River, which flowed through a deep glacial trough, the Mississippi meandered in a shallow channel choked with bars and banks. Recognizing the Mississippi's navigational difficulties, especially in its delta, French leader Pierre Le Moyne d'Iberville (1661–1706) approached the river artery through Lake Pontchartrain (see map 3.4). From forts at Biloxi (1699) and Mobile (1701),

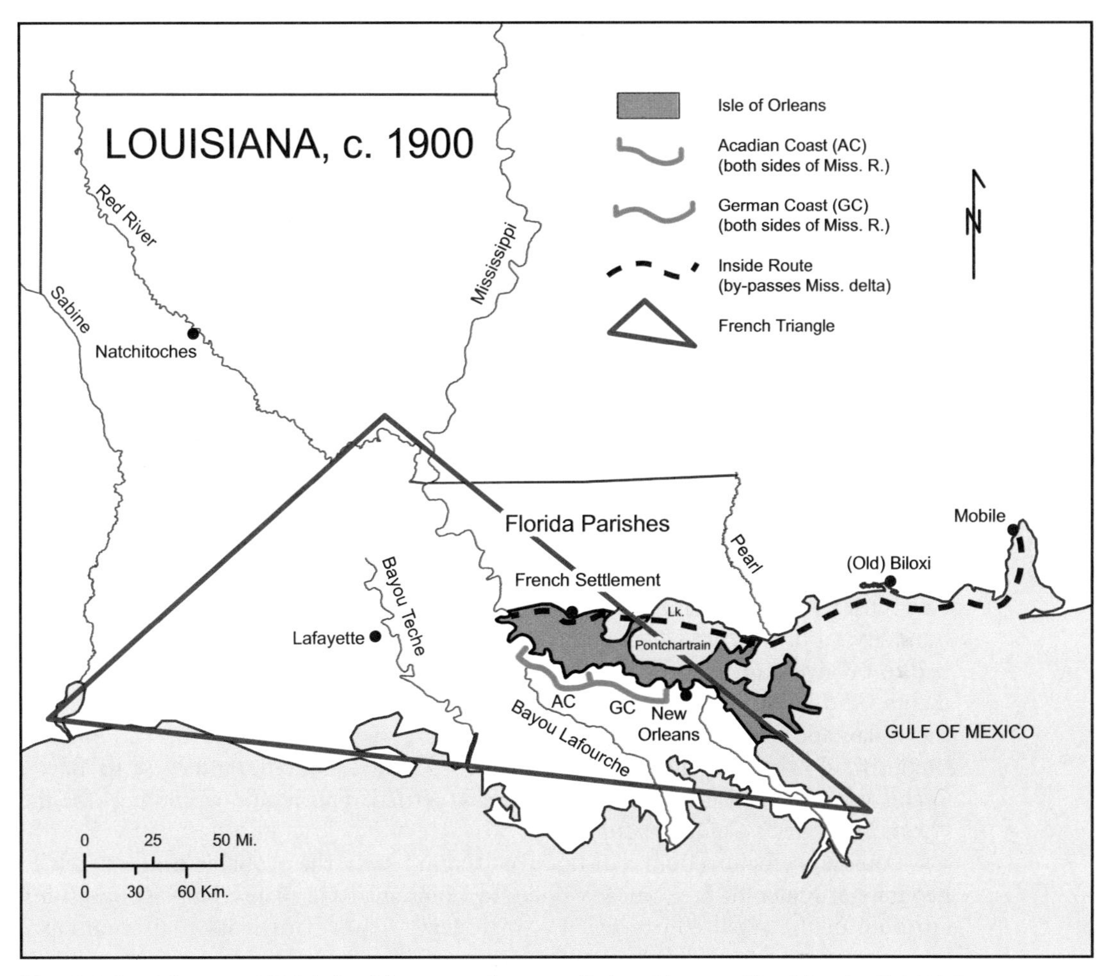

Map 3.4 Louisiana, c. 1900. In Louisiana, counties are called parishes, and Louisiana's official highway map calls the French Triangle its "Acadiana" parishes.

Source: Meinig, *The Shaping of America*, 1:196 and 2:8, for the "inside route" and Isle of Orleans.

Iberville's small boats went behind coastal bars, through Lake Pontchartrain, and up Bayou Manchac to reach the Mississippi, but nothing permanent came of it. Instead, Natchitoches, a post built in 1714 at the foot of the "Great Raft," a log blockage on the Red River, became present-day Louisiana's first permanent settlement. In 1718, on a meander of the Mississippi south of Lake Pontchartrain, New Orleans was founded by a company with commercial monopoly rights and obligations to introduce colonists. New Orleans became Louisiana's capital in 1724. The company, and after 1731 the French government, awarded French settlers land tracts that commonly measured 8 by 40 arpents (1,500 by 8,000 feet—one linear arpent is equal to 192 feet) and stretched at right angles from the Mississippi across its broad natural levees and into the backland swamps. Indigo, tobacco, rice, and cotton, all grown commercially, did well. French people arrived directly from France, Canada, and the West Indies; non-Cajun French born in Louisiana were referred to as Creole. Several thousand slaves were also imported directly from Africa for labor on the evolving plantations.[10]

Louisiana's early settlers also included several hundred Germans whose impact on Louisiana's landscape would be disproportionate to their small numbers. Germans began to arrive in 1721, mainly from the Rhine Valley but also indirectly from failed colonies in the New World. They settled a stretch of the Mississippi upstream from New Orleans that came to be called the Côte des Allemands, or "German Coast" (map 3.4). On small farms located on the slightly elevated natural levees that they had to clear of dense forest, they grew corn, rice, and vegetables that they transported along with dairy products by small boats to a New Orleans market. Adjusting to the oppressively hot and humid conditions of Louisiana had to have been difficult, especially for those who came directly from the cool and moist Rhine Valley. These Germans were eventually dislodged from their small farms by wealthier plantation owners, and they relocated to places like the "French Settlement" east of the Mississippi. In time they were absorbed through intermarriage into the larger French-speaking population. Little that is German apparently remains, with the exception of some surnames—Keller, Montz, Rixner, Weiss, and Wiltz—and an interesting legacy of types of houses and barns adopted and modified by Cajuns.[11]

Some four thousand Cajuns reached Louisiana indirectly from Acadia between the late 1750s and about 1785. Initially these refugees settled an area called the "Acadian Coast" along the Mississippi upstream from the German Coast. Like the Germans, most Cajuns were small farmers (*petits habitants*) who grew crops for a New Orleans market. Like the Germans, they had to adjust to Louisiana's heat and humidity. Cool-weather wheat and apple trees taken from France to the Bay of Fundy had to be given up, as did woolen clothing, which Cajuns replaced with cotton garments. And like the Germans, most Cajuns along the Mississippi were displaced by their plantation-owning neighbors. Cajuns were pushed into the backcountry of sluggishly flowing bayous, notably south to Bayou Lafourche and west to Bayou Teche, where some earlier arrivals had already settled. This region would become the "heart" of today's Cajun subculture.[12]

Louisiana's Cajuns built a distinctive "Cajun" barn, the probable origin of which geographer Malcolm L. Comeaux traces to Louisiana's Germans. The distinguishing attribute of the small (thirty feet by thirty feet) Cajun barn is its central entrance located under a forward-facing gable end (see figure 3.1). This "recessed" opening, which allowed the farmer "to get out of the rain," is flanked by stables on one side for horses or mules and on the other side for cattle. A corncrib and a workspace for shucking corncobs lay behind the six-foot-deep entrance. The Cajun barn is not French; a farmer in France would have had three separate buildings for grain (storage and threshing), cattle, and horses. Germans, on the other hand, kept grain and animals under the same roof, but just where in Germany or in German-speaking Switzerland the prototype of the Cajun barn may have come after 1721 is unclear. Comeaux speculates that Cajuns adopted their barn from those used by Germans living in Louisiana. In Louisiana today few Cajun barns survive, and one's best chance for seeing one is in the French Settlement or in the isolated Upper Bayou Teche country.[13]

Cajuns in Louisiana also built distinctive houses, and unlike the barns, house examples are plentiful. The distinguishing feature of the "Cajun" (sometimes referred to as "Creole") house is its built-in porch (see figure 3.2). The gable ends of this house face the sides, and beneath the roof line a deep front porch is "built in" so as to be an integral part of the structure. Other attributes of the Cajun house type are its heavy wooden understory framework set on elevated footings, its central chimney or sometimes chimneys built on the outsides of the gable walls, and its paired front doors. Not much seems to be known about the origin of the Cajun house, yet Comeaux asserts that, like the Cajun barn, it was borrowed from Louisiana's Germans. In his classic study of Louisiana house types published in 1936, geographer Fred B. Kniffen found this built-in-porch house type to be strongly correlated

Figure 3.1 Cajun Barn, St. Landry Parish, Louisiana, 1982. Note the central recessed entrance. Sheds on both sides make this barn look less square.

Photograph taken (August 3, 1982) by and courtesy of Malcolm L. Comeaux.

Figure 3.2 Cajun House, Thibodeaux, Louisiana, 1986. Note that the built-in porch is structurally part of the house type.

Photograph by the author, taken on the Laurel Valley Plantation, July 23, 1986, on a field trip led by Malcolm Comeaux.

geographically with Louisiana's Mississippi Valley waterways, and to fade into insignificance before it reached northern Louisiana.[14]

The arrival of some four thousand Cajuns beginning in the late 1750s added substantially to the estimated six thousand to ten thousand people, largely slaves and French immigrants, already living along the lower Mississippi. All these people were soon in for some major political realignment. In 1762, just before the end of the Seven Years' War, France transferred Louisiana to its ally, Spain. Thus Louisiana west of the Mississippi and the Isle of Orleans east of the river stayed out of the hands of the British. Then in 1800, in a secret treaty between Spain and France, Spain returned Louisiana to France. And in 1803 Napoleon sold Louisiana to the United States (for the relatively small sum of $15 million). With a "stroke of the pen," and without authorization from Congress, President Thomas Jefferson suddenly doubled the size of his country. In this sale, Spain was upstaged by its French friends: The French buffer between aggressively land-hungry southeastern Americans and Spaniards in Texas was now gone. Just what the United States bought from France in 1803 was not officially defined until the drawing of the Adams-Onís Treaty boundary between the United States and Spain in 1819 (map 2.3).[15]

What became of Louisiana's French people? In 1812 the state of Louisiana was carved out of a corner of the Louisiana Purchase. Its French residents, both Creole and Cajun, remained the state's majority population until about 1830. Meanwhile, both lowland and upland southerners arrived with their "double-pen" houses to settle in the pine hills of the so-called Florida parishes (north of Lake Pontchartrain) and in northern Louisiana west of the Mississippi. Midwesterners from the Corn Belt also took their "I" houses to Louisiana's southwestern prairies. (Dogtrot [a synonym for double-pen] and I-houses are discussed in Chapter 7.) And Louisiana's French people themselves notably moved west into the state's southwestern prairies. By 1900, Louisiana's French still represented 30 percent of the state's population. Most lived in a large triangle with its base along the Gulf of Mexico and its apex near the junction of the Red and Mississippi rivers (map 3.4). Known as the French Triangle, this is where the much-diluted yet still highly distinctive French subculture resides today. Lafayette is recognized as the Cajun "capital."[16]

Illinois Country

Illinois Country referred to the vaguely defined area lying between Louisiana and Canada (see map 3.5). French Canadian fur traders were the first Europeans to penetrate Illinois Country and the continent's vast interior. They established dozens of small fur-trading and agricultural posts along the many waterways that connected the St. Lawrence with the Gulf of Mexico. Political jurisdiction of Illinois Country shifted from Canada to Louisiana in 1717. And by the 1750s some two thousand to three thousand French people—mainly Canadians, some with Native American wives and *métis* offspring, and some owning slaves, as at Kaskaskia—were lightly scattered in this interior. Their posts were the way stations that kept open communication between Canada and Louisiana. But the main significance of these posts was to channel furs and skins back to Montreal.

Montreal was the headquarters for the French fur trade and for the Northwest Company (Nor'westers) that controlled it. Montreal had geographical advantages. It was the jumping-off point for the two easiest routes leading to the interior (map 3.5): The first ascended the St. Lawrence to Lake Ontario, Lake Erie, the base post at Detroit, and then along the Maumee-Wabash portage to Illinois Country. The second went up the Ottawa River and to Lake Huron and the base post at Michilimackinac, from which it branched (1) south through Lake Michigan to the

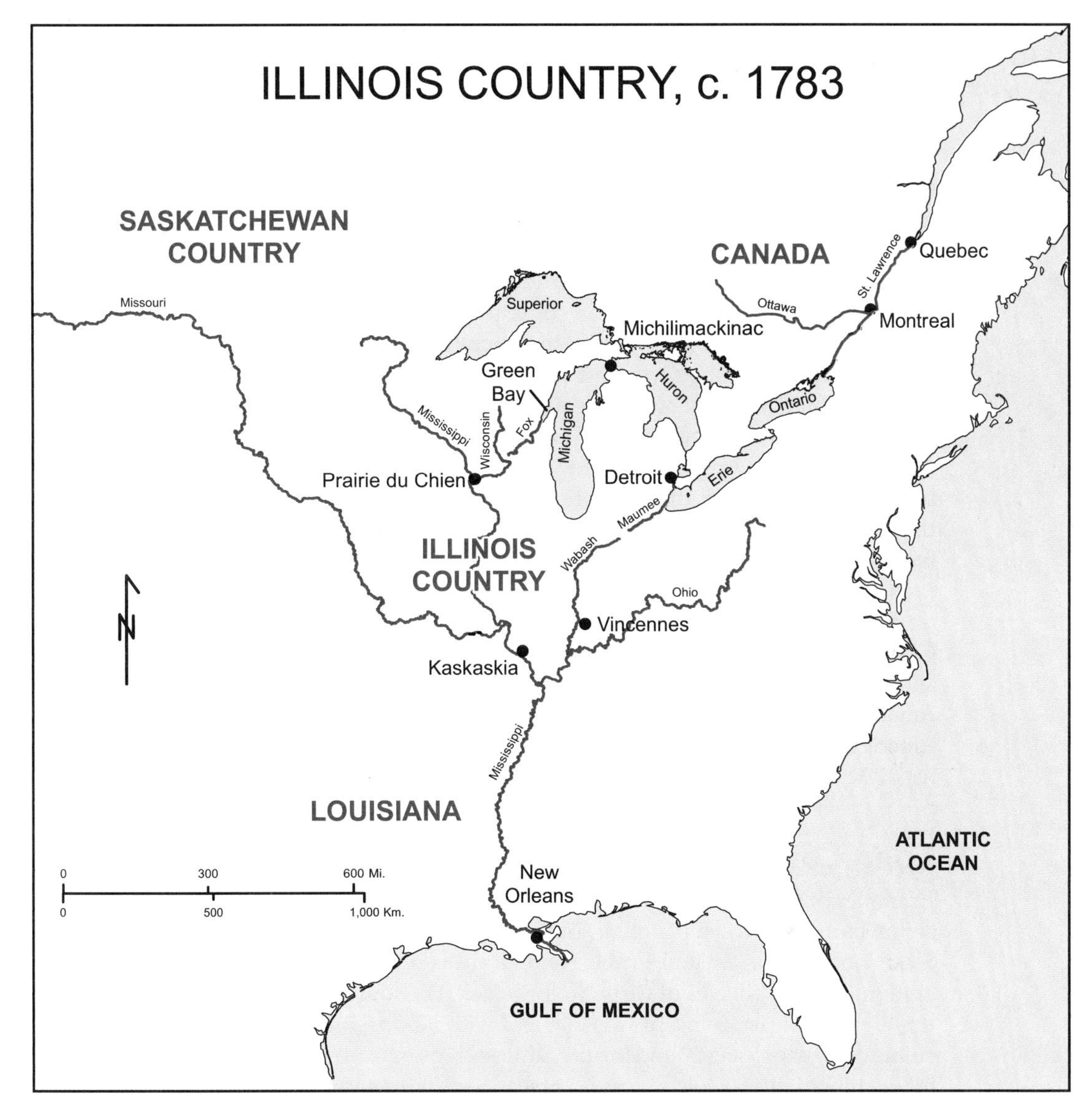

Map 3.5 Illinois Country, c. 1783. From this middle area, furs went northeast to Montreal and agricultural goods flowed down the Mississippi to New Orleans.

Illinois River and Illinois Country, (2) south through Lake Michigan and the Green Bay-Fox-Wisconsin route to Prairie du Chien in Illinois Country, or (3) west through Lake Superior to Saskatchewan Country. The Lachine Rapids on the St. Lawrence (map 3.3) also made Montreal a transfer point between riverboats and the birch-bark canoes (Algonkin technology) used to go inland. In this "peltry trade" French Canadians came together in a successful alliance with Algonkin and Huron Indians. The Indians hunted or trapped deer, beaver, and raccoon and prepared the animals' skins and furs, which they traded at the French posts. Come spring, the French transported these items by canoe to Montreal. And from Montreal, skins and furs went to Quebec for shipment to the single port of La Rochelle, in France.[17]

French Canadians established their first post west of Montreal at Green Bay in 1670 (map 3.5). By 1696 they had built some twenty posts west of Montreal, eight of them in the Mississippi watershed. More posts followed. Each probably began as a fort built of logs stuck vertically in the ground to form a palisade. Within the palisades were buildings made up of squared timbers laid horizontally. The three largest posts, Detroit, Michilimackinac, and Kaskaskia (which began as a Jesuit mission in 1703) had platted streets. Michilimackinac was located on Mackinac (English spelling) Island. Indian lore had it that a "Great Turtle" (Michilimackinac) froze in place to become the island—an image that captures the reality of a place covered with ice for much of the winter (but bustling with tourists in the short summer). The French fur trade reached its zenith in the early 1750s, at which point the French controlled some 80 percent of the North American business. Almost all the French trade passed through Montreal and Quebec en route to La Rochelle.[18]

The French may have engaged in agriculture at all these interior posts. Following tradition, they laid out their fields in long lots so as to give individual habitants riverbank frontage. Wheat was their mainstay crop, and Illinois Country gained a reputation for being the "granary of Louisiana." From posts like Vincennes, located on the Maumee-Wabash portage and at the point where the Wabash was forded on a trail leading west from the Falls of the Ohio to Illinois Country, wheat traveled down the Ohio and Mississippi in boats also loaded with beef, salt pork, and corn. Using bateaux, flat-bottomed boats with flaring sides and raked (inclined) bows and sterns, French people from the lower Mississippi transported tobacco, sugar, and cotton on the more difficult journey upstream along the Mississippi and Ohio. As Illinois Country changed hands politically, becoming British for twenty years (1763–1783) and then American in 1783, the French presence at Vincennes and elsewhere attracted Americans ahead of their normal settlement frontier. Within a generation of the American takeover, most French families at, for example, Vincennes and Kaskaskia, had sold their long lots to Americans and had moved on.[19]

Long Lots

Examples of the French legacy in Illinois Country and Louisiana include place names, architecture, and town planning, but of special interest here is the legacy of long lots. How long lots originated in the New World is not altogether clear. That they were used in the dike villages of eastern Normandy, the source area for perhaps one-fifth of the French people going to Canada, and with whom long lots must have diffused as cultural baggage, seems undeniable, although exactly how this happened awaits careful documentation. Cole Harris is persuaded that an equally strong reason behind the origin of long lots in Canada is the "preference of settlers." Long lots simply suited the New World habitants who wanted their own land to front on a river. That grants of land awarded to seigneurs were themselves oriented to rivers likely encouraged the same alignment for the long lots awarded by seigneurs to individual farmers. But then French people laid out long lots in Louisiana and in Illinois Country, where there were no seigneurs and seigneuries to set the bigger pattern. To be sure, long lots gave habitants a variety of soils and vegetation as they stretched back from a river, and they also insured a reasonable proximity to neighbors, but these two reasons were likely less important in explaining why long lots were used. That long lots could be easily and cheaply surveyed may have had importance.[20]

French surveyors measured long lots in arpents. As noted above, a linear arpent is equal to 192 feet, and a square arpent is about 0.85 acres (one side of a square arpent is 192 feet). The dimensions of these ribbonlike farms varied considerably in New France. Canadian long lots were generally smaller than those in Louisiana and perhaps those in Illinois Country. In his diagram of long lots in a hypothetical seigneurie in Canada, Harris draws farms to measure one-eighth by one and one-quarter miles (one hundred acres; see figure 3.3). These dimensions amount to 3.44 by 34.40 arpents. Long lots of this size conform exactly to the one-to-ten ratio of width to length that Harris suggests was typical. Long lots of this size probably also represent the large end of the spectrum: Harris notes that Canadian long lots were more typically seventy-five to ninety acres in size. In Louisiana and Illinois Country, long lots commonly measured forty arpents (about one and one-half miles) in length, and what varied were their widths. For Louisiana, Kniffen says that long lots during the French regime were commonly eight arpents wide (8 by 40 arpents equals 270 acres). Spanish authorities also used the French arpent as their measure when awarding land in Louisiana, and Kniffen notes that under Spain, long lots were six to eight arpents wide (6 by 40 arpents equals 202 acres). At Vincennes in Illinois Country, geographer R. Louis Gentilcore found long lots to measure two to four arpents wide and forty deep (2 by 40 arpents equals 68 acres).[21]

Clearly, French long lots had different dimensions. That they were all rectangular, however, had implications. On their ribbonlike farms, French people built their houses at one end of their properties, and the result was a row of houses that formed a linear settlement pattern. At first, houses and outbuildings lined up along rivers. When the better land along rivers had been distributed, officials (generally after the French regime) awarded a second *rang* (row or tier) of long lots going inland, and when good land allowed, additional rangs were conceded (figure 3.3). Behind the first rang, houses lined up single file along one side of the parallel concession road, but when tiers three and four were surveyed, houses were now built along both sides of a concession road in a *rang double*. The width of long lots determined how far apart these houses would be: In Canada, where many long lots are one-eighth-mile wide, houses are some 660 feet apart. Thus the French system of rectangular farms insured that farmhouses would be in close proximity—neighborhoods in rows. This contrasted with the square farms used by Americans, which insured that farmhouses would be more dispersed. As we shall point out in Chapter 14, shapes—rectangles versus squares—do make an important difference in rural settlement patterns.

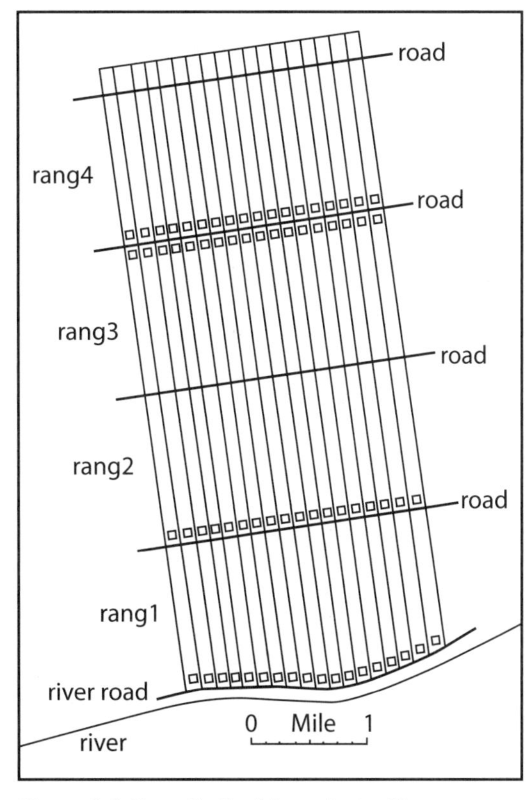

Figure 3.3 Hypothetical Long Lots. Shown are farms that measure 1/8 by 1¼ miles (100 acres). Houses are roughly 660 feet apart. In rang 1, houses front a river and road; in rang 2, they front a concession road; and in rangs 3 and 4, they front both sides of a concession road—a *rang double*. On the ground the pattern is rarely this regular.

Source: Modified and redrawn from figures in Harris and Warkentin, *Canada before Confederation*, 39 and 74.

The legacy of French long lots, then, is a landscape wherein properties are long and narrow and farmhouses and their outbuildings line up in rows. In French Canada and along Louisiana's lower Mississippi Valley, this landscape pattern is most striking. Elsewhere in the United States, long lots are more localized. For example, they persist along the Wabash River at Vincennes and at places like Green Bay and Prairie du Chien in Wisconsin. There are even traces of long lots in the Detroit area. And because of French influence, long lots also exist outside Louisiana and Illinois Country in parts of Texas and especially in northern New Mexico, as noted in Chapter 2. A basic functional difference between French long lots and those in New Mexico needs pointing out: French long lots fronted a river or a road to give people access to transportation, while long lots in New Mexico fronted an irrigation ditch to give people access to water.

Notes

[1] French source areas (Normandy and the La Rochelle hinterland) and numbers (five hundred to Acadia, ten thousand to Canada) are from Harris and Warkentin, *Canada before Confederation*, 19–21. Cole Harris revised downward to some three hundred the number of French going to Acadia in *The Reluctant Land*, 53, his wonderfully thoughtful resynthesis, wherein he draws heavily on the *Historical Atlas of Canada*, which he edited.

[2] Butzer, "French Wetland Agriculture in Atlantic Canada and Its European Roots," 452, for tidal figures; 457–58, for how extended families carried out reclamation; and 456, for how harvests could be abundant as early as the second year.

[3] Ibid., 452, for the fort in 1605; 454–55, 464, for the activities of Razilly and D'Aulnay.

[4] For the Acadian expulsion, see Meinig, *The Shaping of America*, 1: 270–75, quote on 271. For ten thousand Acadians by 1755, see Harris, *The Reluctant Land*, 57. Andrew Hill Clark says that two-thirds of some three hundred thousand Acadians living in the Maritime Provinces in 1968 resided in New Brunswick, with Moncton as their "capital"; see his *Acadia*, viii.

[5] Harris finds that seigneuries in Canada neither expedited colonization nor shaped a way of life, in *The Seigneurial System in Early Canada*, 193. For habitant as an early substitute for censitaire, see ibid., viii.

[6] In "The Extension of France into Rural Canada," 37, Harris notes that a habitant could clear only two arpents a year, which in a lifetime became thirty to forty arpents of cleared land. He provides the phrase "ragged and unkempt" in Harris and Warkentin, *Canada before Confederation*, 49.

[7] Harris discusses frost-free days and wheat in *The Seigneurial System in Early Canada*, 15–16. Harris notes, in Harris and Warkentin, *Canada before Confederation*, 23, differences in January mean temperatures; 23–24, that pear, peach, and walnut trees died in Canada's severe winters; 48, that, owing to the abundance of firewood, Canadians may have suffered less from cold than did peasants in France; 51, that in Canada wheat comprised three-quarters of the land in crops.

[8] See Harris, *The Reluctant Land*, 67–85, for descriptions of Canada's town and country landscapes, esp. 72, for population along the St. Lawrence River.

[9] For archaic given names in New Mexico, see Nostrand, *The Hispano Homeland*, 9. For two-wheeled carts, see Peattie, "The Isolation of the Lower St. Lawrence Valley," 117.

[10] Much here is from Meinig, *The Shaping of America*, 1:193–201.

[11] See Kniffen, *Louisiana*, 124, for the Rhine Valley source, the location of the "German Coast" in the St. Charles and St. John the Baptist parishes, and German surnames. That Germans relocated to the "French Settlement" is central to the thesis developed by Malcolm L. Comeaux, "The Cajun Barn," 53.

[12] Kniffen, *Louisiana*, 127, for the location of the Acadian Coast in the St. James and Ascension parishes upstream from the Germans. Meinig notes the presence of Cajuns in Louisiana beginning in the late 1750s in *The Shaping of America*, 1:283. Lawrence E. Estaville Jr. shows Cajuns in bayous Lafourche and Teche by 1800 in "Mapping the Cajuns," 165.

[13] See Comeaux, "The Cajun Barn," 47–62, esp. 48, for "to get out of the rain," and 53, for Germans as the source for the Cajun house.

[14] See Kniffen, "Louisiana House Types," 183 and 188.

[15] In "Retracing French Landscapes in North America," 88, Harris reports the presence of "perhaps 6,000 (including slaves) along the lower Mississippi" by the early 1750s. Meinig gives a higher figure: ten thousand "mainly French or African origin" people in Louisiana by 1750, in *The Shaping of America*, 1:197, and he provides the "stroke of a pen" quote in ibid., 2:11.

[16] Geographer Lawrence E. Estaville Jr. shows the French Triangle to be emerging by the Civil War, in "Mapping the Louisiana French," 95, 96, 105, 110. He documents Louisiana's French proportional decline in his "The Louisiana French Culture Region," 228, 292, 364, 449.

[17] For the French fur trade, see Harris in Harris and Warkentin, *Canada Before Confederation*, 6–17. For the two main lines of travel leading to "pivots" at Detroit and Michilimackinac, see Meinig, *The Shaping of America*, 1: 112–13.

[18] Harris, "Retracing French Landscapes in North America," 75, for Fort-de-la-Baie-des-Puants on Green Bay in 1670 as the first trading post west of Montreal. Also see Harris, "France in North America," 77, for twenty posts by 1696 and La Rochelle as the French port destination; and 84, for the 1750s as the zenith of the fur trade.

[19] R. Louis Gentilcore notes, in "Vincennes and the French Settlement in the Old Northwest," 285, that Vincennes was located on the trail between the Falls of the Ohio and Illinois Country; 294, that Illinois Country was the "granary of Louisiana"; 291 and 296, that the French at Kaskaskia and Vincennes had sold their lands to Americans within a generation of the American takeover.

[20] Harris, in Harris and Warkentin, *Canada before Confederation*, 40, and Harris, *The Reluctant Land*, 74, for reasons behind the origin of long lots in Canada.

[21] Harris shows farms in a hypothetical seigneurie in similar diagrams in *The Seigneurial System in Early Canada*, 122 and 175. He reproduces the diagram from page 122 in Harris and Warkentin, *Canada before Confederation*, 39, and in "France in North America," 75, where he also notes the seventy-five- to ninety-acre Canadian long lot size. He notes on both pages the one-to-ten ratio. For long-lot dimensions in Louisiana, see Kniffen, *Louisiana*, 122, for French; and 128, for Spanish. For dimensions in Vincennes, see Gentilcore, "Vincennes and the French Settlement in the Old Northwest," 293.

New England 4

New England is an especially distinctive American culture region. Its people speak a regional dialect of English. Many are Congregationalists, the religion that grew out of Puritanism. New Englanders live in political units called towns—counties in the region were an afterthought. And within these towns many live in villages often described as "charming" and "picturesque." In recent decades, however, those eye-catching villages have been more closely scrutinized by revisionist scholars. Using archival evidence, geographers and historians have reinterpreted these Puritan villages and changed our thinking about these and other aspects of colonial New England. We note these new interpretations through three themes: ecology in "The Granite Region," diffusion in "Beachheads to Colonies," and landscape in "The Puritan Village Myth" and "The New England Large."

The Granite Region

In colonial times New England was called just that: "New England"—a name given it by John Smith in 1614. But it was also known as the Granite Region, and New Hampshire today is nicknamed the Granite State. This is because about one-sixth of New England is underlain with granite, and the coarse-grained igneous rock seems to outcrop everywhere. New England's land surface is probably best characterized as a hilly upland. Rising above the hilly upland are two low north-south aligned mountain ranges, New Hampshire's granitic White Mountains and Vermont's Green Mountains, formed from gneiss, which is granite that has been metamorphosed (see map 4.1). In recent geologic times, all of New England was covered by continental ice, and the unfortunate consequence of glaciations for this region was a mantle of thin stony soils that cover almost everything. Large rocks that are transported and deposited by ice are called erratics, and Plymouth Rock, onto which the Pilgrims reportedly stepped from small boats sent ashore from an anchored *Mayflower*, is an erratic—and, predictably, also granite.[1]

Besides its hills and low mountains, New England has three lowlands of significance. The first, Cape Cod, named for its abundance of offshore cod, is a large sandy hook that juts seventy-five miles into the Atlantic. The cape's sandy soils supported little more than grasses and pitch-pine shrubs, and few English chose to settle there. The second lowland, which stretches along New England's coast from Rhode Island to Maine, is simply the smoother margin of the hilly upland to the west. This is where the English first settled. And cutting from north to south through the hilly upland of central New England is the Connecticut River, the region's only major river, which eroded a valley twenty miles wide in places. The alluvial soils and lush grasses that covered the Connecticut's floodplain and terraces were quickly recognized for their agricultural potential, and from the coast the English soon jumped across the hilly upland to the Connecticut Valley. Colonists referred to the Connecticut's floodplain and terraces as "meadows" or "intervals" (meaning intervales or bottomlands), grassland areas much prized because they could be mowed for hay or used as pasture.[2]

Covering New England's uplands and lowlands was a "mixed" forest. Broadleaf deciduous trees such as oak, hickory, and chestnut, which predominate in southern New England, were mixed with needleleaf evergreen trees like white pine, spruce, and hemlock, which predominate in northern New England. Native Americans had much altered this forest by the time the English colonists arrived. For their hunting

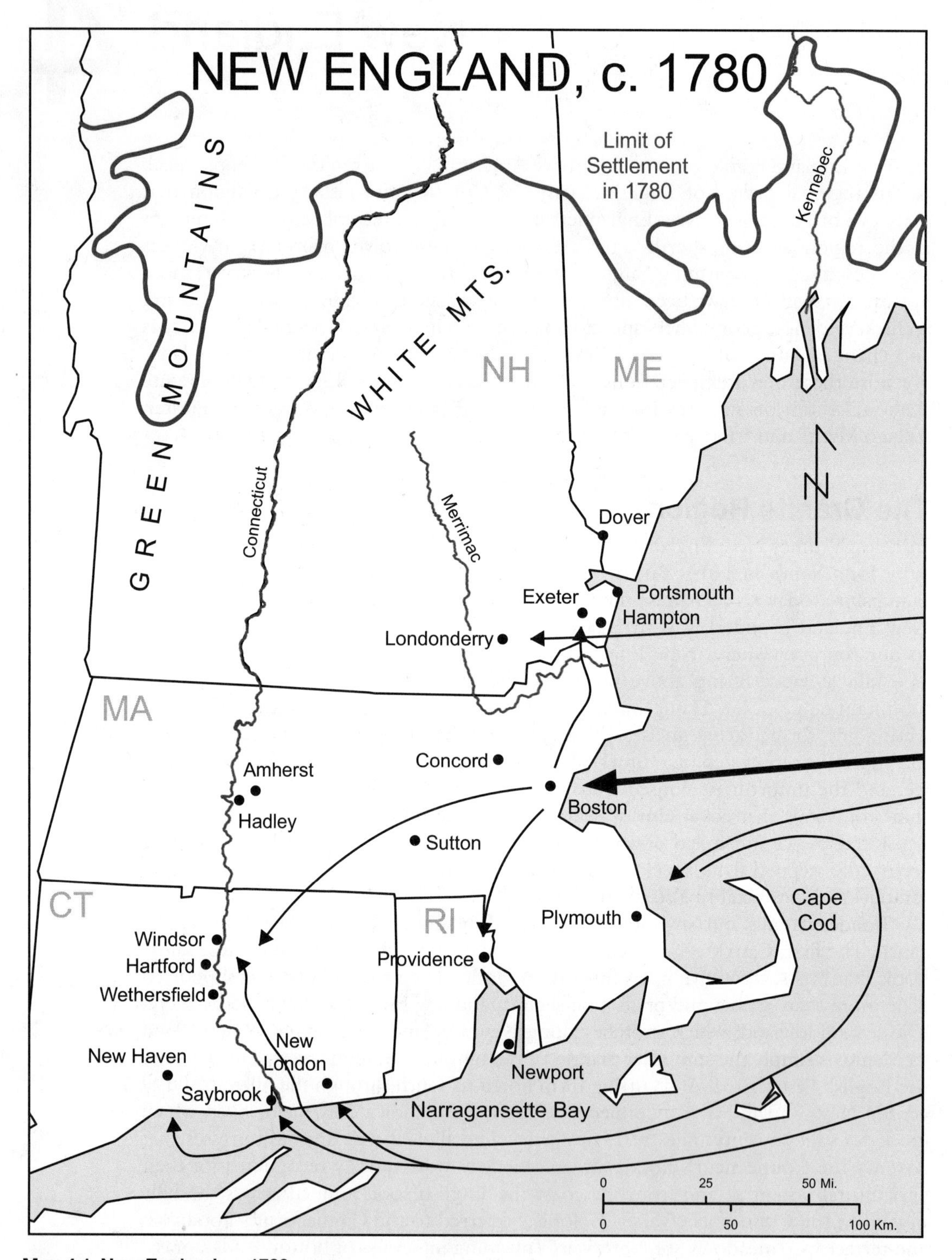

Map 4.1 New England, c. 1780.

Source: Dodge, "The Frontier of New England in the Seventeenth and Eighteenth Centuries and Its Significance in American History," map between 436 and 437, regarding area settled in 1780.

and gathering, each spring and fall the Indians burned off the understory of the forest, especially in southern New England where four-fifths of New England's native population lived. Such burns opened the forest to be, at times, "almost parklike," and this facilitated Indian hunting of white-tailed deer, bear, and turkeys. For their farming, Native Americans made clearings by lighting fire to wood piled around the base of selected trees. These trees would die and eventually topple over, but meanwhile sunlight would strike the forest floor, and in mounds of soil, Indians planted corn, kidney beans, squash, and pumpkins. After some eight or ten years, once soils were locally exhausted (it is apparently a mistaken notion that Indians fertilized soils with fish) and the firewood was gone, the Indians would move their wigwam villages to new clearings, leaving behind "Indian Old Fields." The English were attracted to these old fields and other burned-over sites, and more than fifty early English villages, including Plymouth in 1620, were located in such clearings (map 4.1).[3]

Besides cleared areas, New England's first colonists also sought meadow grasses. The distinction between a meadow and a pasture is that grasses in a meadow are to be mowed for hay while grasses in a pasture are for animals to graze on. With the English came cattle, sheep, oxen, and horses, and over the winter months these livestock had to be fed hay cut in meadows. Rather lush *Spartina* grasses grew in "salt marshes" along New England's coastal lowland, including even Cape Cod, and English colonists located their villages to be near these marshes. Marshes or intervals where hay could be mowed, as noted, attracted settlers to the Connecticut Valley. New England's native grasses, which besides the *Spartina* included broomstraw and wild rye, were less nutritious than grasses in England, and when grazed too closely, the pasture grasses died. Recognizing this, by the 1640s settlers had imported superior grasses including bluegrass and white clover from England, and in the 1700s still more grasses were introduced, including timothy, red clover, and lucerne (alfalfa). Carl Sauer, commenting on how the rich and nutritious grasses of northwestern Europe would "thrive under repeated mowing and grazing," noted with significance, "We may rank the meadow and pasture grasses as the most important plant contribution of this region [northwestern Europe] to the outside world."[4]

The English gradually pushed into the region's hilly upland, and leading the way were lumbermen, who sought, for ship's masts, the white pine that often grew in impressively tall stands on ridge tops. They were also after white and red oak for ship planking and house construction, and white and red cedar for clapboards, shingles, and fence posts. Following the lumbermen were the farmers, who carefully studied the tree species for indications of soil fertility. They found the prime soils—those black with humus—beneath hickories (the wood favored for fuel), maples, ashes, and beaches. Oaks and chestnuts indicated thinner but usable soils. To be avoided were the often acidic and sandy soils found beneath hemlocks, spruces, and white pine. To clear the forest, these English "girdled" trees—that is, they killed the trees by removing the bark from around the entire circumference of the trunk, which saved labor but wasted wood. Labor was always in short supply, but when labor was available, colonists would fell trees with an ax, sell the wood, and burn off what remained, unfortunately including some of the humus. Deforestation continued throughout the colonial period, and the process did not seem to peak until about 1850, when estimates suggest that only one-quarter of southern New England's forests remained.[5]

To their upland clearings the English introduced livestock and crops. As in the lowlands, for hay they sought marshy areas, which in the uplands lay by stream banks and in glacial bogs. Cattle were an important income earner: "Dry cattle" could be driven with relative ease to coastal markets, while "milch cows" were kept for family milk and dairy products. Colonists were less successful in raising sheep, which were vulnerable to predators and whose hooves wore heavy on pasture grasses in the forest. Hogs or swine, by contrast, could reproduce quickly, fend against predators, and, when turned into the forest, grub roots and eat acorns. Hogs were regarded as "filthy cutthroats" by Indians, who especially despised them. Oxen were used to clear the land and plow the fields, and horses provided transportation. The two leading crops, because they did well in colonial New England, were corn and rye. Short growing seasons made New England climatically marginal for growing wheat, the preferred grain. Wheat seems to have grown best in the Connecticut Valley lowland, where the growing season was somewhat longer. But even there, wheat was devastated first by a fungus called black stem rust or the "blast," introduced in the 1660s, and later by the Hessian fly, introduced during the Revolutionary War, supposedly by Hessian mercenaries.[6]

In *Changes in the Land*, historian William Cronon makes many of the foregoing points and reinterprets our understanding of human-environment relations in colonial New England. He contrasts the ecological impacts of Indians and English on the environment between 1600 and 1800, and the contrasts are dramatic: While Indians burned trees and underwood in their seasonal shifting from spring planting to fall hunting, they exploited soils and game only to meet their needs, and their overall impact on the environment was light. The English cut down the forests, overgrazed the pasture grasses, and, in fenced-in fields, exhausted soil nutrients while compacting the ground under the hooves of oxen. All this led to greater surface runoff, flooding, and soil erosion—all measurably destructive to the environment. Meanwhile, those Native Americans who survived the inadvertent introduction of smallpox and other disease-carrying pathogens—estimates have New England's Indian numbers

Spotlight

Martyn J. Bowden (1935–)

Martyn J. Bowden. Taken at Putnam House, Sutton, Massachusetts, April 14, 2008. Photograph by the author.

Born in Bolton in the Manchester area of England, Martyn Bowden took his undergraduate degree at the University of London before arriving in the United States to earn a master's degree at the University of Nebraska (1959) and a PhD at the University of California–Berkeley (1967). After teaching one year at Dartmouth College (1963–1964), Bowden put in a long career in the Graduate School of Geography at Clark University in Worcester, Massachusetts (1964–2004), along the way serving as the major advisor to twelve doctoral students. Bowden is known for his publications on cognition of the Great Plains and the growth of urban downtowns, notably San Francisco, and he has led the way in revisionist scholarship about colonial New England. His fifteen New England beachheads and his five-house solution to Yankee shelter in early Massachusetts Bay Colony are Bowden originals. At Clark University Bowden was a renowned classroom teacher. One of his students told me that Martyn Bowden was, hands down, the best teacher he had ever had, and for every such testimonial one knows empirically there are a hundred more.

at seventy- to one-hundred thousand in 1600 but reduced to twelve thousand in 1800—were forced onto small, marginal tracts of land where shifting seasonally to be both farmers and hunter-gatherers was no longer possible. The English greed for land and profits had triumphed, and their treatment of Native Americans was shameful.[7]

The English in colonial New England did confront long, severe winters, short growing seasons, trees to be felled, and thin stony soils. But the environment was in fact a less formidable challenge than early English colonizers claimed. Geographer Martyn J. Bowden (see spotlight) points out that Puritan leaders in the mid-seventeenth century purposely invented New England's "tough to conquer" image to glorify the accomplishments of their own people as well as to justify their dispossession and displacement of Indians. In sermons and in writings, they characterized New England as a "desart" (deserted place) and a "wildernesse" (uninhabited by humans) in whose forest lived "salvages" (inhuman beasts). This post-settlement manufacture of tough-to-conquer imagery with the aim of glorifying the first European colonists, Bowden argues, would be repeated west of New England in the tallgrass prairies of the Middle West, the shortgrass steppes of the Great Plains, and the Mormon-settled Great Basin Desert.[8]

Beachheads to Colonies

Across the Atlantic, five areas in England contributed significant numbers of emigrants to New England in the seventeenth century. Bowden has identified these source areas using genealogical data for 1620 to 1650 and place-name transfer data for 1620 to 1720. Both sets of data show a clear source-area hierarchy (see map 4.2): The leader was East Anglia and its three counties of Suffolk, Essex, and Norfolk, with the Stour Valley on the Suffolk-Essex border an especially heavy contributor. The Southeast, notably Kent and Hertfordshire counties and London, was second. Third was the West Country, especially Devon, Somerset, Dorset, and Wiltshire counties. The North, notably Yorkshire, was fourth, and the Midlands was fifth.[9]

Well over half the English going to New England came from East Anglia and the Southeast. Relatively heavy numbers arrived during the "Great Migration" from 1630 to 1642; according to D. W. Meinig, nine thousand Puritans had arrived by 1640, and a total of twenty thousand would come by 1660. Most of the English came in families that averaged over six people in size, represented the middle class, and were literate. Heads of households had various occupations including artisans, merchants, tradespeople, and farmers. Fewer than a third were already agriculturalists, the occupation most would follow in New England. East Anglia, England's most populous region, was farther along than other English regions in the enclosure of agricultural lands: They enclosed with hedges ("closes") strips of cultivated land in once open fields where farmers had worked cooperatively and shared tools, and they lived on single-family farmsteads. East Anglia and the Southeast also had the largest numbers of Puritans—members of a religion whose deeply felt Calvinist beliefs made them "Nonconformists" within the Church of England, or the Anglican Church. For many Puritans the desire to reform the Church of England justified their emigration to shape a new England.[10]

In New England English colonizers from four of the five source areas came together in significant numbers to establish fifteen "beachheads," as Bowden called their footholds, mainly along New England's coast. Based on the same sets of genealogical and place-name transfer data, Bowden's beachheads are shown in map 4.3: West Country (6), East Anglia (3), Southeast and London (3), and North (1). Two of the fifteen beachheads located in coastal New Hampshire and in the

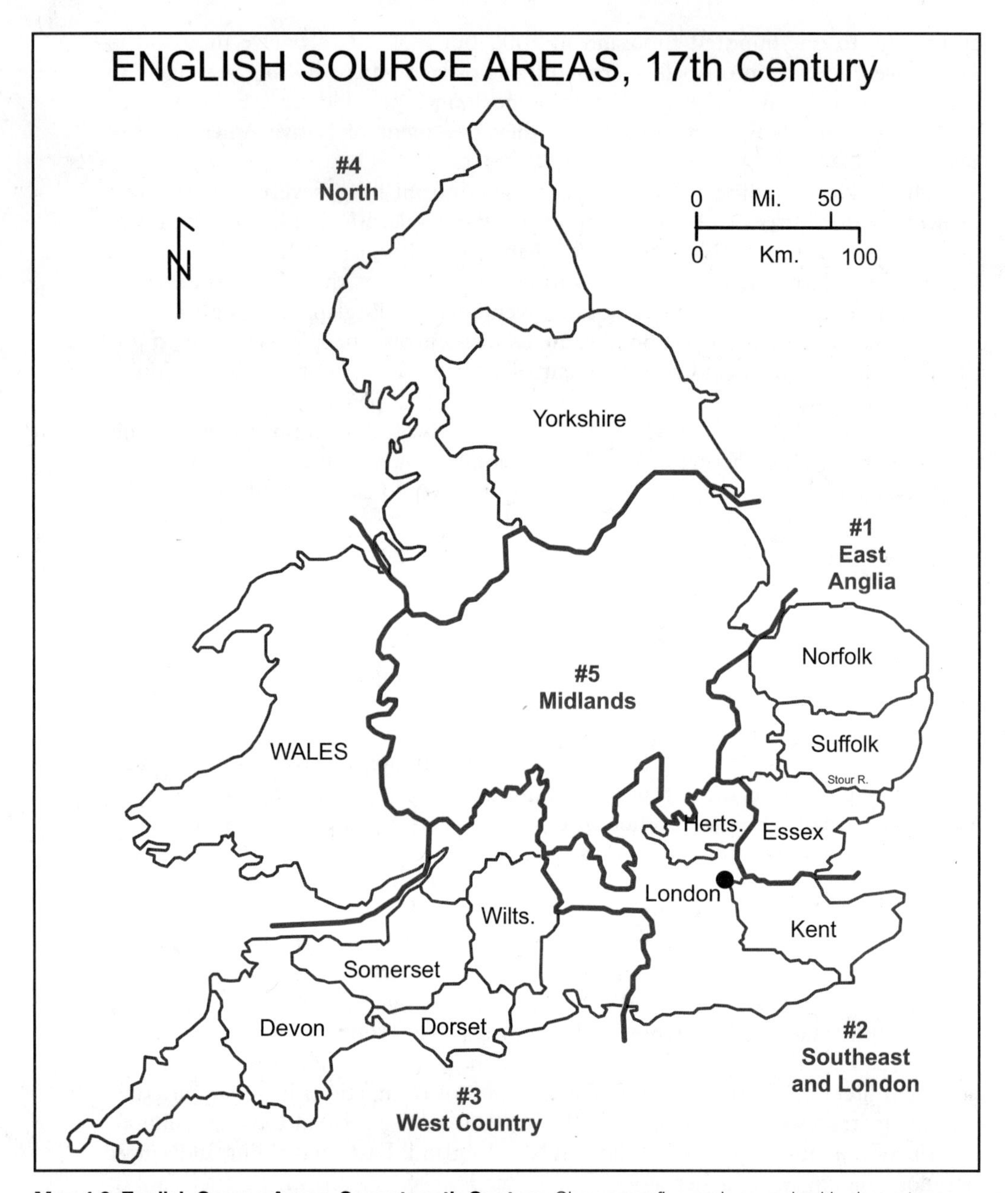

Map 4.2 English Source Areas, Seventeenth Century. Shown are five regions ranked by importance.

Sources: Emigrant numbers (1620–1650) and transfer place names (1620–1720), modified from Bowden, "The New England Yankee Homeland," 3, in Nostrand and Estaville, eds., *Homelands*; reprinted with permission of Johns Hopkins University Press.

Connecticut Valley of Massachusetts could not be confirmed with certainty, but they appear to be East Anglian. Identifying these beachheads and their English source areas, Bowden points out, has implications for what was introduced to New England in terms of building types and construction techniques, agricultural practices and field systems, and settlement patterns. Significantly, these beachheads gave rise to the subregions that became New England's colonies and eventually its six states.[11]

Developing the story of New England's colonization chronologically, Plymouth initiated the region's permanent settlement in 1620 (map 4.1). Its founders were mainly Pilgrims, a religious body known also as Separatists because they had separated

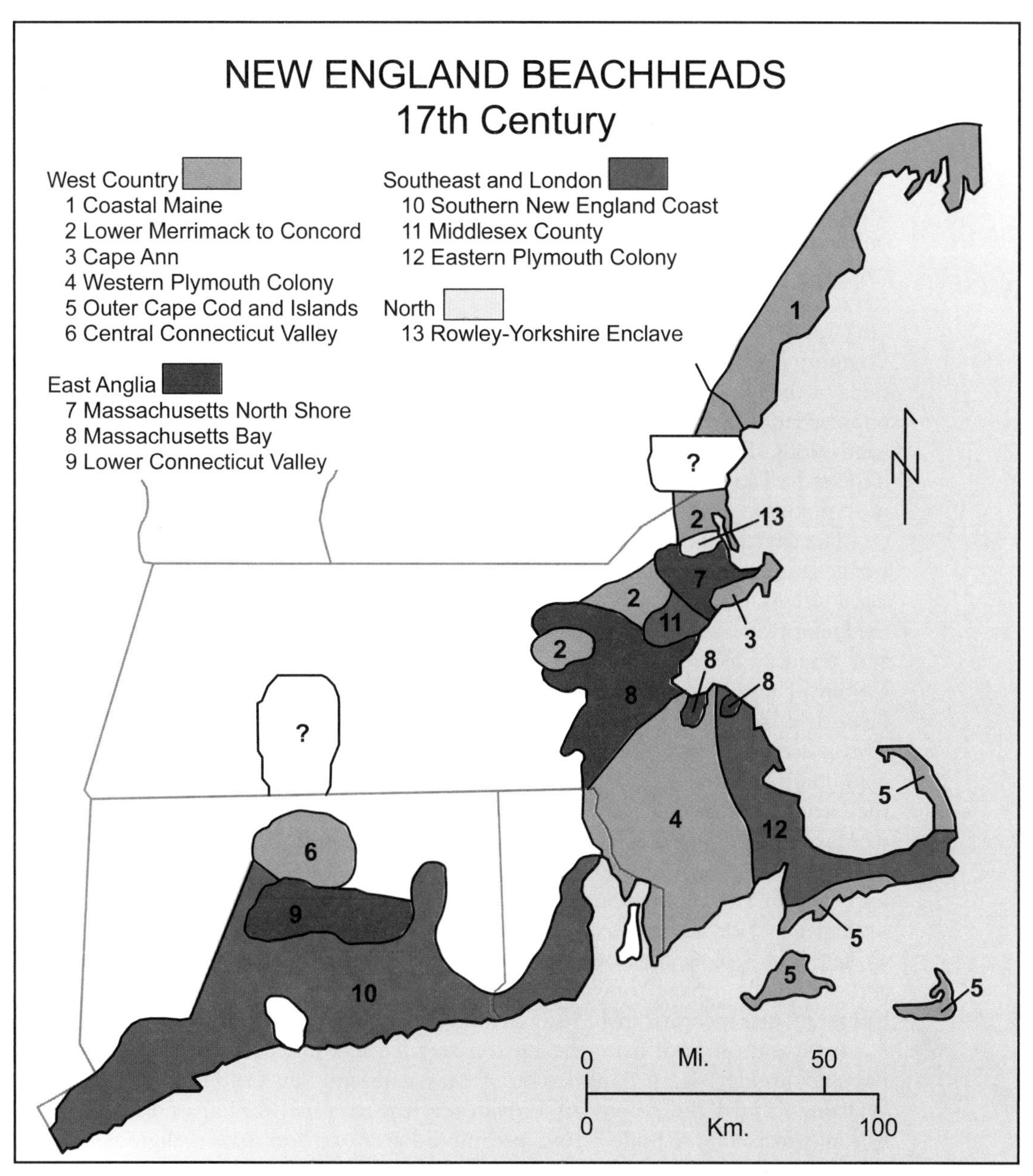

Map 4.3 New England Beachheads, Seventeenth Century. Shown are thirteen confirmed and two unconfirmed beachheads.

Sources: Emigrant numbers (1620–1650) and transfer place names (1620–1720), modified from Bowden, "The New England Yankee Homeland," 7, in Nostrand and Estaville, eds., *Homelands*; reprinted with permission of Johns Hopkins University Press.

from the Church of England. They came seeking religious freedom, and they were far more tolerant of other religious sects than the Puritans were. The initial settlers at Plymouth were from the Southeast, but settlers from West Country, who had a pervasive dislike for Puritans, soon overwhelmed Plymouth. For a decade after 1620, Plymouth Colony was New England's regional focus, yet numbers at Plymouth were small—only some five hundred by 1630. The governor, who lived in the tiny village of Plymouth, oversaw a loose association of settlements that engaged in agriculture, especially the raising of livestock for a Boston-area market. Initially successful attempts to develop fur trade had led to the founding of a post at Windsor in the Connecticut Valley in 1633 (near a Dutch fur post at the site of Hartford), but little came of this effort or of fishing or lumbering. Lacking in vitality and unable to survive as a colony, the now ten thousand English settlers in Plymouth were absorbed in 1691 by their dynamic colony neighbor, Massachusetts Bay.[12]

Massachusetts Bay was settled by Puritans, or Nonconformists, coming largely from East Anglia. Beginning in the mid-1620s, they settled along the coast, including on the narrow Shawmut Peninsula that in 1630 they renamed Boston. Meanwhile, a royal charter for the Massachusetts Bay Colony was secured in 1629. These Puritans came in large numbers, and the emigrant families often arrived as parts of entire congregations with their minister leader. They were well organized and well financed. To quote Meinig, "The sheer scale of this emigration, . . . the character and motivations of the people involved, and the wealth and organization that sustained it were unprecedented." From their capital and focus at Boston, Puritans spread north and south along the coastal lowlands and then inland, and with exceptions they settled in dispersed villages within areas called towns. Massachusetts Bay Colony became the most populous and most influential in the region, and from it sprang offshoots.[13]

The first offshoot went to the Connecticut Valley of Connecticut (map 4.1). Fertile lands certainly pulled settlers there, yet religious issues also pulled these particular Puritans, for the beliefs of those who relocated were more moderate than those of the majority in Massachusetts Bay. Wethersfield was settled first in 1634, and Hartford and Windsor followed in 1635—all three becoming the nucleus of Connecticut Colony. These settlements and their upstream offspring, like Hadley, were of the open field-street village type. In 1634, Puritans coming directly from England had already settled at Saybrook at the mouth of the Connecticut, also home to the port community of New London. These settlements of coastal Connecticut soon united as the New Haven colonies, and their orthodox beliefs made them closest theologically to Massachusetts Bay. In 1638, New Haven was founded by rather large numbers of Puritans who arrived from the home counties adjacent to London. A royal charter issued in 1660 gave Connecticut Colony claim to the whole of the New Haven settlements. Not wishing to risk being absorbed by New York, New Haven joined Connecticut Colony in 1664. Relationships between the English in Connecticut and the Dutch in New York were somewhat antagonistic, and the Dutch-based term "Yankee," first recorded in 1758, was used derisively to refer to the English.[14]

A second offshoot from the Boston area led to the settlement of Narragansett Bay and Rhode Island. Banished from Massachusetts Bay Colony in 1636, Roger Williams formed the colony of Providence, the eventual capital of Rhode Island Colony (map 4.1). Rhode Island became a haven for additional religious dissenters, including English Quakers, Irish Catholics, French Huguenots, and Portuguese Jews. The colony prospered through the raising of sheep, cattle, and hogs, which could rather easily be controlled on the many islands in Narragansett Bay. Newport, founded in 1639, became the "Boston" of Rhode Island Colony, with merchants who had their own connections for trade with South Carolina and the West Indies. Rhode Island's prosperity led to its residents owning a disproportionately large number of servant-slaves—some five thousand of an estimated thirteen thousand in New England. In religion, Rhode Island was clearly New England's major deviant, and its northern and western borders, Bowden notes, were "cultural fault lines" to be crossed at some risk.[15]

Still another offshoot went to coastal New Hampshire in 1639, when John Wheelwright led Puritan exiles from Massachusetts Bay to Exeter (map 4.1). Initiated as a beachhead of settlers from West Country, the area north of the Merrimack River also saw Puritan exiles founding Hampton and Dover. New Hampshire's on-again off-again administrative relationship with Massachusetts Bay lasted until 1741, when merchants at Portsmouth, while reasserting their ties with the Church of England, used their political connections in England to engineer colony status for New Hampshire. Meanwhile, some Scotch-Irish Presbyterians who settled at Londonderry in 1718 became a non-English anomaly in a region that was 95 percent English. Maine continued to be part of Massachusetts Bay Colony and the Commonwealth of Massachusetts until 1820, when it split off to become the twenty-third state. Its long connection with Massachusetts is curious given that Maine was the beachhead of West Country people who, as religiously tolerant people, avoided East Anglian Puritans. Vermont was colonized late in the colonial era, when New Englanders mainly from Connecticut settled east of the Green Mountains and New Yorkers settled west of the Green Mountains at the same time. Both New Hampshire and New York laid claim to Vermont, claims that were not relinquished until 1782 and 1790, respectively. As we shall see in Chapter 8, Vermont became the fourteenth state in 1791.[16]

The Puritan Village Myth

In 1967 I took my first teaching position at the University of Massachusetts (UMass) in Amherst. While a graduate student, I had read about New England villages and towns, and in my new job I looked forward to seeing them. Amherst, I found, was by New England standards a relatively young community, having "hived off" from its parent village of Hadley in 1759. Hadley, founded in 1659, was the early colonial village I would study. So I read and took copious notes on a detailed town history of Hadley written in 1863 by a Hadley resident, Sylvester Judd, and I led my UMass students in historical geography on field trips to observe Hadley's well-preserved village. From this I published "The Colonial New England Town," in 1973. My stated purpose was "to compile a concise geographical summary" of a colonial New England village and town for teachers like me to use in the classroom. The problem is that I was wrong: Hadley, I later learned, did not represent a composite of greater New England's villages and towns.[17]

When I wrote my article, conventional thinking among students of New England, including geographers like Edna Scofield and Glenn Trewartha, went like this: Villages in colonial New England originated as nucleated or compact settlements centered around an open space called a common. Fronting on the common were a meetinghouse (church) and its burying ground, a nearby tavern (inn), perhaps a blacksmith, and the houses of settlers or "proprietors" set on house lots that rimmed the remainder of the common. In several open fields beyond the village, these proprietors owned rectangular strips of land subdivided for planting crops and mowing hay, and they also held in common areas of pasture and woodland. I had read this literature based on conventional thinking, and there I was observing in Hadley what I had read. So what went wrong? Later in the 1970s, after I had written my article, revisionist scholars were finding out that the great majority of New England's villages began not as nucleated affairs but as dispersed settlements. Hadley, it turns out, represented only a subset of nucleated Connecticut Valley "street" villages like Wethersfield and Hartford whence many of Hadley's proprietors had come.[18]

In the 1970s, then, scholars in several disciplines were beginning to question just how colonial New England villages did originate. In this revisionist scholarship in geography, Bowden at Clark University led the way, and he inspired a second principal contributor, Joseph Wood. What Bowden, Wood, and others learned from archival research was that in the colonial period dispersed villages were typical, while nucleated villages like Hadley were uncommon outside Massachusetts's tidewater areas and the Connecticut Valley. Owning land was the primary motive behind most of the English who went to New England. The new villages of these colonists did have a focus composed of a meetinghouse and its burying ground in addition to "nooning" sheds (because Puritan services continued into the afternoon), "Sabbath Day" houses for more remote families, a tavern, and the houses of several proprietors. But to satisfy their desire to live on their own privately held (fee simple) land, most proprietors lived on detached single-family farms that were spread out along lanes and roads that wound away from the meetinghouse focus for up to three or four miles. Geographer Michael Steinitz found that the majority of the houses that seventeenth- or eighteenth-century farmers lived in were one-story cottages, or Cape Cods, of two or three rooms; two-story white clapboards were common only in areas of better soils and greater prosperity, as in Essex County in northeastern Massachusetts and Middlesex County in Connecticut's Connecticut Valley. The loose neighborhoods typical in colonial times were indeed viable affairs, for despite their lack of compactness, they provided the social networks and religious cohesion necessary for a strong sense of community.[19]

According to this revisionist thinking, then, colonial villages, with exceptions, were dispersed neighborhoods with a meetinghouse focus. The myth of the nucleated Puritan village, these academics assert, began when scholars and others looked at what villages had become in the nineteenth century and "projected" their observations to colonial times. Following the colonial era, in the Federal (or National) period, especially in the last quarter of the 1700s and the first quarter of the 1800s, New England evolved from a dominantly agrarian society to a more commercial society. Many men left farming to become tradesmen, artisans, and professionals. In village after village, as Joseph Wood shows in numerous map sequences, the open spaces around seventeenth-century meetinghouses filled in with businesses and houses, and "center villages" emerged. Successful businessmen and artisans built two-story houses, at first in a Federal style (gable ends to the sides), then in Greek Revival style (gable ends facing the street). The one-time common, which grew out of the meetinghouse lot, was transformed into a "green," now beautified with paths and trees. New two-story meetinghouses with steeples on top replaced the one-story churches, and burying grounds came to be enclosed with walls. These nineteenth-century village creations, many assumed (and others argued), were part of the colonial legacy.[20]

The myth of the nucleated colonial village, Wood asserts, was "invented" by romanticists, a literary elite, and regional boosters. Wood cites several dozen authors, including Timothy Dwight, Harriet Beecher Stowe, Ralph Waldo Emerson, and Henry David Thoreau, who created the context for an idealized colonial village by extolling the virtues of village life, pastoral ideology, and New England's "near perfection" politically, culturally, and spiritually. From this, regional boosters, social reformers, and landscape architects collectively invented a geographical past for the nineteenth-century village. Scholars, including geographers Scofield and Trewartha, added to the nucleation argument when citing defense and Puritan ethics as reasons why colonists lived in close quarters. But the copious archival evidence marshaled by Wood and others suggests that we need to change our thinking: Seventeenth-century villages were different from the nineteenth century's bustling clusters of two-story houses located around an open space. Michael Steinitz's research on colonial houses, based on careful use of the Federal Direct Tax Census of 1798, only reinforces the point with his finding that in the last quarter of the eighteenth century, two-thirds of the houses in the central uplands of Massachusetts had only one story. In all this lies a lesson: *Archival evidence is needed alongside observation in the field.*[21]

Revisionist scholarship addressed only villages, not towns. Villages were, of course, the settlements. Towns were the grants of land within which these settlements existed. Town grants were awarded by the legislative assemblies of the several New England colonies to groups of proprietors who had petitioned for the land. Towns were of different sizes, yet thirty-six square miles came to be something of an average by the end of the colonial period. Town boundaries were drawn with straight lines, but the shapes these lines created were often quite irregular. The elected leaders in a given town took charge of dividing up the land, assigning rights to commons, and establishing political and religious regulations. When town populations grew through natural increases or immigration, one of two things usually happened: Either the excess colonists, often the sons and daughters of the original settlers, established satellite villages with meetinghouses within unoccupied town land and, through petition to the colony, then hived off to become new towns, or a group of proprietors would petition the colony for a new grant on the frontier. In these ways towns replicated themselves, and all six New England states came to be carved into hundreds of towns.[22]

Regional variety makes the United States interesting, and use of the term *town* is an example. In New England, the word refers to an area, not a settlement, as it does elsewhere in the United States. I knew this when I moved to Amherst, but I still got tripped up. I remember phoning the telephone company to initiate service in the house I had rented, which was located several miles outside the village of Amherst. When I called I was asked if I lived in the town of Amherst. Remembering that I had driven several miles to get beyond the built-up community, I instinctively answered no. The telephone company employee politely asked, "Then why did you call us for service?" "Oh," I said, "you mean the *town*." As one of my UMass students quipped, in New England, it is impossible "to get out of town."

The New England Large

Once in New England, the English found an abundance of timber for building houses. Oak and chestnut, as hardwoods, were favored for framing; white pine, a softwood, was used for walls; and red or white cedar, light in weight and decay-resistant, were used for exterior shingles, clapboards, and fence posts. The availability of wood had implications: In a general way, it meant that full-timbered houses could replace the half-timbered houses common in England. Lightweight cedar shingles meant that East Anglian carpenters in Massachusetts Bay Colony could build rafter roofs with fewer components; in England, the heavier slate, tile, and thatch materials required more substantial rafter roofs. And then there was the need to adjust to the harsh winters. To better insulate houses, East Anglian carpenters in Massachusetts Bay could now add cedar shingles and clapboards to the pine underboards of walls, a New England innovation. And in a second innovation, in either Massachusetts Bay or Plymouth Colony, English settlers dug the first cellar holes beneath houses to facilitate food storage in both cold winters and warm summers.[23]

Bowden argues that in the 1600s, to adjust to their new environment around Massachusetts Bay, settlers from East Anglia and Kent developed a five-house solution to their needs for shelter. Initially, they built *one-story*, one-room frame houses with rafter roofs that had gable-end chimneys; or they built one-story, two-room houses with central chimneys. Soon, one-story houses grew to two stories, with two rooms down, two rooms up, and a central chimney, a house type that was later named "I-style." The two-story house evolved into a third type, the "saltbox." Along the way there emerged the wide-base "Cape Cod," with five rooms down, two rooms up, and a central chimney—a fourth house type much favored in Plymouth Colony. And the foregoing houses culminated in a fifth form—the central-chimney, five-over-five house known as the New England "Large."[24]

The Large has been called New England's "highest form" for vernacular (commonplace) architecture; in fact, Martyn and Margaret Bowden own an exquisite

Large that they call Putnam House (figure 4.1). Built by Edward Putnam in 1737 atop Putnam Hill, a drumlin in Sutton, Massachusetts, it was the first Large to be built in Sutton. It was purchased in 1972 by the Bowdens who, through their restoration, know its every inch. The massive chimney made from gneiss measures fifteen feet on a side on the cellar dirt floor, tapers to five feet on a side at the top, and has five hearths—three downstairs and two upstairs. The front door, really a double door, opens onto a narrow hall and a triple-run staircase that leads upstairs, with single-run extensions to the cellar and attic. To the right and left as one enters the house are two deep front rooms, above which are similarly sized chambers, all four with hearths of varying widths. Behind the chimney stack downstairs are three small rooms: in the middle, a kitchen with a nine-foot-wide hearth for cooking, to the right in the northeast corner, a buttery or dairy, and to the left in the northwest corner, a pantry. The front of the house faces sixteen degrees east of south, to maximize solar insulation.[25]

Putnam House eventually became a "connecting barn"—that is, lesser buildings were configured to connect the back of the house to a large frame barn sometime after the barn was built in 1876. Connecting barns are houses that are physically linked by workshops or carriage houses or other structures to a general-purpose, wood-frame barn (see figure 4.2). Careful archival research and fieldwork undertaken by architectural historian Thomas C. Hubka have substantially revised what geographer Wilbur Zelinsky reported in 1958 in "The New England Connecting Barn." Hubka found that while connecting farmsteads existed in colonial New England, reconfiguring farm buildings to connect house and barn was most intense in the nineteenth century, peaking after the Civil War. Yankees cited being able to avoid bitter cold winters when tending to livestock as a quite logical reason for connecting their houses and barns, yet Hubka found of greater climatic importance the desire to protect critical "dooryard" space located on the south and east sides of

Figure 4.1 Putnam House (a New England Large), Sutton, Massachusetts, 1998. This example of the regional style was built in 1737.

Source: Nostrand and Estaville, eds., *Homelands: A Geography of Culture and Place across America*. Photograph by the author, March 25, 1998; reprinted by permission of Johns Hopkins University Press.

the connected farm assemblage. In southwestern Maine and adjacent southeastern New Hampshire, Hubka found some 60 percent of farmsteads to be connecting farm buildings; beyond that area percentages decreased to the point that in southern New England and west of Vermont's Green Mountains, connecting barns are quite rare. Thus, connecting barns are highly associated with New England, and they constitute a regional landscape "signature."[26]

Figure 4.2 Connecting Barn Examples, Hadley, Massachusetts, 1973. Houses and barns are connected in various ways. (Top) On Hadley's original Broad Street, January 30. (Bottom) In North Hadley, July 6.

Photographs by the author.

Readers who know *Charlotte's Web*, E. B. White's book for children, will be pleased to know that Mr. Zuckerman's barn was a New England connecting barn. One learns in the story that Mr. Zuckerman had a house, a woodshed, and a barn that had a hayloft and a cellar where the animals lived. In a corner of the cellar entrance, Charlotte, a large gray spider, wove "SOME PIG!" in an open space in her web and thus was able to save Wilbur the pig from Mr. Zuckerman's dinner table. But neither the book's text nor its illustrations make it clear that the house and the barn were connected. In replying to appreciative letters from a fourth-grade class in Australia eight years after *Charlotte's Web* appeared in 1952, E. B. White (1899–1985) wrote of his fifty-acre farm in Brooklin, Maine: "My barn is a good deal like Mr. Zuckerman's . . . [it] is attached to the house through a woodshed, so I can walk out into the barn without going outdoors."[27]

All this might suggest that one day, as you are driving through New England, you can impress your friends by pointing out that those connecting barns you are observing are a regional signature, and those Larges you see—for example, the Louisa May Alcott house built in 1700 in Concord, Massachusetts—represent the highest form of vernacular architecture reached in colonial New England. But as your automobile rounds the hill on the contour into the next valley, remember two things: First, those charming communities with green-facing Congregational churches and two-story white clapboards actually became nucleated villages *after* the colonial era; in colonial times they were loose neighborhoods with scattered one-story houses. Second, the quaint stone fences bordering fields and the region's appealingly manicured ambiance came to exist only with much English sweat and blood. Geographer Peirce Lewis contrasts the New England Yankee, "a long-face fellow . . . who copes with his impossible environment by the exercise of scrooge-like thrift," with the German farmer in rich and genial Pennsylvania, a "jolly, fat and generous" person alongside his "rosy-cheeked wife of ample proportions." Such caricatures, of course, distort and exaggerate, but they also have an element of truth, as we shall see in Chapter 5.[28]

Notes

1 For John Smith's term "New England," see McManis, *European Impressions of the New England Coast 1497–1620*, 111–12. For "Granite Region," see Brown, *Mirror for Americans*, 11, and Brown, *Historical Geography of the United States*, 98. For Plymouth Rock as an erratic, see Lineback, "Pilgrims' Progress," 14.

2 For "intervals," see Cronon, *Changes in the Land*, 31, 114, 148. Speaking of the Connecticut Valley, Ralph H. Brown (*Mirror for Americans*, 187) says that colonists called the floodplain the "first meadow," and above the floodplain built their homes on the "meadow hills."

3 See Cronon, *Changes in the Land*, 25–27, for forest composition and "almost parklike"; 42, for the concentration of Indians in southern New England; 45, for fish not used for fertilizer; 48–49, for burning selected trees seasonally; 90, for fifty villages founded in cleared land.

4 Ibid., 31, 114, 141–42, for salt marshes and intervals; 141–42, 144, for Spartinas and grasses. See Sauer, "European Backgrounds of American Agricultural Settlement," 26, for the "most important plant contribution."

5 Cronon, *Changes in the Land*, 30, 109, 112, for trees sought by lumbermen; 28, 115, for trees and soil fertility; 116–17, for "girdling"; 156, for three-quarters cleared by 1850; 30, for white-pine stands; 131, hickory for firewood.

6 Ibid., 134, 139, 140, for "dry cattle" and "milch cows"; 129, for sheep and hogs; 136, for "filthy cutthroats"; 150, 153, 154, for corn (maize), rye, wheat. See Brown, *Mirror for Americans*, 192–93, for problems with wheat in the Connecticut Valley.

7 Cronon, *Changes in the Land*, 85, for microorganisms including smallpox; 42 and 89, for Indian numbers; 159, for marginal lands.

8 Bowden, "The Invention of American Tradition," esp. 5–12.

9 See Bowden, "The New England Yankee Homeland," 2; and Bowden, "Culture and Place," esp. 74, 84, 110.

10 See Wood, *The New England Village*, 16–20, 29. Puritan emigrant numbers are from Meinig, *The Shaping of America*, 1:90, 92, and McManis, *Colonial New England*, 68.

[11] Bowden, "The New England Yankee Homeland," 2, 6–8, for beachheads; also see Bowden, "Culture and Place," 99, 137, 139.
[12] McManis, *Colonial New England*, 42, for five hundred by 1630; Meinig, *The Shaping of America*, 1:96, for ten thousand in 1691.
[13] Meinig, *The Shaping of America*, 1:90, for the "unprecedented" quote.
[14] Bowden, "The New England Yankee Homeland," 18 and 19, for Puritan differences; 1, for "Yankee," citing H. L. Mencken as his source.
[15] Meinig, *The Shaping of America*, 1:106, for Newport as "Boston"; 1:107, for slave numbers. See Bowden, "The New England Yankee Homeland," 18, for borders as "cultural fault lines."
[16] Meinig, *The Shaping of America*, 1:93, 95, 96–97, 108–9, for New Hampshire; 1:349, 391, for Vermont. See McManis, *Colonial New England*, 70, for New England as 95 percent English.
[17] Judd, *History of Hadley*; Nostrand, "The Colonial New England Town," 45, for "concise geographical summary."
[18] Scofield, "The Origin of Settlement Patterns in Rural New England"; Trewartha, "Types of Rural Settlement in Colonial America."
[19] Wood, *The New England Village*, 55, for owning land as motive to emigrate; 116, 166, for loose focus on meetinghouses; 67, 115, for strong sense of community despite dispersed living; 139, for nucleated village exceptions; 71–87, 167, for the findings of Michael Steinitz. Steinitz earned his PhD under Bowden at Clark University; Wood had much interaction with Bowden, but earned his PhD at Pennsylvania State University.
[20] Wood, *The New England Village*, 71, for "projected"; 89 and 102, for agricultural to commercial transition and new jobs; 103–7, 110, 120–21, for map sequences showing change; 114, for two-story houses, and walls around cemeteries; 129, for commons to greens; 127, for one-story to two-story meetinghouses.
[21] Wood, *The New England Village*, 2, 135, 141, 142, 173, for "invented"; 141–42, 144, 169–70, for contributors to the invented tradition; 170, for "near perfection"; 75, for Steinitz and the 1798 tax; 166, for two-thirds of the houses having just one story.
[22] See Wood, *The New England Village*, 39, 55, 156, for towns not as settlements but as vehicles to encourage settlement; 118, for how "colonial assemblies" awarded towns; 39, 54, for the responsibilities of town leaders; 45, for town growth by replication.
[23] Cronon, *Changes in the Land*, 112, for uses of wood; 119, for substitution of shingle for slate and thatch. See Cummings, "Connecticut and Its Building Tradition," 193, for the rafter roof modification; and Bowden, "The New England Yankee Homeland," 12, for shingles and clapboards added to underboarding, and for cellars.
[24] Bowden, "The New England Yankee Homeland," 10–13, 19–21, and 22–23, for the five-house solution.
[25] The Bowdens have turned Putnam House into a bed-and-breakfast.
[26] See Hubka, *Big House, Little House, Back House, Barn*, 14, to note that connecting farm buildings existed in colonial New England and peaked in popularity after the Civil War; 12, convenient "winter passage" reason; 13, protection of "dooryard" space reason; 20, figure 18 (map), for 60 percent of Maine-New Hampshire farmsteads. See also Zelinsky, "The New England Connecting Barn."
[27] The quote, "My barn . . .," is from a letter by White written October 30, 1960, and reproduced in (Australian teacher) Hummel, "I tore open the envelope—'E. B. White!' I blurted."
[28] Lewis, "When America Was English," 344, for quotes; more lengthy contrasts between New England and Pennsylvania are in his "Americanizing English Landscape Habits," 96–102.

The Middle Colonies 5

Colonists in the "Middle Colonies" knew that their stretch of the Atlantic Seaboard was just that—midway in location between New England and the South. As such, they applied the term quite logically. And they recognized that what made their region different was its cultural diversity. The first to arrive after Native Americans were the Dutch to the Hudson Valley; then came Swedes and Finns to the Delaware Valley, and to Pennsylvania came large numbers of English, Germans, and Scotch-Irish. To put these peoples in their geographical setting, we begin Chapter 5 with a section entitled "Terrain Overview." Sections on "New Netherland and New Sweden" and "Pennsylvania" follow, and we end the chapter with thoughts on why the so-called Pennsylvania Culture Area represents such a distinctive cultural landscape.

Terrain Overview

Starting with an old geography teaching artifice has some merit here: If you were to make landfall on the New Jersey coast and then travel west to follow the southern border of Pennsylvania into Ohio, what landforms would you encounter (see figure 5.1)? First, you would cross the Atlantic Coastal Plain, a flat and sandy area that widens to the south and gets pinched out at Cape Cod to the north. Next, in Pennsylvania, you would traverse a hilly area called the Piedmont, which means literally "foot of mountain." You would then come to the first of the mountains, the northern reaches of the Blue Ridge range that, farther south in North Carolina, becomes the Great Smoky Mountains. After the Blue Ridge you would cross the Great Valley before climbing over a series of northeast to southwest trending ridges and valleys in the Ridge and Valley Province. The ridges stand high because they are composed of less readily eroded sandstone, while the valleys are underlain with soluble and more easily eroded limestone. Finally you would come onto the highly dissected Appalachian Plateau, which in Pennsylvania begins at the Allegheny Front; the Appalachian Plateau ends in central Ohio as it grades into the Interior Lowlands.

During colonial times European colonists were familiar with these physiographic divisions, but they knew them by different labels. The Atlantic Coastal Plain was the "Region of Sea Sand"—the first area to be reached, but only in certain places the first to be settled, because the land was not especially good for agriculture. The Piedmont was called the "Upper Country" because it was higher in elevation than the Region of Sea Sand. New England, whose hilly upland was similar to the Piedmont, was rightly considered as a separate physiographic unit. The Blue Ridge was called just that in colonial times, while the Ridge and Valley Province was known to some as the "Endless Mountains," a term somehow derived from Indians who clearly appreciated the seemingly unending series of ridges and valleys. Finally, the Appalachian Plateau was called just that in colonial times, although in Kentucky and Tennessee the subregional name was Cumberland Plateau.[1]

Separating the Upper Country and the Region of Sea Sand was the *fall line*. At the break in slope between the hills and the low country, water in streams occasionally became turbulent with rapids, cascades, and, from time to time, a waterfall. This fall line can be traced from Trenton, New Jersey, on the Delaware River; south through Philadelphia on the Schuylkill River; to Georgetown (within the District of Columbia), on the Potomac; Richmond, on the James; Augusta, Georgia, on

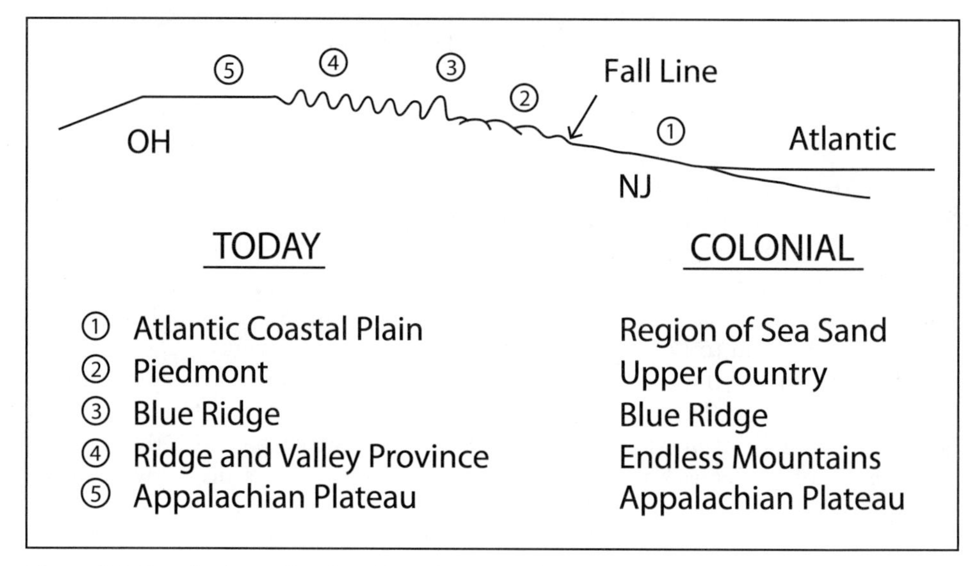

Figure 5.1 Terrain Cross Section, New Jersey to Ohio. Not drawn to scale.

the Savannah; and Columbus, Georgia, on the Chattahoochee (see map 5.1 inset). Generations of historians and geographers have written that communities came to exist along the fall line for two reasons: Oceangoing sailing ships or lesser craft could go no farther upstream because stream turbulence marked a "head of navigation"; and the water flowing over the occasional falls was a source of power for mills. But the role played by the fall line in the selection of sites for new communities seems to have been overstated. In 1964 geographer Roy Merrens found that the fall line played no role in community site selection in North Carolina. On the other hand, geographers Ronald Grim and Gerald Holder found that the head of navigation factor was significant in initial community site selection in Virginia and Georgia, respectively.[2]

Crossing the Middle Colonies from north to south were three major rivers along which much of the settlement action unfolded (see map 5.1). The most northerly was the Hudson. At its mouth, islands and sandbars obstructed navigation, but beyond New Amsterdam, oceangoing sailing ships could ascend with ease. The Mohawk River flows into the Hudson from the west about 170 miles above present-day New York City. Ralph Brown points out that on the Hudson, the head of *sloop* (smaller sailing craft) navigation was at Fort Orange (Albany) some ten miles below the mouth of the Mohawk, while the head of *ship* navigation was at the "Overslaugh," where after 1784 the city of Hudson was founded thirty-six miles below Albany. The second major river, the Delaware, could be ascended only to falls at present-day Trenton, the head of sloop navigation. And the third major river, the wide but disappointingly shallow Susquehanna, was rarely used for ship navigation. Owing to land subsidence around its banks, the Chesapeake Bay itself is actually a drowned estuary of the Susquehanna. Sailing ships could easily penetrate to the vicinity of Baltimore.[3]

The Middle Colonies, then, stretched from the Mohawk south to the upper Chesapeake Bay. By about 1700 some of its Region of Sea Sand and much of its Upper Country had been effectively settled, and by 1740 colonists had penetrated the fertile limestone valleys well into the Endless Mountains. By 1810 the Atlantic Coastal Plain, because of its infertile sandy soils, would already have gone through a "forest to forest" cycle, especially south of the Middle Colonies in Maryland and Virginia, because settlers had already abandoned their farms and moved west. In the Upper Country, the best soils—those with a top layer of humus known to colonists as

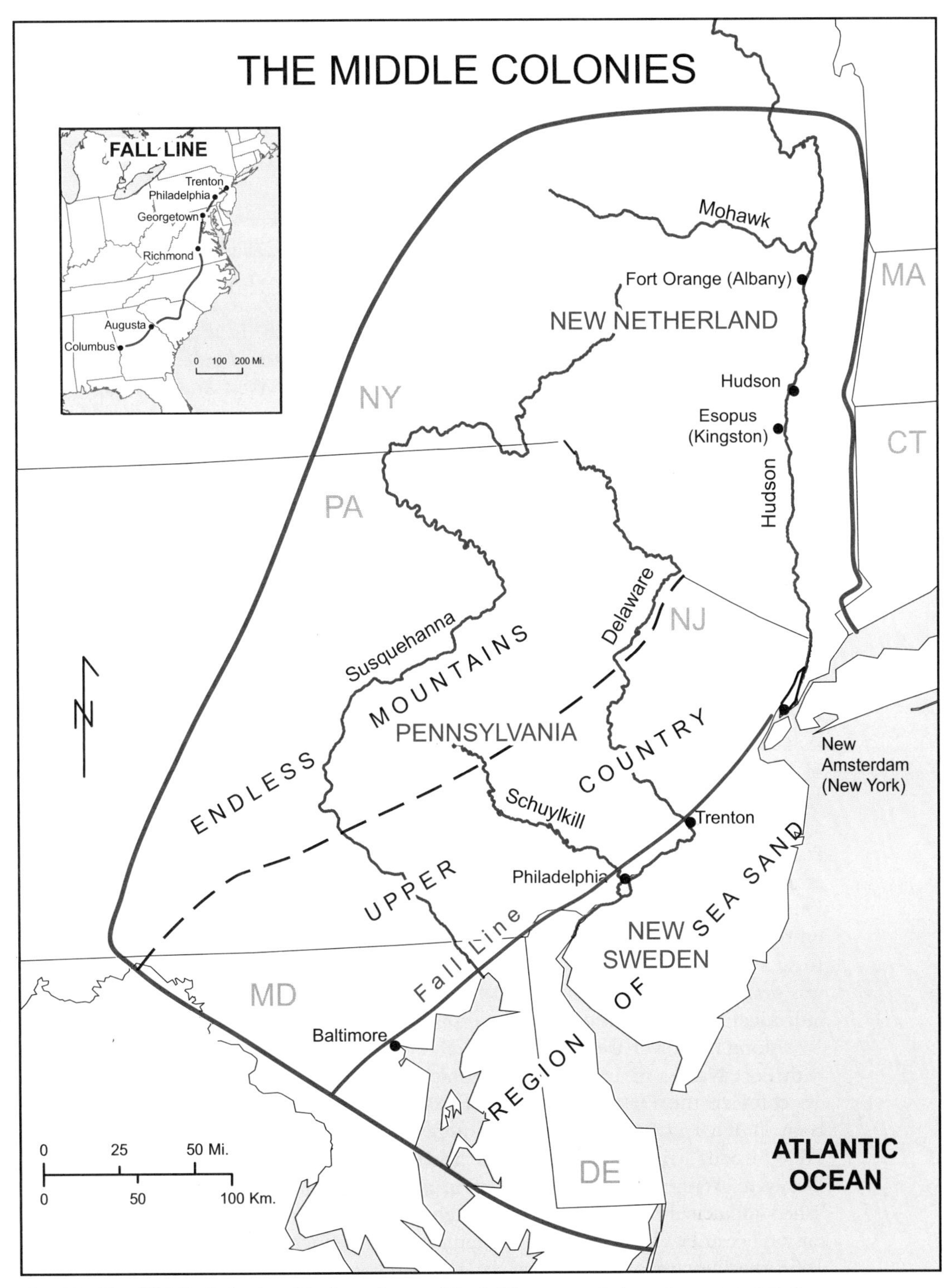

Map 5.1 The Middle Colonies. The importance of the fall line seems to have been overstated.

"black mold"—were identified by their groves of nut-bearing trees. The Dutch and the Swedes found reasonably good land to farm along the Hudson and the Delaware, and the most productive mixed farming area would emerge in the Piedmont of southeastern Pennsylvania west of Philadelphia.[4]

New Netherland and New Sweden

In 1609 Henry Hudson sailed into the Hudson River, known to the Dutch as the North River, and the Delaware River, which the Dutch called the South River (see map 5.2). Although he was an English mariner, Hudson had been hired by the Dutch East India Company to look for a westward passage to Asia, and accordingly he claimed the areas adjacent to both rivers for the Dutch. In this New Netherland, the Dutch engaged in fur trade, primarily along the Hudson. Under the United New Netherland Company, replaced in 1621 by the Dutch West India Company, the Dutch built Fort Orange near Albany in 1614; at a midway location up the Hudson, they founded what later became Esopus, near Kingston, in 1615; and in about 1614 or 1615 they built warehouses at the southern tip of Manhattan Island. At first the Dutch traded with the local Algonquians of the Hudson Valley, but an alliance with the Mohawk, the easternmost of the five Iroquois Nations, allowed them to tap lands that were far richer for furs and stretched west to the Great Lakes. Fort Orange, near Albany, became their major fur entrepôt; indeed, Albany on the Hudson became the Dutch counterpart to Montreal on the St. Lawrence for the French.[5]

For the Dutch (and the French), the role of the home government was of less importance than was that of the company chartered by the home government. Thus it was the Dutch West India Company that introduced colonists, and among the first were Walloon and Flemish Protestants who came as religious refugees. Some located in New Amsterdam, which by the 1620s had become a small community. A wall built in 1653 across the tip of Manhattan Island between the East River and the Hudson gave protection to those in New Amsterdam. More Dutch arrived, and with them came blacks, mainly slaves from the Caribbean. Against the wishes of Governor Peter Stuyvesant, a small group of Sephardic Jews, who in 1654 reached New Amsterdam as exiles from Brazil, were allowed to stay. Meanwhile, to promote agriculture, the Dutch West India Company awarded estates to "patroons" or proprietors who took on the obligation of introducing a minimum number of colonists. One patroon was Kiliaen Van Rensselaer, a director of the Dutch West India Company, who in 1630 was granted land on both sides of the Hudson at Albany. Awarding small tracts to individual farmers proved more successful than did the estates given to patroons.[6]

Along the lower Delaware the Dutch built additional fur trading posts beginning with Fort Nassau in 1623, but the company rather neglected the South River area. To colonize the Delaware, Dutch officials visited Sweden and negotiated to have both Dutch and Swedish financial interests back what became a New Sweden. In 1638 Swedes arrived to found Fort Christina on the site of present-day Wilmington, Delaware. With the Swedes came perhaps as many Finns, and together they established additional fur trading posts and channeled their efforts to growing wheat and raising livestock. These Swedes and Finns also introduced the log cabin. However, in 1655 the Dutch reasserted their claim to the lower Delaware, expelled the Swedish officials, and set up their headquarters at Fort Casimir (New Castle) south of Fort Christina. This ended New Sweden.[7]

Nine years later (1664) Dutch rule also ended when the English sailed into New Amsterdam harbor and told Governor Stuyvesant to step aside. Except for a brief return to Dutch control from 1672 to 1674, New Netherland became English. Interestingly, the possibility that Dutch and not English might have become the

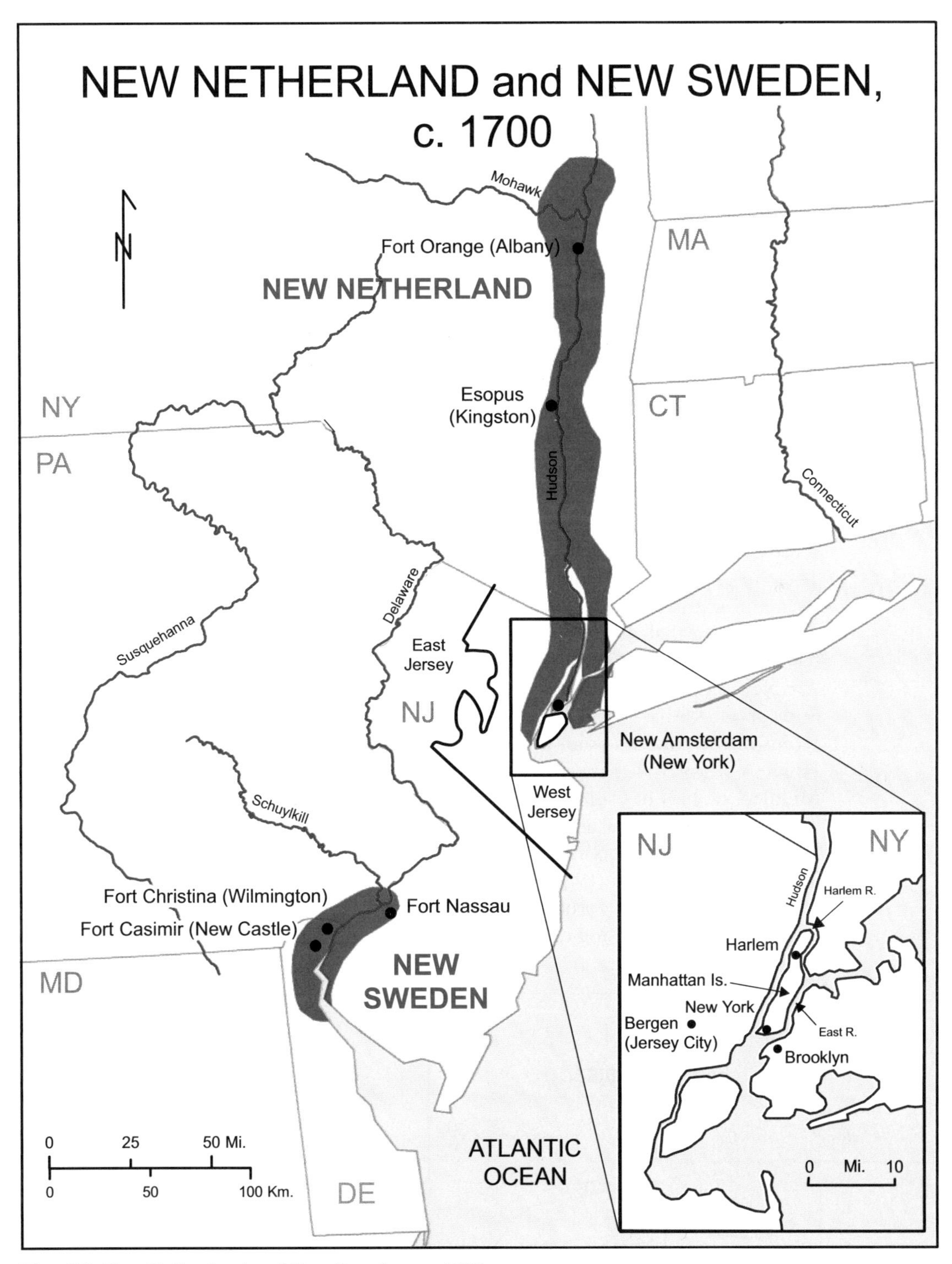

Map 5.2 New Netherland and New Sweden, c. 1700.

Sources: Modified from Meinig, *The Shaping of America*, 1:127, 132 (maps); and Wacker, *Land and People*, 122 (map), for the East-West Jersey boundary.

colonial mainstream language ended with this English takeover. New Amsterdam was renamed New York, Esopus became Kingston, and Fort Orange became Albany. By 1664 some ten thousand Europeans, mostly Dutch, lived in New Netherland, and most of them would stay. Their numbers were greatest along the lower Hudson. By 1700 New York had breached the wall that would become part of Wall Street, Harlem was a village on the north end of Manhattan Island, across the East River was Brooklyn on Long Island, and across the Hudson was Bergen, now Jersey City (see map 5.2 inset). As one went up the Hudson, the percentage of people who were Dutch increased, and for decades after 1700, Albany remained a Dutch stronghold.[8]

After their takeover of New Netherland, the English created New Jersey, which they subdivided into East Jersey and West Jersey. The two Jerseys were united into a single colony in 1702, yet as geographer Peter O. Wacker underscores, the colony's diverse population—in East Jersey, Dutch, blacks, and New England–derived English; and in West Jersey, Quaker English, Germans, Scotch-Irish, and old-settled Swedes and Finns—and its position between dynamic New England and Pennsylvania left New Jersey with a weak identity. In East Jersey and much more importantly in the Hudson Valley, a Dutch legacy persists through two folk house examples. One is the Flemish cottage introduced by Flems (not Dutch) to New Netherland, a one-and-one-half story, one- or two-room deep house with a shallow-pitched gambrel roof and overhanging eaves possessing curves that are similar to flanges on bells. The so-called Dutch Colonial style that was especially popular in the first half of the twentieth century found its antecedents in this Flemish cottage. And the second is a type of house that displays corbie-step gables, or gable ends that look like stairsteps, reportedly once plentiful in Albany (see figure 5.2).[9]

New Sweden's legacy became the log cabin, a simple one-room, one-story, one-door abode with a pitched roof that protected the horizontally laid log walls. Log cabins in the colonies seem first to have been introduced to New Sweden, likely by Finns. A scholarly debate over the European antecedents of log construction in the Middle Colonies has some scholars giving greater importance to Swedes and Finns on the lower Delaware, and others to Germans in Pennsylvania. After systematic field research in the Baltic countries, especially Finland, Sweden, and Norway, geographer Terry G. Jordan in the 1980s "resurrected" the Fenno-Scandian origin of log architecture and carpentry. Jordan concluded that the log house was largely Germanic in carpentry, but that the log cabin was Fenno-Scandian. He found that Fenno-Scandian

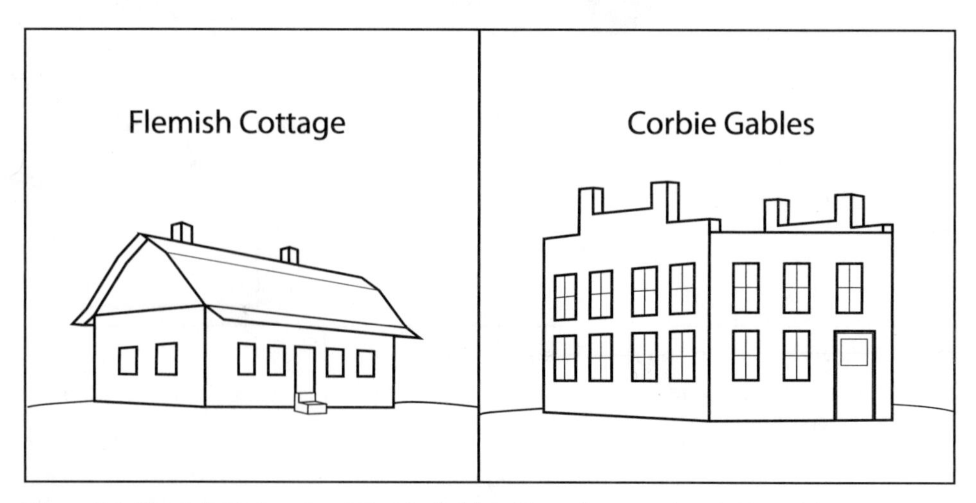

Figure 5.2 Flemish Cottage and Corbie Gables. Few relic examples of either of these landscape features in New Netherland have survived.

carpentry was "simpler and cruder" and "better suited to frontier conditions." As colonists moved west in the forested East, then, those log cabins—built with ax and adz, and for many a pioneer family the first solution to housing—diffused initially from New Sweden.[10]

Pennsylvania

Pennsylvania came to exist as a colony in a most amazing way. In March 1681 William Penn received an enormous grant of land lying west of the Delaware River. This grant was repayment for a debt King Charles II of England owed Penn's deceased father. Penn called his grant Pennsylvania, or "Penn's woods." Penn was a Quaker, a member of a faith that believed in religious tolerance, and his goal was to create a model colony open not just to Quakers but to peoples of all faiths. Penn was also wealthy, and with his resources he could plan a model city and launch his colony's settlement in an immediate and major way. Indeed, Penn himself came to America to take charge.

Penn called his model city Philadelphia, or "city of brotherly love." His instructions to a party of agents whom he sent in advance were to identify a site on the Delaware River that was "high, dry, and healthy" and would avoid lighterage—meaning the use of lighters, or flat-bottom barges, to transport people and goods from ship to shore. In the summer of 1682, such a deepwater site was found on the east side of a neck of land that lay between the Delaware and the Schuylkill rivers (see map 5.3 inset). Penn's agents surveyed a rectangular grid that would eventually stretch two miles to the Schuylkill. Penn arrived in October 1682 to oversee the inception of his preplanned community, which at first was very much clustered along the Delaware. Philadelphia's grid filled in only gradually: By the end of the colonial period, the city had not expanded much beyond Washington Square and the Pennsylvania State House (later, Independence Hall), and construction in Philadelphia would not reach the Schuylkill until 1836 (see figure 5.3).[11]

In numbers of people, Penn's colony grew quickly; twenty-three vessels arrived in the first year alone. English people, mainly Quakers coming largely from the North Midlands, made up the majority group, but there were also Quakers from Wales who settled west of Philadelphia and Quakers from the lower Rhineland who settled in Germantown, north of Philadelphia. Inland from Philadelphia, as lands were surveyed into farms, English settlers did well growing wheat and raising cattle, and they came to dominate the first tier of counties next to the Delaware—Chester, Montgomery, and Bucks. About 1710, at the invitation of the Penn family, Germans began to arrive in large numbers, and they dominated the second tier of counties—Cumberland, York, Lancaster, Berks, and Northampton. They also prospered at this mixed grain-livestock agriculture in the fertile, rolling Piedmont. At inland centers that developed at York, Lancaster, and Reading, visitors commented on how the Germans were "clanish." For generations, these so-called Pennsylvania Dutch spoke German at home and conducted their Lutheran, Reformed (Calvinist), and Mennonite church services in German. To this day the Amish and Mennonites conduct services in German (see below).[12]

As Germans arrived there came a third wave of people, the Scotch-Irish. The Scotch-Irish were Scots who in the early 1600s left lowland Scotland for Northern Ireland or Ulster. They were Presbyterians, and religious strife with Irish Catholics in Ulster prompted their emigration to Pennsylvania beginning about 1725. In Pennsylvania they settled in every major district, and they were especially well represented beyond the Germans in the frontier Ridge and Valley Province. They also grew wheat, raised livestock, and prospered. Geographer James T. Lemon, in a

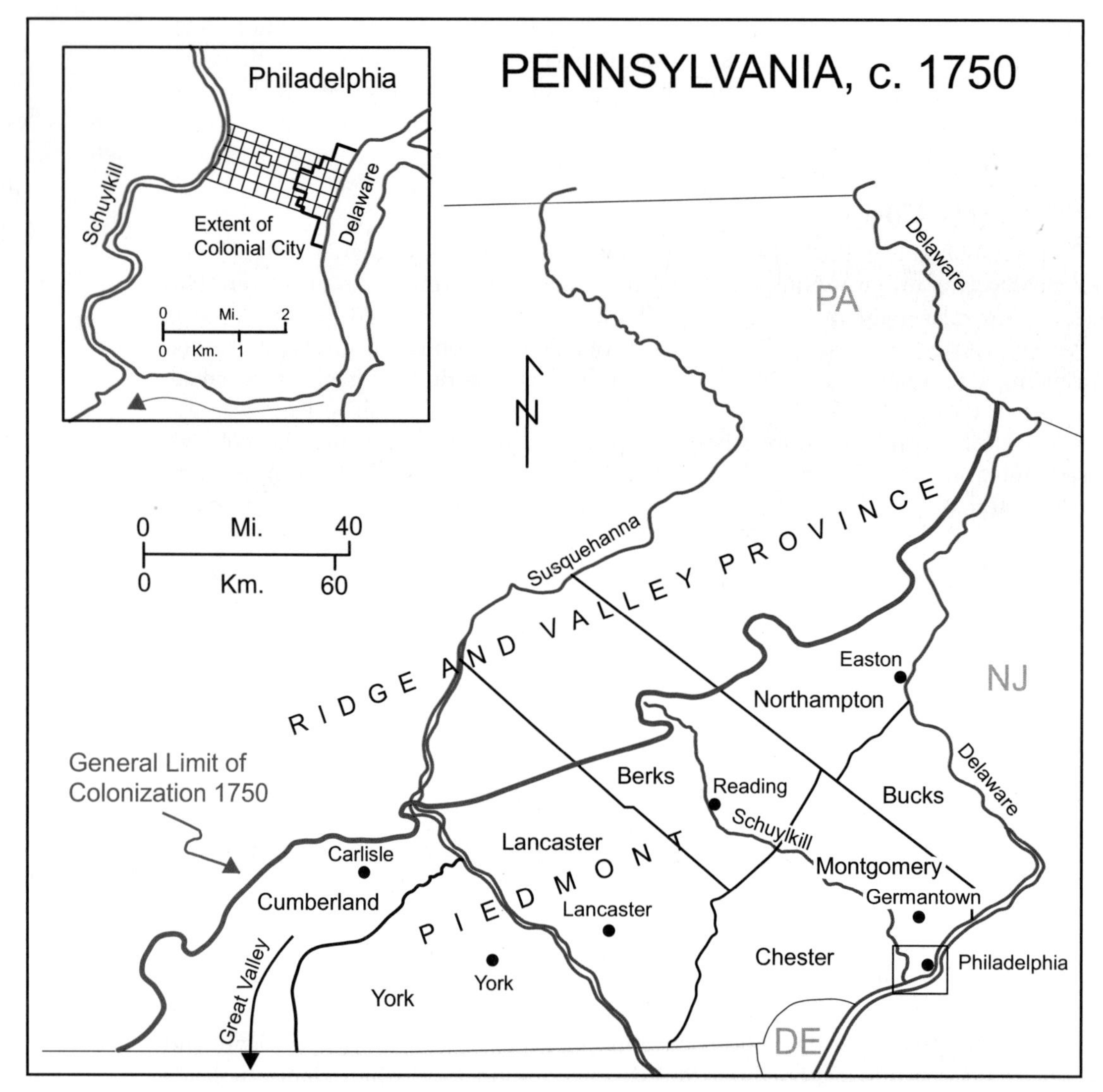

Map 5.3 Pennsylvania, c. 1750.

Source: Meinig, *The Shaping of America*, 1:138 (map), for county boundaries and "general limit of colonization 1750."

careful analysis of how a group's national origins correlated with agricultural success in colonial southeastern Pennsylvania, has invalidated the stereotype that Germans preempted the best limestone soils and were better farmers than were their English and Scotch-Irish neighbors. In this densely inhabited and bountiful "bread" colony west of Philadelphia, everyone seemed to prosper. Meanwhile, on Philadelphia's skyline, the steeples of Presbyterian churches were added to those of the Lutherans and Anglicans. But there were no steeples or spires on the modest meetinghouses of the Quakers who had initiated the colony.[13]

For his Philadelphia, William Penn wanted a central square with a distinctive layout. In the middle of Philadelphia's grid at the intersection of its two major streets, High (later, Market) and Broad, Penn created a central square carved out of the inside corners of four adjacent blocks (see figure 5.4). Soho Square and others like it in the grid pattern of London, implemented after the Great Fire of 1666, must have been Penn's inspiration. Within his central square Penn envisioned small buildings in each corner and apparently a somewhat larger structure in the middle, but the

Figure 5.3 Philadelphia's Red Brick Streetscapes, 1976. Much modified since colonial times, these persist within a few blocks of the Delaware River. (A) Residences on Delancey Street. (B) Retailing on the ground floor and residences above, on Second Street.

Photographs by the author, April 18, 1976.

square's center, perhaps its original corners, appears to have been left open. This same distinctive Philadelphia Square, known to many today as the "Pennsylvania diamond," diffused as part of the grid pattern to Lancaster, which was founded in 1731 and soon became Pennsylvania's largest inland community. As the county seat, Lancaster came to have a courthouse in the middle of its square. The literature on central courthouse squares notes that the "Lancaster Square" has a courthouse in the square's center while the "Philadelphia Square" does not. Ironically, the two-story brick courthouse built in 1739 in Lancaster's square burned to the ground in 1784, thus leaving the square open. On Philadelphia's square stands City Hall, which was built beginning in 1871.[14] City Hall surrounds, in atrium fashion, City Hall Courtyard, an open area used for tree-lighting and increasingly for other events.

The Lancaster Square and its Philadelphia prototype are of seminal importance. They made southeastern Pennsylvania the American source area for the idea of the central courthouse square, argues geographer Edward T. Price. For transacting political business, the original thirteen colonies and the states that followed were carved into counties—today some three thousand across the country—and a grid pattern was often favored as the layout for new county seats. "Pioneer pride," suggests Price, may explain why many county seats have central squares for their courthouses. Price shows that the Lancaster Square diffused with importance within Pennsylvania before the mid-1790s, and between about 1780 and 1840 it spread west and south from Pennsylvania, but only rarely to the Lowland South or the Trans-Mississippi West. About 1780 an apparent variant of the Lancaster Square appeared in Harrisonburg, Virginia, and a square with its name then diffused with some importance to Ohio, Georgia, and elsewhere (figure 5.4). About 1810 a courthouse was placed in the middle of a single block of the grid in Shelbyville, Tennessee; this so-called Shelbyville Square became the most frequently used model for the location of new courthouses until about 1900, when the founding of new counties had pretty much run its

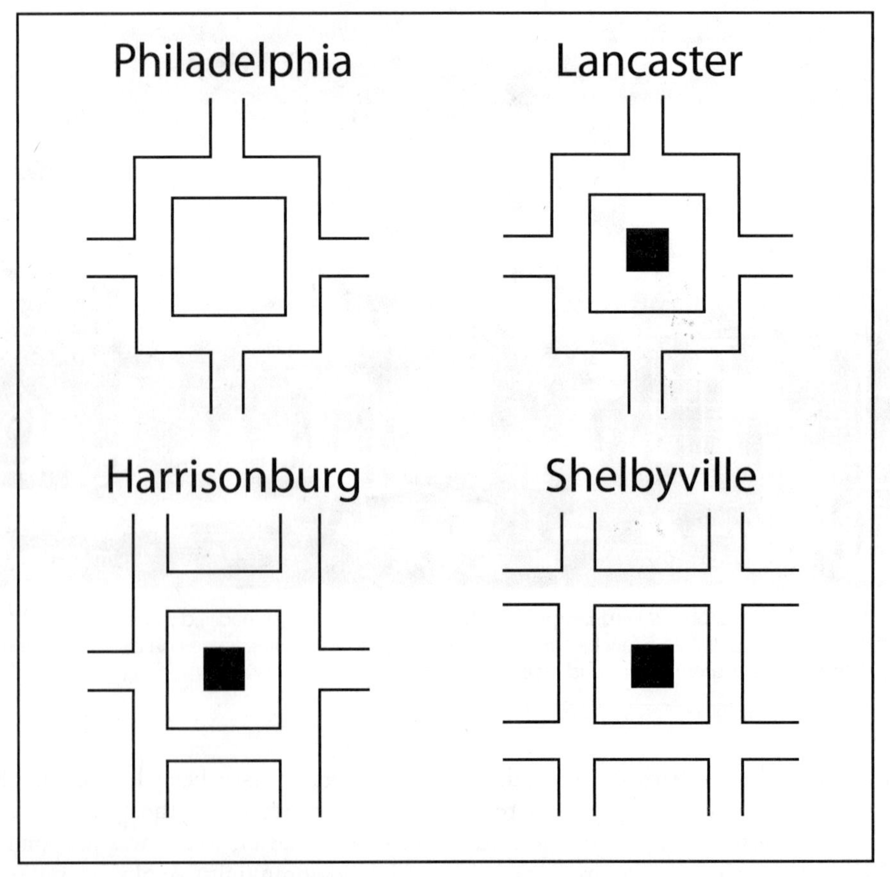

Figure 5.4 Courthouse Squares. Black squares indicate courthouses.

Source: Price, "The Central Courthouse Square in the American County Seat," 30.

course. Many counties in the United States have no central courthouse squares in their county seat communities, but where such squares do exist, they are prominent landscape features and, especially in counties that have remained rural, continue to play important roles politically as well as economically and socially.[15]

Southeastern Pennsylvania became the source area for a second landscape feature, the so-called bank barn (see figure 5.5). Apparently introduced from the German part of Switzerland to Pennsylvania in the early eighteenth century, this large barn built of stone, brick, or wood has two levels: a basement where livestock are kept in stables and an upper level used for threshing grain (above which is a loft for storing hay). What makes the bank barn distinctive is a "forebay" that overhangs the basement level and provides shelter at the barn's livestock entrance. A bank of earth or a ramp gives access for wagons to unload grain to the upper level, and for this reason bank barns are often built into a hill. That bank barns are large reduces the need for a farmer to have toolsheds or chicken houses or other outbuildings, and Pennsylvania farmsteads are characterized by the absence of such outbuildings. Bank barns, also known as Sweitzer barns or simply as Pennsylvania barns, diffused south beyond Pennsylvania notably to the Shenandoah Valley of Virginia. Their diffusion west to Ohio and beyond is somewhat correlated with the Amish.[16]

The Amish, Pennsylvania's most distinguished German group, broke away from the less conservative Swiss Mennonites in Europe in the 1690s. They arrived in Pennsylvania beginning in 1717, and they have from the beginning chosen to live in separate colonies. After the colonial period the Amish spread to most of the northeastern states, but their core states are Pennsylvania, Ohio, and Indiana. Living in colonies, speaking German in the home (but also being fluent in English), and dressing simply are aspects that distinguish these people. They have their own grade schools, and because their education ends with the eighth grade, Amish men are usually farmers and craftsmen (increasingly also factory workers), not professionals such as doctors and lawyers. The Amish have no churches, preferring to meet biweekly in private homes, where services are conducted in German. They are not opposed to new technology like electricity and automobiles, but to avoid contact with the "English," as outsiders are known, they use propane for fuel and horse-drawn buggies for transportation. Lancaster County's Amish attract thousands of tourists to Pennsylvania annually, but they are most numerous in Ohio. Holmes County in east central Ohio has the largest single Amish colony, and Millersburg, its county seat, is where the Amish manufacture their buggies.[17]

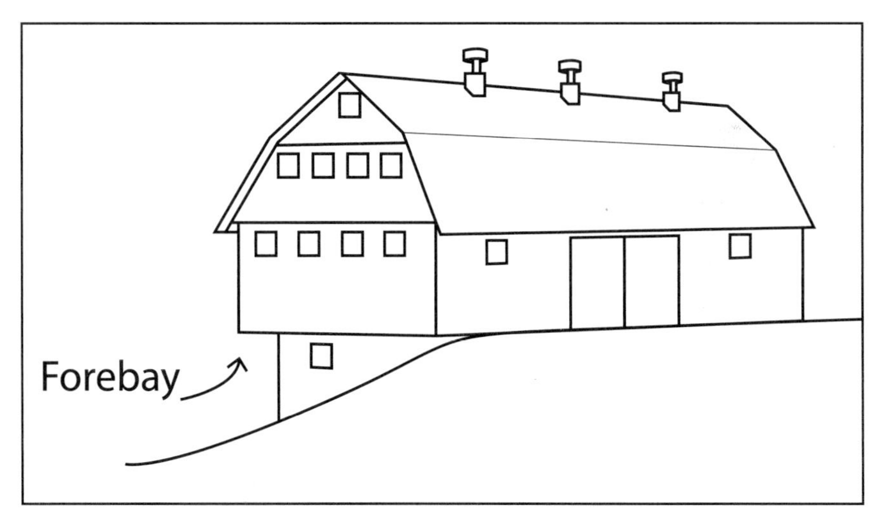

Figure 5.5 Pennsylvania or Bank Barn. Also known as the Sweitzer barn.

Pennsylvania Culture Area

In a delightful contribution to geography's literature, Wilbur Zelinsky (see spotlight) in 1977 analyzed the "townscapes" of what he called the Pennsylvania Culture Area or PCA. From his home base at Pennsylvania State University, located in State College on the far western edge of the PCA, Zelinsky crisscrossed southeastern Pennsylvania many times, and from his systematic field observations he came up with a list of attributes shared by 234 towns in the PCA. These attributes account only for pre-1870 features, and seven that stand out are summarized here:

- Compactness of settlement features such as row houses that abut sidewalks;
- Use of red brick for the construction of buildings, often sidewalks, and sometimes streets;
- Rectilinear streets lined with curbside shade trees;
- Streets having arboreal names like Maple and Walnut;
- Alleys that are paved, well lighted, and intensively used for shops, restaurants, and dwellings;
- A "scrambling" of land-use functions including residential, retail, professional, government, religious, and recreational; and
- A community focus in a central square or in what locals sometimes call a "diamond." Zelinsky notes that some central squares are elongated to look more like diamonds than squares.[18]

Zelinsky's PCA, with its "bumpily ovoid, or amoebic" shape, occupies a significant part of the southeastern corner of Pennsylvania (see map 5.4). Indeed, the PCA

Spotlight

Wilbur Zelinsky (1921–2013)

Wilbur Zelinsky in front of the Crossroads Cathedral on Shields Boulevard, Oklahoma City, Oklahoma. Photograph by the author, April 12, 1981.

All of Zelinsky's publications are a delight to read. And there are seemingly few geographical topics about his "beloved country" that escaped Zelinsky's curious gaze. The range spans from connecting barns to Pennsylvania towns, from cemetery names to vernacular regions, and from the season of marriage to women in geography. The book Zelinsky was proudest of, he once told me, is his *Nation into State* (1988), which he thought had yet to receive full critical appreciation. Zelinsky was born in Chicago, where he was his high school's valedictorian in 1939. He earned both his undergraduate degree (with honors in 1944) and his doctorate (1953) at the University of California–Berkeley. He taught at three universities and worked as an industrial analyst for the Chesapeake and Ohio Railway before ending a long career at Pennsylvania State University (1963–1987). From 1972 to 1973, he served as president of the Association of American Geographers. A cultural geographer who approached his research historically, Zelinsky's fame will rest on his unique creativity, his alluring and clever prose, and the array of significant topics he tirelessly pursued.

stretches beyond Pennsylvania into the Piedmont of Maryland and to small areas of West Virginia and Virginia in the lower Shenandoah Valley. Zelinsky postulates that the townscape morphology that defines the PCA came to cover this large area simply through inertia: It was the urban style in vogue during a time of vigorous population expansion in the eighteenth century. The unanswered question is why there is a "gap" of some twenty to fifty miles that separates the inland PCA from the Delaware River and its source of inspiration at Philadelphia. Zelinsky notes that his PCA correlates nicely with the "Pennsylvania Barn Region" as mapped in 1971 by geographer Joseph W. Glass (map 5.4). However, unlike Zelinsky's PCA, there is no

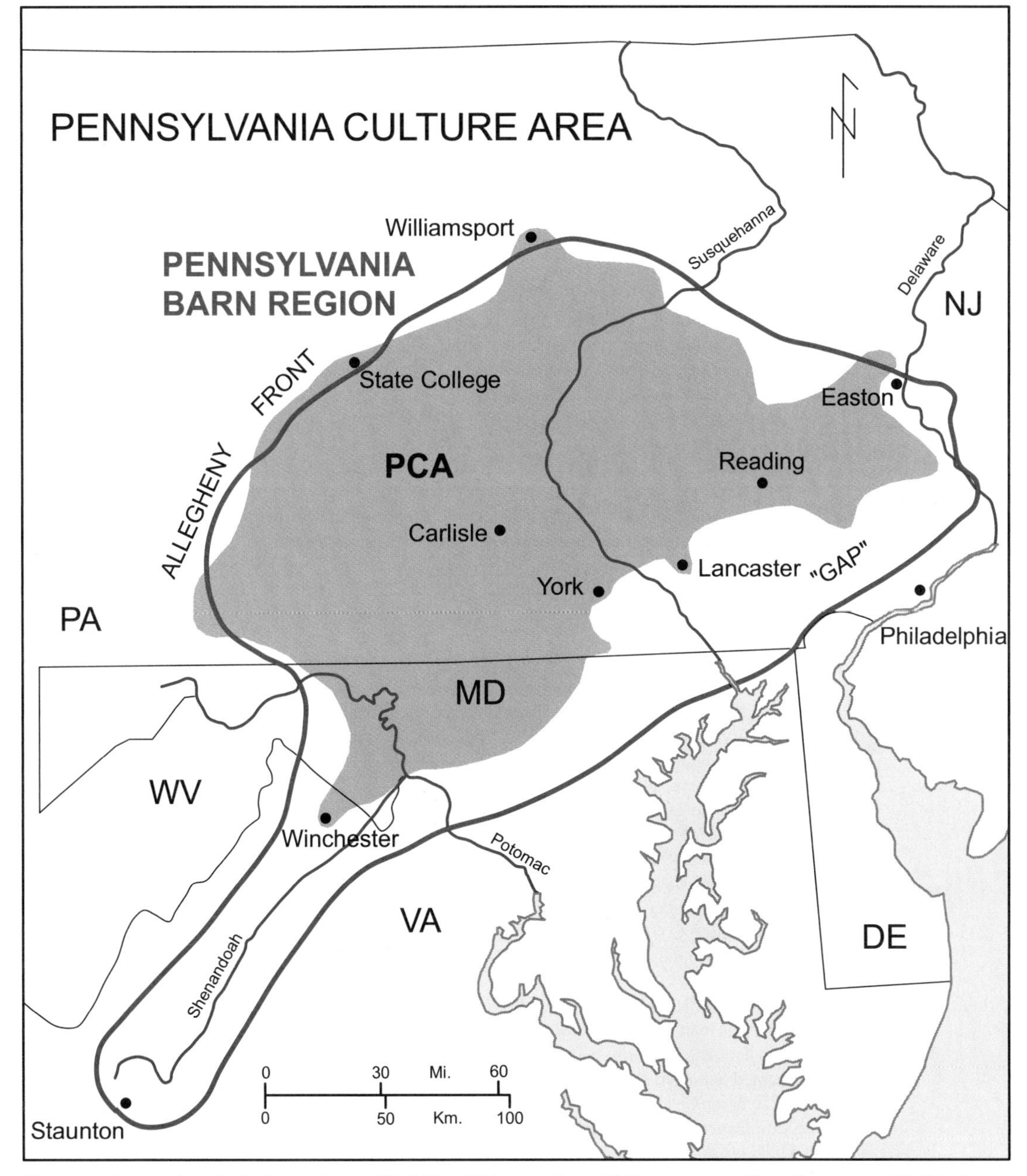

Map 5.4 Pennsylvania Culture Area, 1977. The PCA contains 234 "townscapes" that display pre-1870 features.

Source: Zelinsky, "The Pennsylvania Town," map 130, for the PCA; maps 142 and 143, for reproductions of Joseph W. Glass's Pennsylvania Barn Region, shown here.

gap between Philadelphia and the bank barns mapped by Glass. Clearly, there is need for someone to explain the gap.[19]

As aesthetically pleasing as they are, the attributes of the PCA—compactness, red brick, intense use of alleys—did not move west as a complex. However, as we have seen, one item that did move west out of southeastern Pennsylvania was the central courthouse square. And with or without a courthouse in the middle of it, this focus on squares and community came to be associated with business. A community focus with a business orientation was an American innovation. The plaza in the middle of a Spanish civil community had a social role but was dominated by the Roman Catholic Church—that is, by religion. So, too, did religion dominate the open common in the middle of the New England village. But in Pennsylvania, with exceptions, central squares were for periodic markets and commerce, and churches were located at non-central points elsewhere in the grid. As Americans moved west, it was the business-oriented community center out of southeastern Pennsylvania that triumphed. Peirce Lewis, Zelinsky's colleague in geography at Penn State, captured nicely the importance of southeastern Pennsylvania when he wrote that this tiny part of the colonial Atlantic Seaboard may well have been "the very headwaters of America's cultural mainstream."[20]

Notes

[1] In *Mirror for Americans*, 7–11, Brown also notes "Maritime Plains" and "Low Country" as terms used for the Region of Sea Sand, and "Upland Country" for Upper Country. See also Brown, *Historical Geography of the United States*, 97–98. Wacker, *Land and People*, 127 (map), shows New Jersey's Atlantic Coastal Plain as still unsettled in 1765.

[2] Brown, *Mirror for Americans*, 9, lists fourteen fall line communities. In "Fall Line," *Encyclopedia of Southern History*, 420, Gary S. Dunbar questions the fall line's "causal role." Merrens notes the absence of urban centers along the fall line in his *Colonial North Carolina in the Eighteenth Century*, 171, 178–79. In "Historical Geography and Early American History," 533–38, Merrens argues that along its entire length, the fall line played only a minor role in the origin of urban centers. Nevertheless, Grim and Holder give significance to the head-of-navigation reason in Virginia and Georgia. See Grim, "The Origins and Early Development of the Virginia Fall-Line Towns," and Holder, "The Fall Zone Towns of Georgia," 274–77 (Holder cites 3 and 275 as pages in Grim, whose thesis is unavailable).

[3] Brown, *Mirror for Americans*, 176, 208, discusses sloop and ship heads of navigation on the Hudson and Delaware, respectively.

[4] Brown, *Historical Geography of the United States*, 98, for black mold and nut-bearing trees, and the "forest-to-forest" cycle. Detailed dot maps constructed by geographer Herman Friis, in "A Series of Population Maps of the Colonies and the United States, 1625–1790," show the limestone valleys of Pennsylvania and the Shenandoah Valley of Virginia to be settled between 1720 and 1740.

[5] Meinig, *The Shaping of America*, 1:126 and 128, for the Dutch-French comparison; 1:41 for the Dutch East (1602) and West (1621) India companies.

[6] When torn down in 1699, New Amsterdam's wall became Wall Street. See Nelson, "Walled Cities of the United States," 8–10.

[7] Meinig, *The Shaping of America*, 1:129, for Dutch entrepreneurs visiting Sweden.

[8] Ibid., 1:119, for ten thousand Europeans in 1664.

[9] For the Flemish cottage, Wacker, "New Jersey's Cultural Landscape before 1800," 47 and 49; also Stump, "The Dutch Colonial House and the Colonial Revival," 45–46.

[10] See Jordan, "A Reappraisal of Fenno-Scandian Antecedents for Midland American Log Construction," 59 and 94.

[11] See urban historian John W. Reps, *The Making of Urban America*, 160, for "high, dry, and healthy." Philadelphia historian Jeffrey P. Roberts, in his lecture, "Historical Geography of Philadelphia," noted 1836 as the date construction in Philadelphia reached the Schuylkill. *Kill* means "stream" in Dutch; a channel named Schuylen Kill became known as the Schuylkill River.

[12] Meinig, *The Shaping of America*, 1:131, for twenty-three vessels arriving in 1682; 1:138–39, for English dominance in the first tier of counties and German in the second. David H. Fischer notes the North Midlands as the source area for "Friends" or Quakers; see his *Albion's Seed*, 438 and 440 (map). See Brown, *Mirror for Americans*, 202–3, for visitors to heavily German York and Reading reporting that Germans there spoke no English.

[13] For the sameness in agricultural success, Lemon, *The Best Poor Man's Country*, xvi, 63, 107, 216; also see Lemon, "The Agricultural Practices of National Groups in Eighteenth-Century Southeastern Pennsylvania," 467–96.

[14] Reps, *The Making of Urban America*, 163, for city planning in London after 1666 as an influence on Penn; 161, for Penn's plan for buildings in each "angle" of the central square; 166, for a map of Philadelphia circa 1720 showing a building that was projected for the middle of the square but was apparently never built.

[15] Price, "The Central Courthouse Square in the American County Seat," 41 and 43, for southeastern Pennsylvania source area; 60, for "pioneer pride"; 35 (map), and 42 and 44 (dates), for Lancaster Square diffusion; 51, for Harrisonburg Square; 44 (map), 48, 49, 51, for Shelbyville Square.

[16] See geographer Robert F. Ensminger's comprehensive study, *The Pennsylvania Barn*, 148 and 261, for maps that show the barn's diffusion; ibid., 163, notes that the Amish built bank barns in Holmes County, Ohio, and Elkhart County, Indiana.

[17] See geographer William K. Crowley's "Old Order Amish Settlement," 251, for arrival in 1717; 261, for three "core states," Ohio as the leading state in Amish numbers, and Holmes County as home to the largest colony.

[18] For a discussion of these attributes, see Zelinsky, "The Pennsylvania Town," 131–38.

[19] Zelinsky, "The Pennsylvania Town," 139, for "bumpily ovoid, or amoebic"; 145, for the origin of the PCA; 146, for the unanswered question of the Philadelphia to PCA gap. Zelinsky, 142 and 143, reproduces the regional boundary of the "Pennsylvania Barn Region" as delimited in Glass, "The Pennsylvania Culture Region." Richard Pillsbury notes the curious absence of Philadelphia's rectangular grid in the same gap area identified by Zelinsky, in "The Urban Street Pattern as a Culture Indicator," 439, 440, 442. Pillsbury found communities in the gap to have irregular or linear layouts.

[20] For the "very headwaters" quote, see Lewis, "Small Town in Pennsylvania," 330.

The South 6

The two Tidewater subregions of the colonial South differed in major ways from New England and the Middle Colonies (see map 6.1). In "Tidewater Chesapeake," a subregion that lacked any large urban centers, we will observe the English's use of slaves to grow tobacco on plantations. George Washington's "Mount Vernon," although atypical because of its size, is discussed as a plantation example. "Tidewater Carolina" will show that the English grew rice and indigo as commercial crops on their coastal plantations, and their imported slaves far outnumbered the area's population of white people. Charles Town (Charleston after 1783) quickly emerged to be the regional focus. Lying west of both Tidewater subregions, "The Upper Country" represented a mixture of westward-moving English and slaves in addition to Scotch-Irish and Germans who moved south from Pennsylvania.

Tidewater Chesapeake

For Tidewater Chesapeake Bay, imagine a section of the coast where the land has subsided relative to the offshore ocean (see map 6.2). Rivers like the James and the Potomac are literally drowned with saltwater as they flow into Chesapeake Bay; indeed, the bay itself is the huge drowned estuary of the Susquehanna River, as noted earlier. As a consequence of this drowning, the water level of rivers rises and falls twice daily with the tides, and the shorelines of rivers are indented with bays and necks of land. In colonial times, the bays provided anchorage for small, oceangoing sailing vessels, while the necks provided defensible sites for early settlements. One such neck of land, found forty miles up the James River, was selected in 1607 as the site for James Towne (later, Jamestown).

Much is known about Jamestown. Three ships carrying 144 Englishmen (males, literally) sailed from Blackwell near London on December 20, 1606, reprovisioned in the West Indies, and landed at the Jamestown neck of land on May 14, 1607. Mainly swampy lowland, this neck contained some 1,600 acres and, depending on the season, was an island at high tide and a peninsula at low tide. On a high riverbank next to deepwater anchorage, the neck's highest ground and the site of an Indian Old Field, the Englishmen in a short nineteen days built a triangle-shaped fort that enclosed about an acre. Erecting the fort's walls with timbers set upright in trenches in so short a time was, according to archaeologist William M. Kelso, a truly "Herculean" effort. But the events that followed brought the colony to the brink of collapse. The men, without much leadership but bent on finding minerals like gold or some resource of value to satisfy their sponsor, the Virginia Company of London, scattered and had to be forced to remain and work at James Fort. In their exploits these men alienated the local Powhatan Indians. Meanwhile, they were dying in large numbers, primarily from typhoid and dysentery. Of the 104 men left at Jamestown when the ships returned to England in June 1607, only 38 remained alive when reinforcements came back from England in January 1608.[1]

But as we know, the tiny settlement survived to become the first permanent English foothold in the present-day United States. More reinforcements came, including the arrival of the first women, in September 1608. Soon the men built houses beyond the walls of the fort, the Virginia Company transferred land to proprietors, and with the introduction of West Indian tobacco in 1612, for which John Rolfe is credited, the colony had a resource of value to export besides naval stores

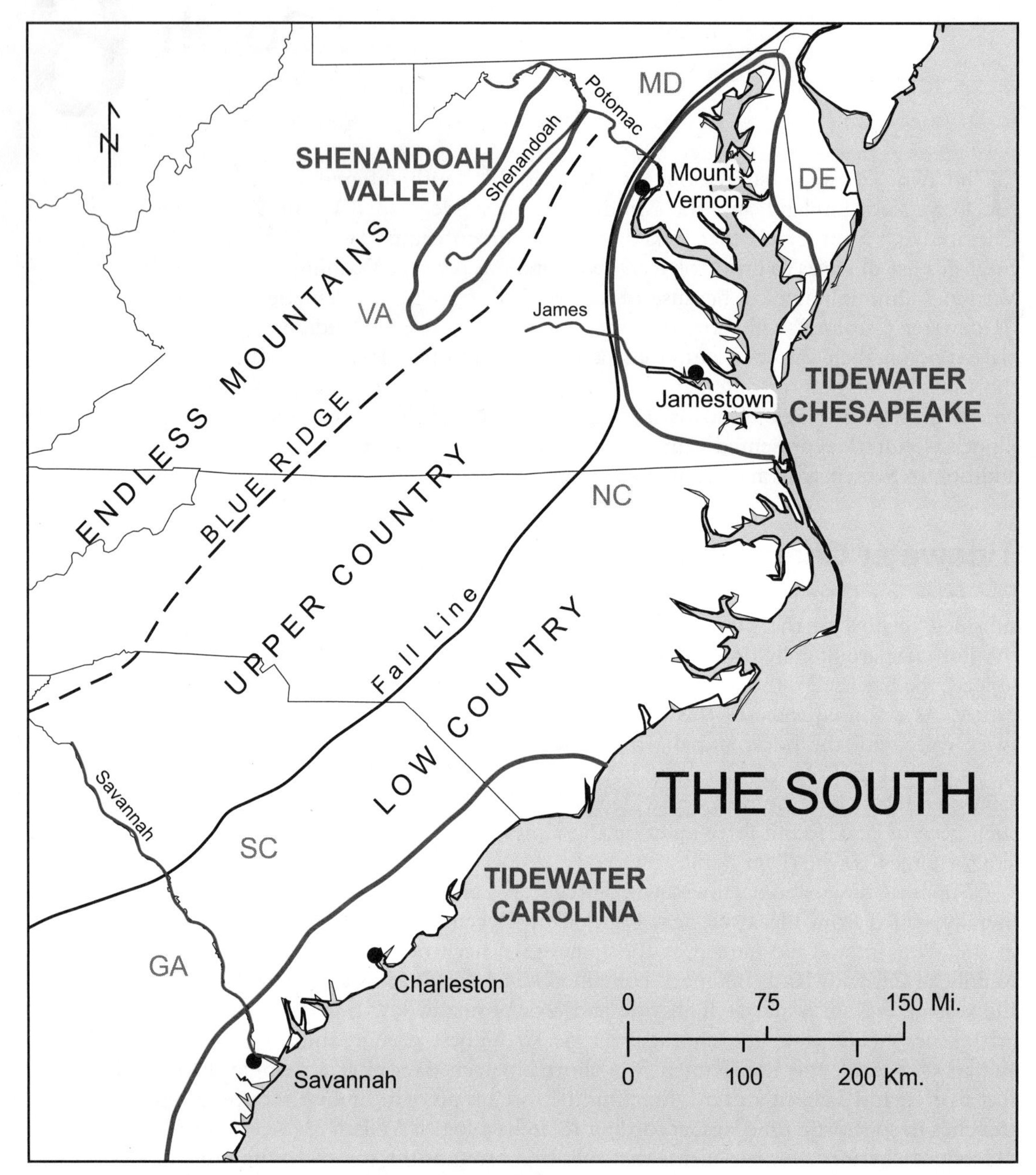

Map 6.1 The South. The colonial South covered half the Atlantic Seaboard from the upper Chesapeake to Georgia.

(tar and pitch) and sassafras (a medicinal cure for syphilis obtained from sassafras trees). The English continued to arrive, primarily from England's sixteen southern counties. Most were single males indentured to landowners—that is, the landowners paid for their passage from England, for which the new arrivals paid the landowners with some six years of labor. Landowners, meanwhile, gained additional land through "headrights"—they received fifty acres per individual introduced. In 1619 the colony received its first slaves—some twenty Africans from Angola who had been captured from a Spanish ship. Under Virginia Company control, the English spread up and down the James on tobacco-growing plantations and in small new communities

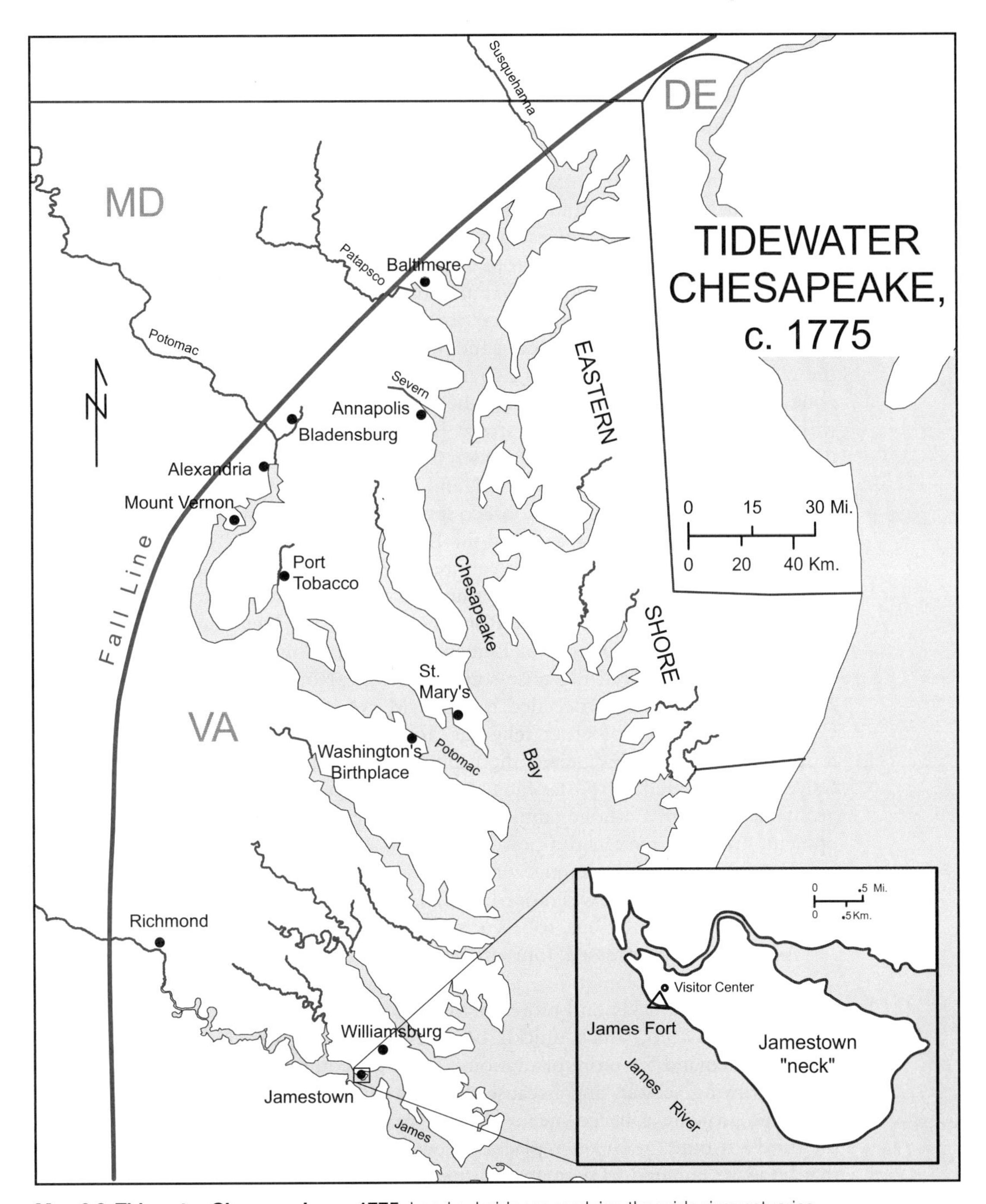

Map 6.2 Tidewater Chesapeake, c. 1775. Land subsidence explains the wide river estuaries.

called colonies or hundreds. But problems, primarily high mortality rates, plagued company efforts, and in 1624 English authorities revoked the Virginia Company's charter and made Virginia a Crown colony. Jamestown was named the colony's capital, and it remained so until 1699, when officials moved it to Williamsburg.[2]

Throughout the 1600s Jamestown floundered because of its unhealthy location. Its low-lying, swampy site seems to have been less the cause for disease than was its situation midway up the James River. Geographer Carville Earle argues that the

English in early Virginia died mainly from typhoid ("burning fevers") and dysentery ("bloudie flixes") and less so from salt poisoning. Mortality rates from typhoid and dysentery were highest in the middle (oligahaline) James River zone, where saltwater and freshwater came together, less in the lower saltwater zone, and least in the upper freshwater zone. In the middle zone at Jamestown, something of a "salt trap" caused river water to be contaminated with fecal material, especially in the late summer months, causing typhoid and dysentery to reach epidemic levels. John Smith, Jamestown's perceptive early leader, understood that unfit river water explained why the Indians, after planting their crops at the river in spring, disbanded to the hills and freshwater springs in summer and returned in fall to harvest their plantings. Smith wisely dispersed Jamestown's settlers in the summer of 1609, but that October, Smith returned to England, ending the policy. Then came perceptive Thomas Dale, the colony's governor from 1611 to 1616, who also dispersed settlers to healthier zones in summer, presumably because he understood the problem of unfit water. But with these exceptions, mortality rates at Jamestown were high. After 1624 a general dispersal to plantations occurred under Crown colony control; epidemics declined simply because of population dispersal, and Virginia slowly became a healthier place.[3]

In 1632, as plantations and tobacco spread in Virginia, King Charles I of England created Maryland. The king awarded the Eastern Shore, as it was called, and the land east of the Potomac on the western shore, to George Calvert, whose title was Lord Baltimore (map 6.2). George Calvert and his several sons, who took charge of the grant when their father died, were Roman Catholics, and their having been awarded land by a king whose Church of England (Anglican Church) had broken with the Roman Catholic Church did raise some eyebrows. Nonetheless, the Calvert family received the grant and proceeded to make Maryland a refuge for English Roman Catholics and people of other religions. In 1634 Leonard Calvert took charge of founding St. Marys, designated the capital, on a small inlet of the Potomac. The Calverts encouraged mixed farming, but their colonists quickly turned to growing profitable tobacco. Catholic families that arrived early controlled Maryland's development, and they remained a powerful faction even after they came to be far outnumbered by Protestant immigrants. Like Jamestown, St. Marys stagnated, and in 1694, Maryland's capital was moved to Annapolis. And like Virginia, Maryland came to be a colony of plantations, tobacco, and slaves.[4]

As the 1600s progressed, four attributes made Maryland and Virginia very much alike. The first was plantations, the privately owned and relatively large landholdings that spread quickly and rather chaotically around the indented shorelines. The second was tobacco, which quickly became the leading commercial crop and propelled the demand for more plantations. Tobacco required high inputs of labor over a long growing season, and because the crop rather quickly depleted soils of their nutrients, growing tobacco meant clearing new land in the winter months. Slaves, the third attribute, gradually replaced indentured servants after 1650; the reality was that bondage equated to permanency, and labor was needed year-round. Already by 1700, of a total 108,000 people living in Tidewater Chesapeake, some 15 percent were slaves in both Virginia and Maryland.[5]

The fourth attribute common to the subregion was its lack of urban centers, which is an enigma. Jamestown and St. Marys, like Boston and Philadelphia, were founded to be capitals and colony foci, but unlike their northern counterparts, neither grew. Their successor colonial capitals, Williamsburg and Annapolis, did not grow either. Earle offers a possible explanation: In New England and Pennsylvania, immigrants came as families; families constituted fully two-thirds of those going to New England and over half of those going to Pennsylvania. By contrast, the preponderance of those arriving in Tidewater Chesapeake were young, single males, first as company employees and later as indentured servants. And young single males, unlike families, had limited needs for the goods and services provided in urban places. Other reasons offered for the lack of urbanization include liberal land-granting policies coupled with the demand for tobacco-producing land, and the relative self-sufficiency of the larger plantations, which precluded many urban needs. Whatever the explanation, the lack of urban centers certainly did not hold back population growth in the area. By about 1775, estimates put the total population of Virginia and Maryland at 780,000, of which 500,000, now almost 45 percent black, lived in the Tidewater area.[6]

And so tobacco, which in 1700 accounted for a whooping four-fifths in value of all exports from the colonies, was shipped in barrels called hogsheads from hundreds of collection points, including plantation wharves (sometimes requiring literage by barges to ships), rolling stores ("rolling" because the hogsheads were rolled to them), and a few small port communities. One such small community was Port Tobacco, located on the Port Tobacco River, a tributary to the Potomac (map 6.2). As pine trees were removed and land for tobacco was plowed in the local river basin, eroded sandy topsoils gradually filled the Port Tobacco River with sediment. Poor Port Tobacco, once the county seat of Charles County, with a courthouse, eighty houses, a tobacco warehouse, and a landing from which hogsheads were loaded onto "large vessels," was left high and dry (see figure 6.1). Already in decline by 1796, Port Tobacco by 1862 had mudflats that put the community some 3,000 feet from the "head of tidewater," by 1882 it was some 4,500 feet from water, and by 1945 it

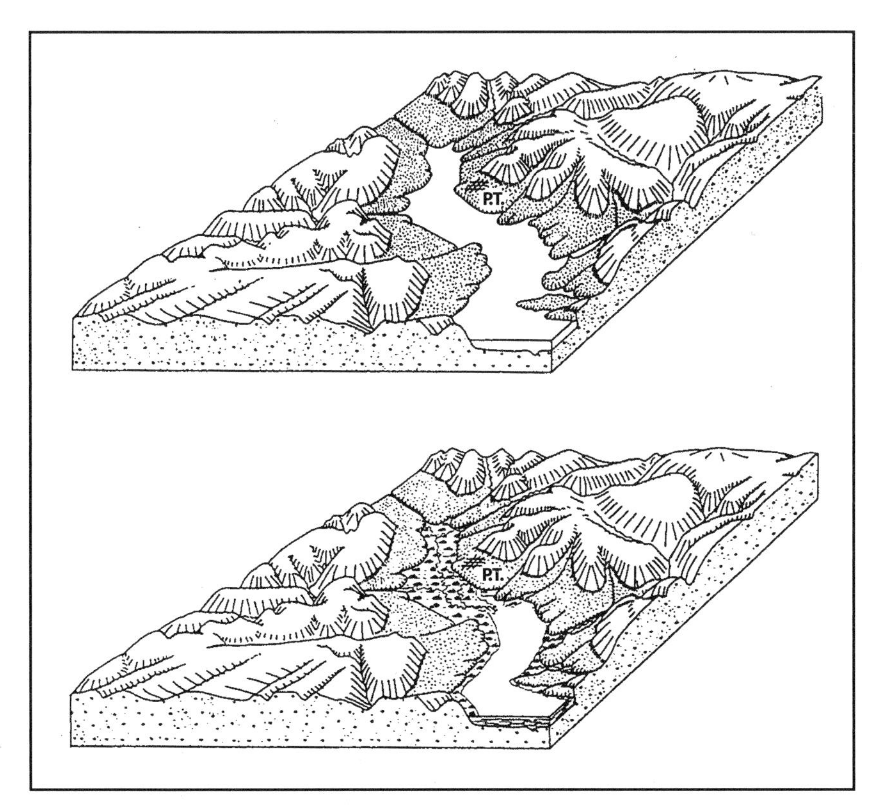

Figure 6.1 Port Tobacco, Maryland, 1945. These two block diagrams illustrate how sediment in the Port Tobacco River made Port Tobacco (indicated by "PT" and grid) inaccessible to ships.

Source: Gottschalk, "Effects of Soil Erosion on Navigation in Upper Chesapeake Bay," 232. Courtesy of *Geographical Review*.

was a mile away from water. L. C. Gottschalk documents how sedimentation caused the demise of a dozen such small colonial ports on the upper Chesapeake, including Bladensburg at the head of tidewater eight miles up the Anacostia River from present-day Washington, DC. Depletion of soil nutrients and soil erosion found Tidewater planters seeking new ground above the fall line even before 1700, and by 1775 half of the tobacco exported from Tidewater Chesapeake came from the Upper Country.[7]

Annapolis, because of its deepwater harbor at the mouth of the Severn River, survived as a port community despite upstream problems with silting. It remained Maryland's capital, and in 1845 authorities chose Annapolis as the location of the United States Naval Academy. By 1800, however, Baltimore, with its more ample harbor at the head of Patapsco Bay, had already surpassed Annapolis in trade (and in time Baltimore would become the seat of the Roman Catholic Church in the United States). Williamsburg had a different outcome. When Virginia moved its capital for a third time from Williamsburg to Richmond in 1779, the small College of William and Mary kept Williamsburg alive, but barely so. Indeed, the houses and buildings along this fine baroque community's Duke of Gloucester Street fell into such disrepair that in 1926 a local Episcopal minister, W. A. R. Goodwin, prevailed on wealthy John D. Rockefeller Jr. to financially support Williamsburg's restoration. Undertaken over several decades, the restoration was carried out with such meticulous accuracy that Williamsburg has become a showcase for historical preservation and a living outdoor museum that is high on the list of destinations for US and international visitors.[8]

Jamestown fared less well than Williamsburg did. After authorities moved Virginia's capital to Williamsburg in 1699, Jamestown was abandoned and its buildings disappeared. Much of the 1,600-acre neck of land went under the plow. The Association for the Preservation of Virginia Antiquities (APVA), formed in 1889, acquired the Jamestown site, and for a tricentennial celebration, in 1907, it erected a monument next to the only remaining above-ground artifact: a brick church tower. For Jamestown's 350th-year celebration, in 1957, interpretive buildings that included a glassmaking factory were on-site, and visitors were told that the original James Fort had been eroded away by the James River. Amazingly, as preparations began for the quadricentennial observance in 2007, William Kelso, APVA's head archaeologist, in April 1994 found what turned out to be two of James Fort's original walls. The third wall had succumbed to erosion (see map 6.2 inset). Artifacts, especially coins and clay pipe bowls, were used to date the stains of the two walls' rolled-out circular timbers that had stood upright in trenches, and the forty-six-degree angle formed by the two walls where they joined at a circular bastion matched precisely a 1610 description of the original fort's design. Inside the fort was evidence of buildings constructed using "mud and stud," and artifacts, including an Indian cooking pot, that suggest that Native Americans spent significantly more time at the fort than has previously been thought.[9]

Mount Vernon

Unlike in New England, where "plantation" meant the planting of people, as at Plymouth Plantation, in the South a plantation meant the planting of crops. This southern meaning—land used for agriculture—originated early in the colonial period, likely on the peninsula between the James and the York rivers, according to Ralph Brown. In the South a plantation referred to a privately held parcel of some size, with the majority around the Chesapeake Bay seemingly between five hundred and one thousand acres. These parcels had a cluster of buildings that included the owner's manor house with its detached kitchen wing, perhaps the quarters of an overseer, and buildings such as a smokehouse, a carriage house, and barns. The slaves' quarters were usually a short distance away, and the complex in its entirety constituted a somewhat self-sufficient "little village." In fields cleared from the pine forest, slaves planted crops of wheat, corn, rye, oats, and most importantly, tobacco. Plantations with Tidewater locations had a wharf, from which hogsheads of tobacco would be exported, and through which came items like rum, molasses, tea, and, at Mount Vernon, undoubtedly oranges, George Washington's favorite fruit.[10]

The foregoing attributes of a plantation describe Mount Vernon. The Washington family acquired Mount Vernon in 1674. George Washington was actually born on the lower Potomac on Pope's Creek Plantation in Westmoreland County in 1732, but the family moved to Mount Vernon on the upper Potomac in 1735 (map 6.2). In 1754, when Washington acquired Mount Vernon at the young age of twenty-two, the plantation had some two thousand acres, thirty-six slaves, and a one-story mansion house. In 1759 Washington married Martha Custis, a widow with two children, and Mount Vernon became their residence until they died (George in 1799 and Martha in 1802). Washington loved his estate, and through hands-on management, he orchestrated Mount Vernon's growth to nearly 8,000 acres, 316 slaves, and the addition to the mansion house of a second story—complete with a high-columned piazza on the Potomac-facing front. When Washington mapped his own property in 1793, the eight thousand acres were carved into five farms (see map 6.3). Four had overseers, and Washington himself ran the Mansion House Farm. George and Martha Washington were part of society's wealthy elite.[11]

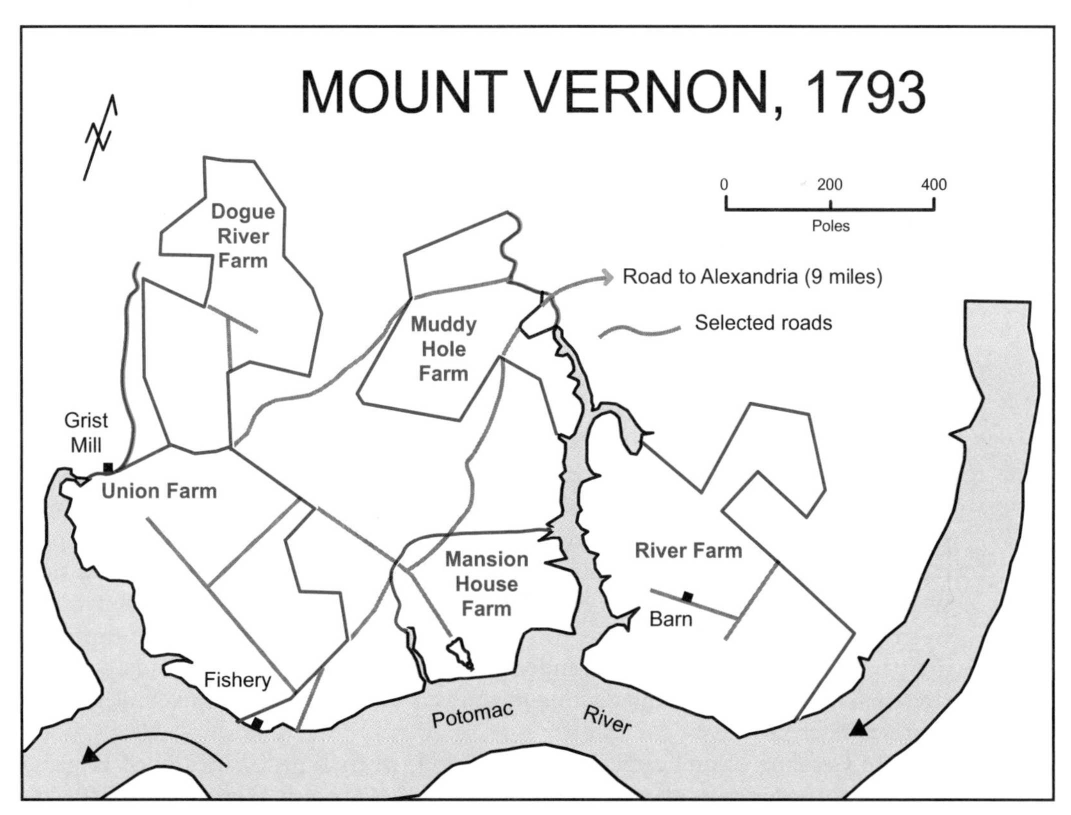

Map 6.3 Mount Vernon, 1793. Washington's own map of his plantation is redrawn and modified to show more clearly his five farms. Scale is in poles: one pole equals 16.5 feet or about 5 meters.

Source: Rees, Mount Vernon, 122–23.

Only some three thousand of Mount Vernon's eight thousand acres were actually cultivated. When he took charge of the plantation, Washington hoped to carry on as a tobacco planter, but he soon realized that tobacco had depleted and eroded the soils, that tobacco's demands for labor would take the full attention of his slaves, and that tobacco merchants in England were taking advantage of him. In the 1760s Washington purposefully diversified to mixed grains, particularly wheat, and for markets he sought local merchants in places like Alexandria. Wheat became his cash crop, but he also grew corn, rye, hemp, flax, potatoes, and clover—altogether some sixty crops and a variety of livestock. Washington was well-read on recent agricultural innovations, and he put his knowledge into practice. To restore his soils, he fertilized liberally with barnyard manure, and he experimented with spreading mud that was scraped from the Potomac River and brought up by slaves in carts. He devised a seven-year rotation plan: After a field lay fallow (unseeded) in the first year, he would plant wheat, peas, barley, soy, oats, and pasture in the next six years. To spread out wheat's high labor demands at harvest, Washington practiced "crop scheduling" to stagger its planting. Slaves, of course, provided the labor, preparing fields, planting and harvesting crops, and constructing the post-and-rail fences that enclosed Washington's rather straight-lined fields. Those who were skilled as carpenters, weavers, tailors, and wheelwrights fulfilled many of Washington's other needs.[12]

Washington appreciated the slaves who worked at Mount Vernon, and in his will he arranged that they be gradually emancipated after the death of Martha. Unfortunately, Washington's heirs rather mismanaged the estate they inherited, and to keep Mount Vernon going, they sold off parcels of land until finally, in the 1850s, only some two hundred acres of the Mansion House Farm remained. In a well-publicized story, Ann Pamela Cunningham, a resident of South Carolina, incensed that the home of the first president might be sold to private property, established the Mount Vernon Ladies' Association (MVLA) and raised the money to buy what remained of the land in 1858. The MVLA runs Mount Vernon as a national shrine today.

Tidewater Carolina

Down the Atlantic coast lay Carolina. Carolina's Low Country stretched inland from the Atlantic to the fall line, which meant a zone some 120 to 150 miles wide (see map 6.4). The first forty miles going inland were especially flat and featureless. Crossing this Low Country were numerous rivers, including Carolina's three "great" rivers, the Pee Dee, Santee, and Savannah. Tides would ascend rivers of slight current to thirty-five miles, but they ascended the Santee, a river of strong current, only fifteen miles. When the water in a river rose, freshwater, which floats on top of the denser saltwater, overflowed the river's natural levees to flood the tidal swamps, as the floodplains that bordered rivers one-half to three miles back were called. Because planters sought them for their rice plantations, tidal swamps were of great importance in colonial Carolina. And just where plantations were located within these tidal swamps had rather well-defined limits: The downstream limit was that point where water overflowing the levees was too brackish for crop growth, and the upstream limit was where higher ground ended a river's tidal effect. Usable tidal swamps stretched for some twenty miles along most rivers that began four or five miles inland from seawater.[13]

The Carolina colony came to exist in 1663, when King Charles II of England awarded to proprietors the coast from Virginia to Spanish Florida. In 1670 colony leaders founded Charles Town. Several years later they moved the community to a neck of land that jutted into salt water between the Ashley and Cooper rivers, a site that was somewhat healthier because of the adjacent salt water, but one

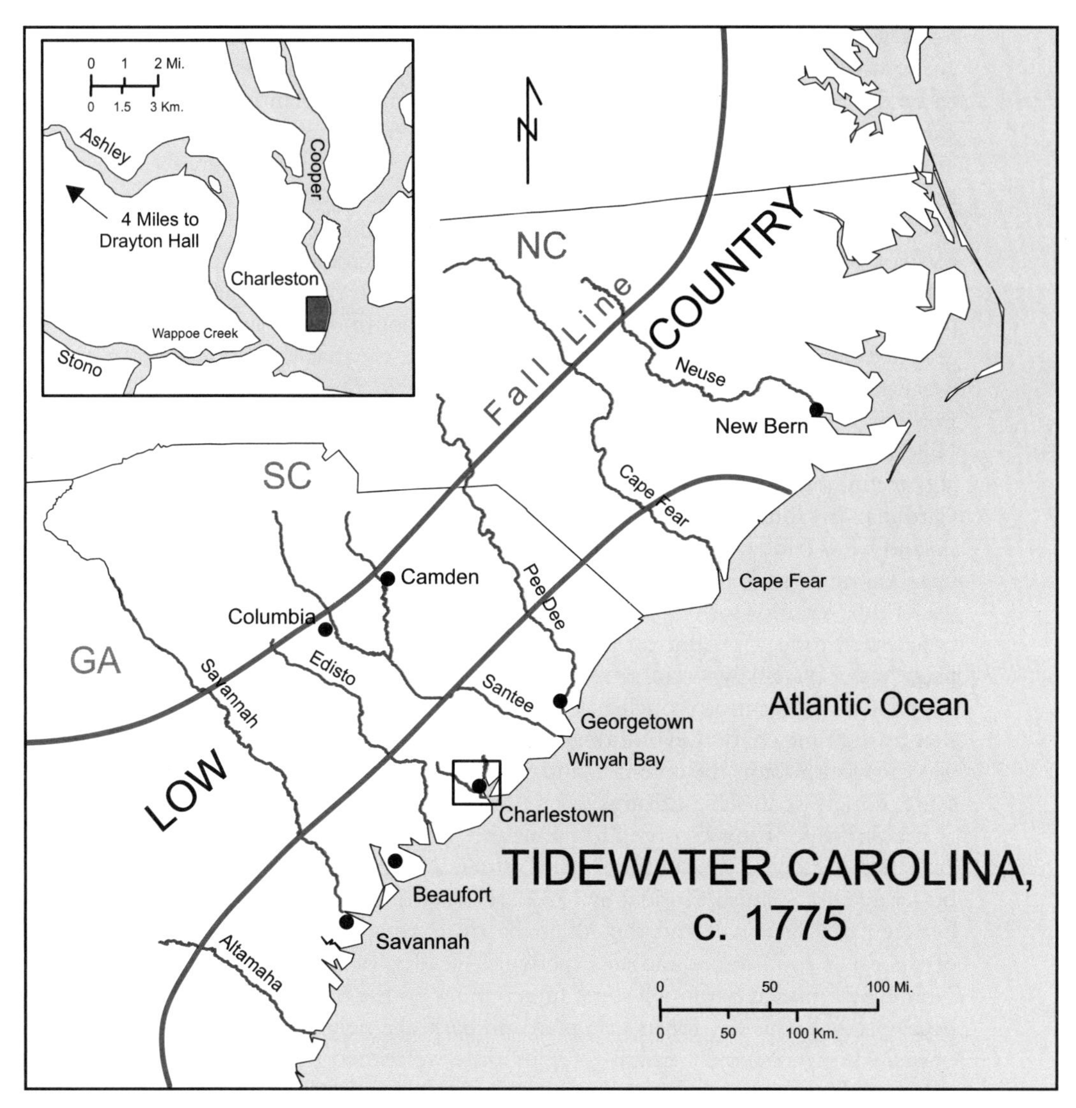

Map 6.4 Tidewater Carolina, c. 1775. The Low Country stretched inland 120 to 150 miles to the fall line, but 90 percent of the population lived in the narrow tidewater zone shown.

whose highest point was still only sixteen feet above sea level (see map 6.4 inset). Designated Carolina's capital, Charles Town (later, Charleston) soon became the colony's economic focus and the funnel through which naval stores, cattle, and rice were exported and slaves were imported. Gradually, English families arrived from England and indirectly from Barbados. Their plantations averaged five hundred to one thousand acres, and rice grown by slaves became the leading crop. By the 1730s, of some thirty thousand people in Carolina, blacks, of whom most if not all were slaves, accounted for twenty thousand, a ratio that haunted leaders, who correctly feared an insurrection. Meanwhile, the northern half of the Carolina grant, largely a neglected backwater, was made the separate colony of North Carolina in 1712, with New Bern, founded in 1710, its future de facto capital.[14]

Introduced perhaps as early as the 1680s, rice became Carolina's main staple export crop in the 1700s. In tidal swamps, slaves dug wide ditches around squared fields, and with the dirt they constructed embankments or dikes. Openings in the dikes, called trunks, were shut by use of floodgates that controlled the flooding of fields at times of high tide. Only after the ditches and dikes were completed did slaves cut down the trees and, in time, remove the stumps from the fields. The many tasks to be done—from leveling fields to controlling water to planting and weeding and harvesting the rice to removing by hand the husks from the grain—all required heavy inputs of labor. In June, after the "first flooding" of the crop, many wealthy planters moved from their rural mansions to town houses in Charles Town to escape the unhealthy "pestiferous effluvia" of the swamps. By the mid-1700s, then, in a patchwork of favored tidal swamp locations along Carolina's extreme coast, rice plantations were in place, and according to geographer Sam B. Hilliard, the "premier" rice-producing part of the coast was around Winyah Bay and the mouth of the Santee.[15]

Geographer Judith A. Carney argues persuasively that slaves taken from coastal western Africa supplied much of the know-how in South Carolina's rice culture. For centuries rice had been grown in West Africa between present-day Senegal and Nigeria, the source area for South Carolina's slaves. Carney notes that microclimates and techniques for water control were "identical" between West Africa and South Carolina. In South Carolina, Carney says, rice was first grown in about 1700 on upland fields (under natural rainfall) in rotation with cattle pasture before shifting to inland swamps and, by the 1750s, finally to the tidal swamps. The first documented use of tidal swamps for rice came from Winyah Bay in 1738. In 1739 a slave rebellion called the Stono Uprising prompted planters to shift from using slaves in "gangs" to a "task" system wherein slaves had free time once their daily assignments were completed. In this more conciliatory environment, rice prospered, and Carney asserts that by the time of the Revolutionary War, the South Carolina rice planter aristocracy was the wealthiest in the colonies, and South Carolina's slave culture was, in a relative sense, closest of all slave cultures to its African roots.[16]

Meanwhile, King George II of England, disappointed that the Carolina proprietors had neglected the southern part of their colony and eager to have a settlement buffer against Spanish Florida, in 1732 created the colony of Georgia from the land between the Savannah and the Altamaha rivers (map 6.4). James Oglethorpe, one of Georgia's proprietors and an experienced leader, personally took charge: For the capital, he founded Savannah some fifteen miles up the Savannah River, a community noteworthy to this day for its grid plan and park-like neighborhood squares. For colonists, who were mostly English, Oglethorpe welcomed debtors and the religiously persecuted, experimented with producing subtropical products such as silk and wine, and imposed strict rules that prohibited slaves and rum. During the 1740s, however, Georgia's "propinquity" to South Carolina, where planters were clearly profiting from rice grown by slaves and where consumption of rum was quite legal, proved too great to overcome. In 1751 the Georgia proprietors terminated their contract and Georgia, turned into a Crown colony, became much like South Carolina—a riverine society of rice plantations carved out of tidal swamps.[17]

Carolina's settlement frontier moved inland with the introduction of a new crop: indigo. Indigo is a tropical and subtropical plant that is grown for the blue dye extracted from its leaves. In 1739 Eliza Lucas was among the first to plant indigo seeds (obtained from her father in Antigua) along Wappoe Creek across the Ashley River west of Charleston (map 6.4 inset). During the second half of the eighteenth century, indigo rose meteorically to become Tidewater Carolina's second-leading staple after rice, only to fall and all but disappear as a crop by 1800, due to East India competition and the rise of upland cotton. Unlike rice, which remained anchored in the tidal swamps, indigo spread inland at least to the fall line. Geographer John J. Winberry points out that because planters tried to spread their risks by growing several crops, few if any grew indigo exclusively. One example was wealthy John Drayton, owner of a seven-hundred-acre plantation located ten miles up the Ashley River from Charleston, whose eight hundred slaves grew three main crops: rice, indigo, and cotton. Drayton's gorgeous Drayton Hall is unique as the only antebellum mansion house along the Ashley River to survive the Civil War (see figure 6.2).[18]

By 1750 some seventy-five thousand people, more than fifty thousand of them slaves, lived between the Altamaha and Cape Fear rivers (North Carolina's lower Cape Fear district was but an extension of South Carolina's people and culture). Roughly 90 percent of them lived in the narrow Tidewater rice zone. Charleston, with its heavily English and slave populations and a sizable minority of Anglicized French Huguenots, in 1800 ranked fifth in size among the Atlantic Seaboard's cities. Along the coast, two additional small, urban foci had developed at Georgetown and Beaufort, and as the English and their slaves pushed inland along rivers to the Upper Country, two more small, urban foci developed at Camden and Columbia (map 6.4). By 1800 upland cotton was on the rise in the Upper Country, and in the Tidewater area rice would continue as the leading crop until the Civil War, after which important rice growing would shift to Arkansas, Louisiana, and Texas.[19]

The Upper Country

Above the often imperceptible fall line lay the Upper Country. Its distance from coastal population centers and markets would be a problem for Upper Country pioneers, yet fertile soils and available land pulled settlers inland. Two distinct streams of colonists settled the Upper Country. Moving west up river valleys went the English with their slaves and plantations and commercial crops. In Virginia these settlements were made above the fall line by 1700, and in Carolina they reached the Upper Country some two decades later. Meanwhile, pushing south out of Pennsylvania and the Middle Colonies by the 1720s came the second stream, mostly Scotch-Irish and Germans with their grain and livestock. The gradual mixing of these two streams set the Upper Country apart from the two Tidewater subregions. Already settled by Tidewater planters, Virginia's northern Upper Country did not attract any significant influx of colonists from Pennsylvania or the Middle Colonies. But west of the Blue Ridge in Virginia's Shenandoah Valley, which was studied carefully by geographer Robert D. Mitchell (see spotlight), significant mixing occurred, and what happened is instructive.

Figure 6.2 Drayton Hall, located ten miles from Charleston, South Carolina. Built by John Drayton between 1738 and 1742, this exquisite example of Georgian Palladian architecture, elevated for greater ventilation in the hot and humid summers, survived Union troop destruction because Charles Drayton displayed yellow flags, signifying smallpox, at the plantation's entrance. The National Trust for Historic Preservation acquired the property in 1969.

Photograph from a National Trust brochure.

Spotlight

Robert D. Mitchell (1940–)

Robert D. Mitchell at Britannia Beach, British Columbia, August 19, 1992. Photograph by the author.

Bob Mitchell and I were graduate students together at UCLA between 1962 and 1964. A Scot by birth, Bob came to UCLA straight from earning his master's degree in geography at the University of Glasgow. He delighted one and all with his marvelous accent, and I admired the detailed map he hung above his desk on which he plotted his every worldly road trip. In 1964 Bob left UCLA to work on a doctorate (completed in 1969) under Andrew Clark at the University of Wisconsin–Madison. Bob put in a thirty-year career at the University of Maryland–College Park (1969–1999). An expert on the geography of the colonial Eastern Seaboard, Bob is perhaps best known for his careful analysis of the Shenandoah Valley in the eighteenth century, synthesized in 1977 in *Commercialism and Frontier: Perspectives on the Early Shenandoah Valley*. The words *commercialism* and *frontier* capture his revisionist argument that even frontier societies sought to be commercially engaged in their agricultural marketing. For eight years (1982–1989), Bob coedited the *Journal of Historical Geography*. And in 1987, with his Maryland colleague Paul A. Groves, he coedited *North America: The Historical Geography of a Changing Continent*, the first synthesis of the historical geography of the United States literature since Ralph Brown's book in 1948.

The Shenandoah River, a tributary to the Potomac, flows northeast for some 175 miles. The valley itself displays variations in three sections that are identified in a general way by counties (see map 6.5). Settled first in the 1720s, the lower third attracted English from Virginia, who entered the Shenandoah through several wind gaps in the Blue Ridge, and Scotch-Irish and Germans from Pennsylvania, who reached the Shenandoah by funneling south through the Great Valley along the so-called Great Wagon Road. The English grew tobacco and other crops on plantations with slave labor; the Germans, who were rather strongly opposed to slavery, specialized in grains and livestock; and the Scotch-Irish, also grain and livestock farmers, found a market for hemp as well. By 1775 each ethnic group represented about a third of the population in the lower (and most heavily populated) valley section, and Winchester, with market ties across the Blue Ridge to Alexandria and Falmouth, had become the focus. In the middle section of the Shenandoah, all three ethnic groups were present, yet Germans, who, as in Pennsylvania, for decades clung to their language (at home, at church, and in parochial schools) and chose to be socially cohesive, constituted about half of the settlers by 1775. The grain and livestock exports of the middle third flowed north and then east, because Massanutten Mountain blocked easy access to gaps in the Blue Ridge that led directly east. In the upper third of the Shenandoah, Scotch-Irish represented well over half of all the settlers in 1775. Hemp, used in making rope and sacks, was the major export, traveling from Staunton to Richmond, through Rock Fish Gap (1,850 feet), and north by road to Philadelphia. By 1800 eighty-four thousand people lived in the Shenandoah Valley, each ethnic group represented a third of the total, and the rather intense social separation of each group was in decline.[20]

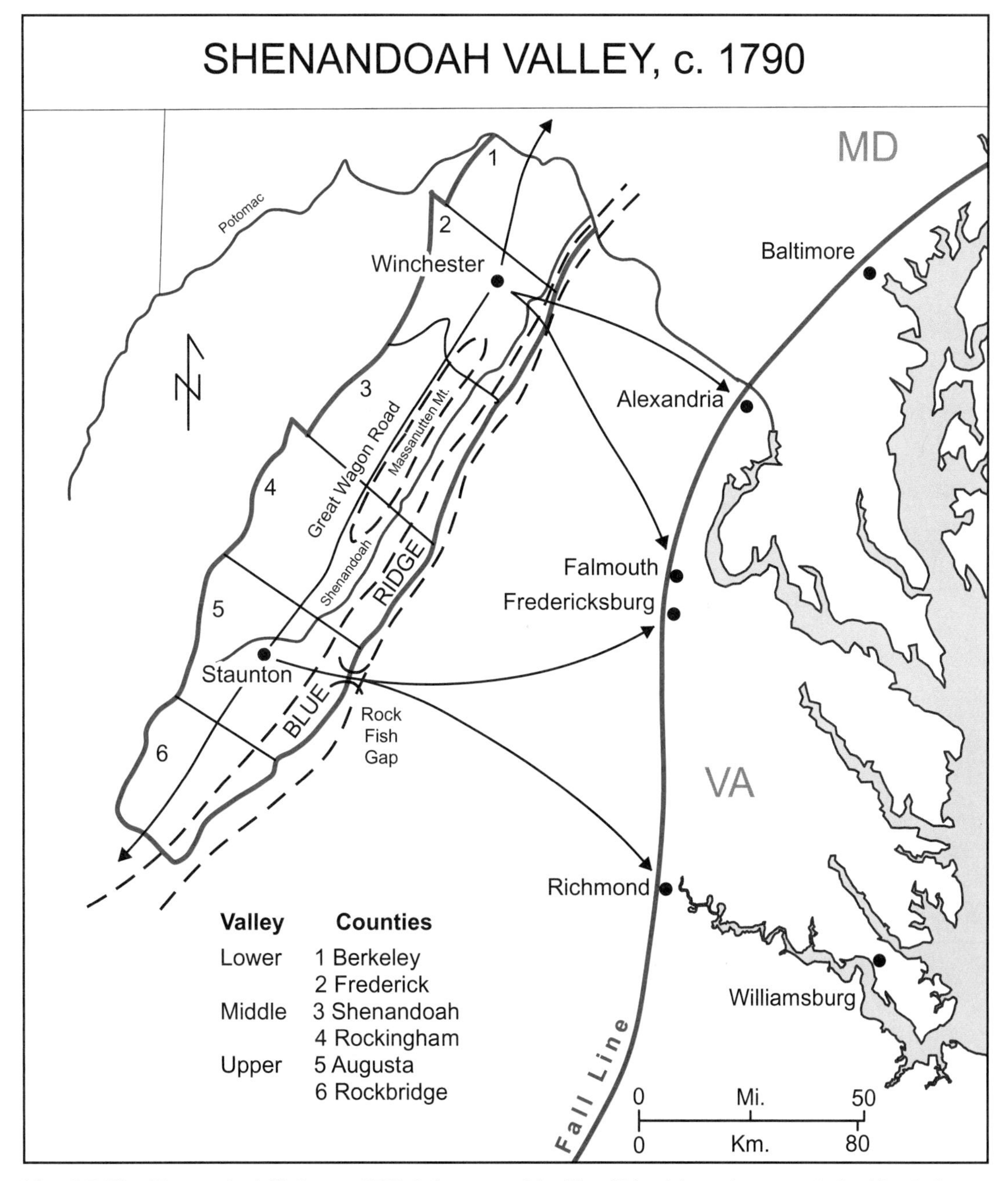

Map 6.5 The Shenandoah Valley, c. 1790. Lying west of the Blue Ridge (shown between dashed lines), the valley is subdivided into the six counties shown.

Source: Redrawn from several of Robert D. Mitchell's maps.

In Carolina's Upper Country toward the end of the colonial period, cattle raising had some importance. Cattle from Virginia and elsewhere reached coastal Carolina with the first settlers in the 1670s. Raised in the coastal swamps and savannas, cattle diffused inland with the English and black populations to the inner coastal plain and then to the Upper Country. Geographer Gary S. Dunbar describes the activity, which had its own terminology: "Cowpens" in colonial South Carolina were ranches, "cowpen keepers" were the white ranch managers, and "cattle hunters" were the black cowboys. On a few hundred cleared acres in the Upper Country, makeshift enclosures would be constructed for keeping cattle, hogs, and the horses used by the cattle hunters. Dunbar notes that such cowpens diffused from South Carolina to North Carolina and to Georgia, but he does not connect this cattle-raising tradition with the Great Plains. Geographer Terry G. Jordan, in his *Trails to Texas*, is quite explicit about the connection between South Carolina's cowpens and the Great Plains Cattle Kingdom, as we shall see in Chapter 19.[21]

Notes

1 Kelso, *Jamestown*, 27, for Blackwell as the port of departure; 173, for "Herculean" and the fort's site as the neck's "highest ground."

2 Fischer, *Albion's Seed*, 236 and 238 (map), for the sixteen southern counties as Virginia's source area.

3 Earle, "Environment, Disease and Mortality in Early Virginia," 368, for typhoid and dysentery; 372, for "salt trap"; 374–75, for John Smith; 377–80, for Thomas Dale; 382, for "oligahaline"; and 385, for Crown colony dispersal.

4 Brown, *Mirror for Americans*, 215, for colonial period use of the "Eastern Shore."

5 Meinig, *The Shaping of America*, 1:144, 153, for the use of "Greater Virginia" to underscore the Tidewater subregion's uniformity, which colonists "recognized as such at the time"; 1:158, for Maryland's Roman Catholic subsociety as the exception to this uniformity. See Mitchell, "American Origins and Regional Institutions," 414, with regard to 108,000 in total population, among them 15 to 16 percent black. Brown, *Mirror for Americans*, 224, notes that one slave in one season could attend to two and one-half acres of tobacco with per-acre yields of one thousand pounds.

6 Earle, "The First English Towns of North America," esp. 44 and 48. Also see Mitchell, "The Colonial Tidewater," 2, for population figures circa 1775.

7 See Gottschalk, "Effects of Soil Erosion on Navigation in Upper Chesapeake Bay," 228–29, for Bladensburg; 231–33, for Port Tobacco. Subsequent research suggests the greater importance of coastal erosion in producing sedimentation; see Marcus and Kearney, "Upland and Coastal Sediment Sources in a Chesapeake Bay Estuary," 408–24.

8 Gottschalk, "Effects of Soil Erosion," 235, for Annapolis; 236, for Baltimore. Visitors at Williamsburg are told that only 88 of some 410 buildings are originals.

9 See Kelso, *Jamestown*.

10 Brown, *Mirror for Americans*, 219, for "plantation" as land used for crops possibly originating between the James and York rivers, and that most plantations were five hundred to one thousand acres; 220, for travelers likening clusters of buildings to "little villages."

11 Fusonie and Fusonie, *George Washington*, 6, for slave numbers. Plantation houses like Washington's could best be seen from water, not roads, as noted in Brown, *Mirror for Americans*, 215.

12 Fusonie and Fusonie, *George Washington*, 8, for problems continuing with tobacco; 9, for diversifying to wheat and seeking Alexandria merchants; 9 and 22, for the use of river mud; 15, for sixty total crops and a seven-year crop rotation plan; 25, for three thousand of eight thousand acres, and the use of animal manure. For "crop scheduling," see Earle, "A Staple Interpretation of Slavery and Free Labor," 58. Washington's post-and-rail fences built as straight-line field borders is noted in Brown, *Mirror for Americans*, 229.

13 See Brown, *Mirror for Americans*, 233, for Carolina's three "great" rivers, tides ascending rivers of slight current to thirty-five miles but the Santee to only fifteen miles; 235 and 240, for tidal swamps and the limits of their linear stretches.

14 Meinig, *The Shaping of America*, 1:177, for the creation of North Carolina in 1712; 1:178, for the 1730s population figures in Carolina; 1:190, for New Bern as the de facto capital of North Carolina. For the five-hundred- to one-thousand-acre plantation sizes, see Brown, *Mirror for Americans*, 240.

15 In "The Tidewater Rice Plantation," 57–66, Hilliard describes in detail the use of tidal swamps and that Winyah Bay was the "premier" stretch of coast. "Pestiferous effluvia" and removal to Charleston after the first flooding are in Brown, *Mirror for Americans*, 237.

[16] See Carney, "From Hands to Tutors," 21, for 1738 at Winyah Bay as the first reference to tidal rice; 25, for implementation of the "task" system after the Stono Uprising. Carney's complete synthesis is *Black Rice.*

[17] See Meinig, *The Shaping of America*, 1:180–82 ("propinquity" on 182).

[18] See Winberry, "Indigo in South Carolina," 92, for Eliza Lucas; 98, for few if any planters growing indigo exclusively. See Meinig, *The Shaping of America*, 1:186, for 1742 as the date for indigo's first successful cultivation.

[19] Meinig, *The Shaping of America*, 1:183, 185, for 90 percent of Carolina's seventy-five thousand living in the narrow tidewater zone; 1:190, for the lower Cape Fear district as an extension of South Carolina. Hilliard, "The Tidewater Rice Plantation," 57 and 65, notes that rice production in South Carolina increased until the Civil War, after which it shifted west.

[20] This valley analysis is from Mitchell, "The Shenandoah Valley Frontier," 461–86 (the presence of English and Welsh from Pennsylvania on 469).

[21] See Dunbar, "Colonial Carolina Cowpens," 125–31, esp. 130 for "no demonstrable connection" between cattle raising in South Carolina and the Great Plains. For contradictory evidence, see Jordan, *Trails to Texas*, especially Chapter 2.

Colonial America 7

Generalizing about the colonial period, in "Comparative Frontiers," we recognize that the Spanish and French had frontiers that were fundamentally different from the frontier comprising the largely British and slave populations. The outcome would be that the dynamic British and African frontier would overrun the static Spanish and French frontiers and, in the process, would exclude Native Americans. In "Culture Hearths," we see that four centers of innovation along the Atlantic Seaboard became the springboards for a westward movement, yet much of the observable material culture that streamed west from them amounted to cultural baggage from Europe rather than innovations developed in those four centers. We then follow westward streams in "Folk Houses Move West," using those structures to represent all that was heading cross-country. And finally, in "Port Cities," we look at the land-use zones that evolved in the major colonial cities that, as points of contact with the Old World, looked east as well as west.

Comparative Frontiers

What happened during the colonial period, then, is that basically three peoples, the Spanish, the French, and the largely British, penetrated what is now the United States and intruded upon its native people. With the largely British group came also a fourth people, Africans, in their tragic forced migration. This convergence of peoples from across the Atlantic was, of course, a major event in world history. Map 7.1 presents a time-space summary: source areas, destinations, directions of penetration, and key early settlements (with times of first permanent colonization and dates marking the ends of the several respective colonial eras, in the caption). In geographical terms, the outcome of this convergence was three identifiable settlement frontiers, one Spanish, one French, and one largely British but including blacks. Two fundamental differences characterized these three settlement frontiers.

The first difference had to do with how the Indians were regarded. For the Spanish and the French, both strongly Roman Catholic peoples, converting natives to Christianity was a major New World goal. Although the Spanish and the French lived, for the most part, separately from the natives, miscegenation was common, and a large mestizo, especially Spanish-Indian, population came to exist. The Spanish and French had what can be called frontiers of *inclusion*. By contrast, making religious conversions was not a major goal for the largely British group. Driven by a basic contempt for natives and a desire to take their land, they either pushed the Indians west or they annihilated them. Examples of racial mixing were rare. The largely British group had what can be thought of as a frontier of *exclusion*.[1]

The second fundamental difference concerned population numbers. The Spanish and the French numbered only in the thousands. Canada in 1763 had some seventy thousand French, far more than Louisiana (ten thousand) or Illinois Country (three thousand), and New Mexico in 1821, with some thirty thousand Spanish, was by far the largest concentration of Spanish people in the borderland. With few reinforcements coming from France or Spain, and with minimal numbers coming north from New Spain or Mexico, the Spanish and French frontiers can be characterized as *static*. By contrast, the population of the largely British frontier along the Atlantic Seaboard numbered in the hundreds of thousands. Population estimates for 1750 were as follows: New England, 400,000; the Middle Colonies, 330,000;

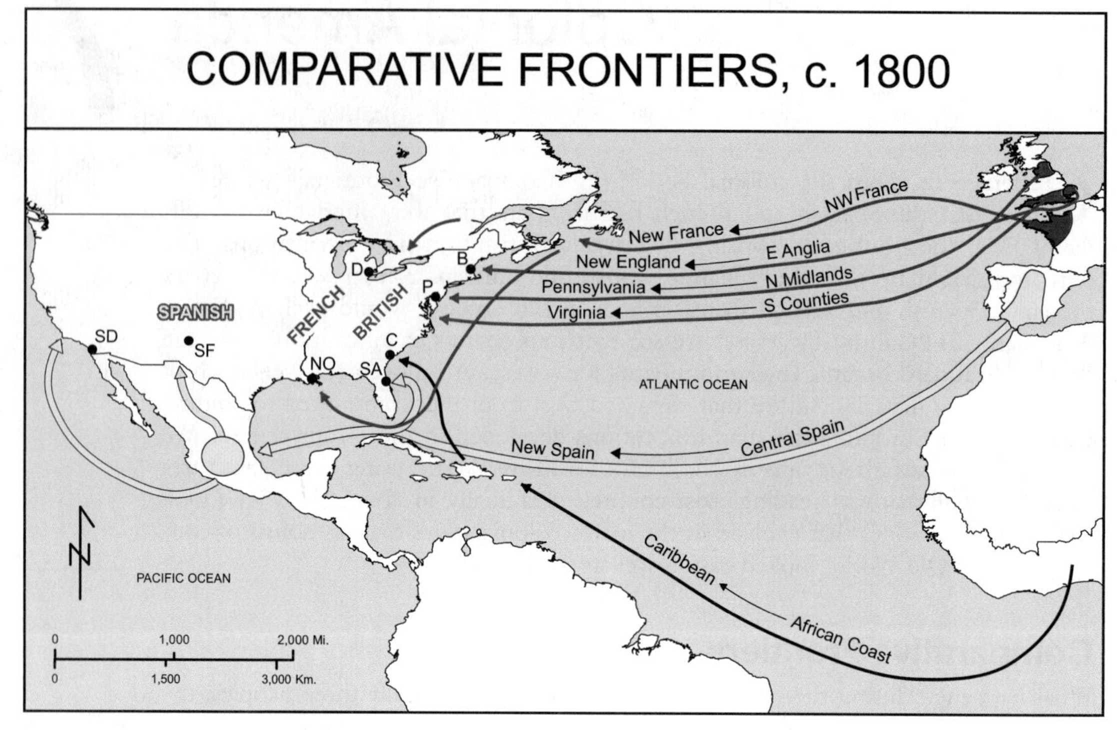

Map 7.1 Comparative Frontiers, c. 1800. The Spanish move north from the Caribbean to Florida (St. Augustine, 1565) and from New Spain to the Southwest (Santa Fe, 1610; San Diego, 1769). The colonial eras end in 1819 for Florida and 1821 for the Southwest. The French penetrate the interior, settling Illinois Country (Detroit, 1701) and Louisiana (New Orleans, 1718). Colonial eras end in 1763 for Illinois Country and 1803 for Louisiana. Meanwhile, several waves of people settle the three Atlantic Seaboard regions of New England (Boston, 1630), the Middle Colonies (Philadelphia, 1682), and the South (Charleston, 1670). Fischer, *Albion's Seed*, identifies the East Anglia (New England), North Midlands (Pennsylvania), and Southern Counties (Virginia) source areas. The colonies declare their independence in 1776, which is confirmed by treaty in 1783.

and the South, 480,000; for a total 1.2 million. And this population was growing rapidly: The census of 1790 recorded some 3.9 million, and the census of 1800 saw an increase to 5.3 million.[2] Sheer numbers along the Atlantic coast translated into a frontier that was already clearly *dynamic.*

For the big picture about 1800, then, a sizable, largely British population living in three regions along the Atlantic coast would easily sustain a dynamic frontier. And this dynamic frontier—the westward movement—becomes the main theme in America's changing geography. As this dynamic frontier moves west, it overruns the static frontiers of first the French in Illinois Country and Louisiana, and eventually the Spanish in the borderland. Meanwhile, the native peoples who have survived beyond the colonial period are excluded socially and politically by the largely British people, and in time their dwindling numbers are confined to reservations. Of this incipient westward movement, Meinig wrote, "The basic dynamics of American expansion were unprecedented. Never had so many people acting in their own private interest under conditions of great political freedom had access to such a large area of fertile lands."[3]

Culture Hearths

If we make a list of those items that diffused to the West from the three dynamic Seaboard regions, the concept of a "culture hearth" comes into play. Meinig defines a culture hearth as "an area wherein new basic cultural systems and configurations are developed and nurtured before spreading vigorously outward to alter the character of much larger areas." Geographers have identified culture hearths for each of the

Seaboard regions; indeed, for the South, they identify both tidewater areas as culture hearths (see map 7.2). Geographer John C. Hudson, moreover, has mapped in great detail the Yankee, Midland, and Upland Southern streams of migration that mixed in a general way in a somewhat vaguely defined Middle West. All of this leads us to ask, what were the new items that developed in these culture hearths, and what paths did they take moving west? The list of newly developed items of culture such as technological innovations, business methods, social mores, and political procedures must certainly be long. But the list of new examples of observable material culture that geographers delight in mapping—importantly, houses, barns, and fences—appears to be rather short. Such mappable items that moved west seem instead to have come mainly as cultural baggage from Europe.[4]

Let us consider the three Seaboard regions individually. In New England, Boston was the urban focus, and its surrounding lowlands were the region's core. Basically this

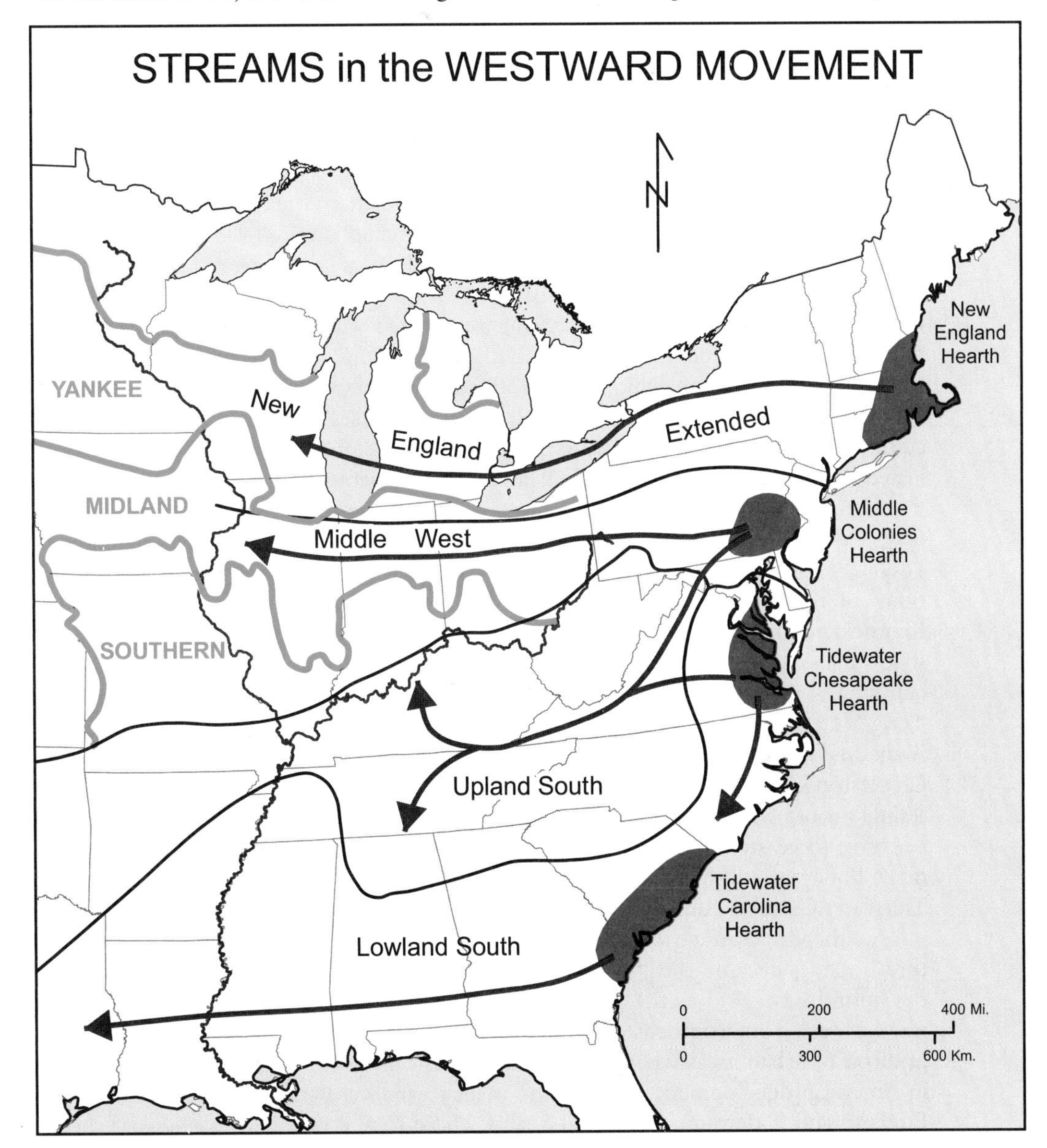

Map 7.2 Streams in the Westward Movement.

Sources: Glassie, *Pattern in the Material Folk Culture of the Eastern United States*, 37 and 39, for culture hearths and regions; Hudson, "North American Origins of Middlewestern Frontier Populations," 411, for streams (Yankee, Midland, Southern).

core, with all its vibrant activity, is usually recognized to have been the region's culture hearth. In this culture hearth, among English people who had come importantly from East Anglia, a New England culture developed, with its own dialect and the values of a strong work ethic and thrift derived from Puritanism. From this culture hearth, or perhaps from New England more generally, also came something tangible that can be mapped—the innovation of building houses (made of wood) that were free-standing on spacious lots and set back from a road. But the houses themselves—recall Martyn Bowden's five-house solution to shelter that ranged upward from a cabin to the I-house to the saltbox to the Cape Cod to the Large—all came as cultural baggage from England. So, too, did the English barn, the wooden fences, the idea of connecting the house and the barn, and the "town" geographical unit in which all these things came to exist. One searches for mappable examples of material culture that evolved in New England. Hudson identifies quite clearly the Upstate New York to Great Lakes path beyond New England, along which these Yankee traits diffused (map 7.2).

The New England generalizations seem also to fit the Middle Colony hearth in Pennsylvania. Philadelphia was Pennsylvania's urban center, and around it in southeastern Pennsylvania there developed a core that is commonly recognized to have been the Middle Colony culture hearth. Attracted to southeastern Pennsylvania were the English Quakers importantly from the North Midlands, the Germans from the Upper Rhineland, and the Scotch-Irish from lowland Scotland via Northern Ireland. Examples of their mappable material culture, as noted in Chapter 5, included the Philadelphia grid and its distinctive central square, townscapes having row houses built with red brick, the large Pennsylvania or bank barn with a forebay, and the quite prevalent I-house. Except for the row-house townscapes, which pretty much remained within the Pennsylvania Culture Area, these items all moved west from southeastern Pennsylvania in what Hudson calls a Midland stream through Ohio, Indiana, Illinois, and Iowa (map 7.2). They also moved west in a major way to the Upland South.[5] These examples all came initially as European cultural baggage and, again, new items of mappable material culture that may have originated in the Middle Colony culture hearth are not obvious.

For the colonial South, historical geographers identify two culture hearths. The first, Tidewater Chesapeake, had no urban focus (until the rise of Baltimore before 1800). Among the settlers from England's southern counties and the blacks, plantations and tobacco culture moved west from the Chesapeake to the Shenandoah Valley, Kentucky, and Tennessee (map 7.2). With them went the symmetrical, Georgian-style "hall-and-parlor" house and undoubtedly other items of mappable material culture. From the second culture hearth, Tidewater Carolina, after cotton became "king" (i.e., after 1800), the English (who came initially from England and then from Barbados) moved their plantations west, clear to East Texas, taking slaves with them. Charleston's "single" house—whose front facing the street was (for tax purposes) only a single room wide but whose interior dimensions were much more expansive, even reaching three stories tall to catch the sea breezes—and cowpens, as complexes of ranch buildings and livestock enclosures were known, just may have originated in the Tidewater Carolina culture hearth. To what degree they diffused west seems unclear.

So the three Atlantic Seaboard regions, and more particularly their four culture hearths, became the "springboards" for the colonization of half the nation—the humid East. And as the pattern unfolded, there is general agreement that what moved west from southeastern Pennsylvania—for example, central squares in the gridiron of urban areas where the emphasis was on business—defined "mainstream" in American development. Joining the mainstream were strong currents from New England and Tidewater Chesapeake, and where they converged, particularly north of the Ohio River in Ohio, Indiana, and Illinois, these three currents mixed. Out of this mixing, in the basin of the Ohio River, it can be argued, an "American" culture developed, yet geographers stop short of calling the area lying west of Pennsylvania a "national core." Surely, probing the possibility of a national core as a spatial concept deserves the attention of an upcoming scholar.[6]

Folk Houses Move West

When sorting out one current from another in the westward flow of people and their cultures, folk houses become an important diagnostic tool. Americans today share a *popular culture*, which for house types means a sameness about their mass-produced dwellings—one finds California ranch houses in New England and New England Cape Cods in Oklahoma. But during colonial times, people belonged to *folk cultures*, their houses identified these folk cultures, and as the several folk societies moved west, their members continued to build their folk houses down to about 1850. What follows is a discussion of house types that are generally recognized to characterize the folk societies of New England, to some degree the Pennsylvania culture hearth of the Middle Colonies, and the Lowland South. Geographer Fred Kniffen (see spotlight) and anthropologist Henry Glassie are drawn on here, and an important distinction pointed out by Glassie needs to be highlighted: Glassie notes that for *folk* cultures, variations from place to place are major but variations through time are minor. The three Seaboard regions were quite different, and change was slow. The opposite is true of *popular* culture (which grew out of the folk cultures). In popular culture, variations over space are minor yet variations through time are major. America today has a sameness from place to place, and change is rapid.[7]

Spotlight

Fred B. Kniffen (1900–1993)

Fred Kniffen. Taken in October 1965, this photograph appeared facing page 59 in a special diamond anniversary issue of the *Annals of the Association of American Geographers* 69, no. 1 (1979). Editor John C. Hudson credits John B. Rehder for the photograph. Courtesy of *Annals of the Association of American Geographers*.

For most historical geographers, research begins in the archives. Fred Kniffen certainly knew the archives, and of course, he developed his studies historically, but *his* research began in the field. And his field-based findings made him a role model for others, among them Terry G. Jordan and his own twenty-eight doctoral advisees. Born in Michigan, Kniffen earned a bachelor's degree in geology at the University of Michigan in 1922. Fortuitously, he met Carl Sauer and John Leighly while an undergraduate in Ann Arbor, and several years later he followed them to the University of California–Berkeley, where Kniffen in 1929 earned a PhD under Sauer (with lots of work under anthropologist Alfred Kroeber). That same year Kniffen was hired at Louisiana State University, the institution where he helped shape the Department of Geography and Anthropology and spent his entire career. Kniffen had many interests, but his giant reputation rested heavily on his pioneering studies of folk houses; indeed, Henry Glassie, a student of folk cultures and a Kniffen collaborator, acknowledged that his "greatest debt" was to Kniffen. Kniffen's seminal "Folk Housing: Key to Diffusion" was given as his honorary presidential address to the Association of American Geographers in 1965.

We might begin with the elements one should have in mind when describing a house type. Using the example of the I-house makes this easy. Despite the fact that the I-house is "by far the most widely distributed of all old folk types," it seems not to have been named until 1936, when Kniffen, in his study of Louisiana house types, used an "I," in recognition of the Indiana-Illinois-Iowa source of many of its builders, to identify this Midwestern type in Louisiana's southwestern prairie region. The I-house is tall and shallow, a "two over two," as they say (see figure 7.1). These are its descriptive elements:

- Number of stories: two
- Number of rooms: one room deep, two (or more) long
- Gable ends: face to the sides
- Front door: usually centrally located in the broad side
- Roof: pitched
- Chimney: as drawn in figure 7.1, central, but could be located at one or both gable ends, either inside or outside the house
- Material: wood, brick, stone, even log or adobe, or a combination of these
- Placement in lot: also as drawn in figure 7.1, centrally located and set back from the street[8]

In colonial times, I-houses, brought as cultural baggage from England, were found along the full length of the Atlantic Seaboard. They appear to have been the most common folk house in Pennsylvania, and they were certainly the most common of folk houses beyond the Middle Atlantic source area in the Upland South, but their widespread distribution makes them less diagnostic of a single cultural group moving west. Nonetheless, the placement of chimneys did vary along the coast and inland and is a diagnostic factor (see figure 7.2). In Pennsylvania, chimneys were built inside the house at the gable ends. In Tidewater Chesapeake, chimneys were placed outside the house at the gable ends. Chimneys placed outside the house meant less heat radiating inside, which made sense in the South. Glassie points out that external gable-end chimneys came to "bind" both the Upland South and the Lowland South as a regional house-type attribute. For maximum radiation of chimney heat in cold New England, chimneys in I-houses were positioned near the center. But as New Englanders moved west in the early 1800s, central halls replaced the central chimney, and chimneys were now built as internal gable-end features.[9]

Figure 7.1 The I-House. So named because of its prevalence in Indiana, Illinois, and Iowa, this house type is tall and shallow, a "two-over-two," with gable ends to the sides. It is the most common American folk house.

In New England, the I-house sometimes had a shed or a lean-to added to the rear of the house, a "characteristically British" phenomenon, according to Glassie. The result was a saltbox, a house type with two stories, symmetrical gable ends, a central chimney, but now two rooms deep (figure 7.2). Because these frame dwellings had two stories in the front but only one behind, the roof had a long slope to the rear, and the resemblance of this roofline to the lid on a small saltbox for table use seems to have prompted its name. And so, relic examples of

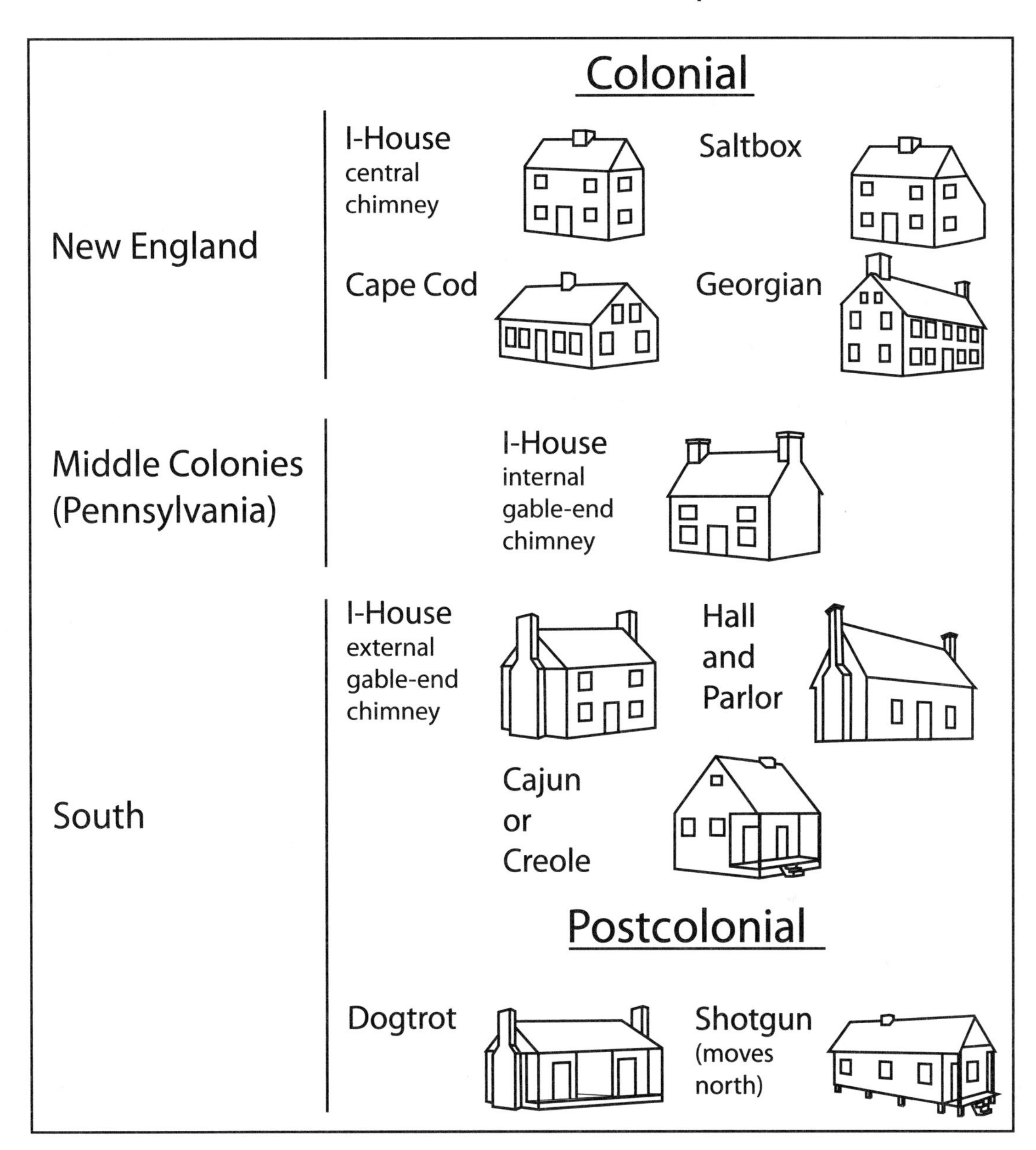

Figure 7.2 Folk Houses Move West.

Sources: Glassie, *Pattern in the Material Folk Culture of the Eastern United States*; Kniffen, "Folk Housing."

saltboxes are diagnostic of Yankee culture, but the saltbox seems not to have diffused west from New England, certainly not to the degree that Cape Cods did.[10]

Cape Cod house types flourished among the English in the old Plymouth Colony, on eastern Long Island, and of course on Cape Cod. Kniffen suggests that their popularity actually rose a notch when New Englanders reached upstate New York about 1790. A Cape Cod can be described as a one-and-one-half-story house with gable ends facing to the sides and a central chimney (figure 7.2). In its most usual form, the floor plan of a Cape Cod, which Glassie says was its "folk essence," had two large front rooms with a chimney in between and three or more back rooms. There were two rooms in the half-story above. Cape Cods were, of course, diagnostic of where New Englanders settled, but in time modifications occurred. Glassie notes that along its path of diffusion from central New York to Ohio and south around the Great Lakes, a "descendent" of the Cape Cod found the central chimney give way to gable-end flues for stoves while the pitch of the roof became shallower and the roof itself higher, but the floor plan remained largely intact.[11]

Also diagnostic of the New England stream is the "Large" or Georgian house—architectural historians make fine points of distinction between the two terms. Bowden characterized this five-over-five symmetrical house as representing colonial New England's highest form of vernacular architecture. About 1700, the Large, exemplified by the house Bowden himself restored (see Chapter 4), had one massive central chimney. This central-chimney house, Kniffen finds, did not diffuse beyond New England. What did diffuse, with decreasing popularity as it reached upstate New York about 1790 and Ohio about 1810, was the Large with smaller chimney stacks built on the insides of each of the gable-end walls (figure 7.2). This symmetrical two-over-two house with a broad central hall seems to fulfill what is meant by "Georgian."[12]

Georgian architecture, a style in vogue during the reigns of the several Georges in Great Britain in the 1700s, is characterized by strict symmetry. Found the length of the Atlantic Seaboard, Georgian architecture in the Tidewater Chesapeake area took the form of a smaller and simpler hall-and-parlor house. On the ground floor, this frame or brick house had two rooms on either side of a central hall and parlor, and there were two or three rooms in the half-story above (figure 7.2). Chimneys were commonly built outside at the gable ends, as was typical in the South. The hall-and-parlor house became the "most usual house" in the Tidewater Chesapeake about 1700, and it became one of the most common house types in the Lowland South. Kniffen shows how, for better ventilation, the hall-and-parlor house, in its evolution, was raised as much as a full story above the ground, either on piers or on a brick foundation.[13]

As the hall-and-parlor house diffused to the west, it evolved into the "dogtrot" house (figure 7.2), asserts Glassie. This happened when someone made the broad central hall into an open "trot" for improved air circulation; indeed, people who live in dogtrot houses refer to the breezeway as a "hall." Glassie speculates that the dogtrot house came to exist in the Tennessee Valley about 1825. It diffused primarily within in the Lowland South. So, too, did the "shotgun" house diffuse largely within the Lowland South after its apparent introduction by Haitians to New Orleans in the early 1800s. One could shoot a bullet clean through the chain of three or four rooms of this one-story, one-room-wide house (when all the doors were open) from the gable-end front door to the gable-end back door, presumably the explanation for its name. Both the dogtrot and the shotgun houses came to exist after the colonial era. The Cajun or Creole house, of course, had been popularized by the French in their colonial period, yet it diffused but little beyond Louisiana. Thus, folk houses have their own geography, and knowing something about this can make traveling in the United States much more rewarding.[14]

Geographer Deryck Holdsworth would likely object to this discussion of house types and their westward movement. In his thoughtful review essay titled "Revaluing the House," Holdsworth was quite critical of Kniffen (but less so of Glassie), whose paths of diffusion leading west from culture hearths, Holdsworth argued, were but superficial arrows untested by detailed fieldwork. Holdsworth pointed out that the choices people made in building their houses, and the roles played by builders in a given local housing market, were indeed complex decisions and needed to be analyzed in full appreciation of their economic, political, and cultural contexts. Holdsworth seemed even to question the validity of the notion that folk houses existed to identify regional differences. In short, Holdsworth was not kind to Kniffen and Glassie in their representation of folk houses and how they may have diffused west in preindustrial America. His plea for additional research is well taken, but until new findings are available, it seems premature to declare the research of Kniffen and Glassie to be invalid and outmoded.[15]

Port Cities

A look at the common morphology of early port cities has its own rewards. In 1800 all fifteen of the largest cities in the newly constituted United States were seaports. Each had at least five thousand people. In rank order, the five largest were the following:

Philadelphia	69,403
New York	60,489
Baltimore	26,114
Boston	24,837
Charleston	20,473

Put in perspective, the total population of these fifteen cities represented only 7 percent of the country's total 5.3 million people. Yet these seaports, especially the five largest, were of great importance as the "connecting points" between Europe and the developing American port city hinterlands. Moreover, Philadelphia and New York were the major foci of a developing American core that stretched north to include Boston and south to include Baltimore. To facilitate their shipping functions, these port cities shared certain likenesses in their land-use layouts.[16]

Drawing on studies by geographers of Boston and Baltimore, one can identify at least four port city land-use zones circa 1800 (see figure 7.3):

1. *Waterfront*
   Along the waterfront were the wooden wharves jutting from the land where vessels conducting long-distance trade docked to discharge and receive cargo and passengers. Facing the wharves were the storehouses used for warehousing merchandise, and the countinghouses used for keeping books and transacting business. Driven by land values to the peripheries of the waterfront (see arrows in figure 7.3) were the shipyards where vessels were repaired and new ships built. Also at the peripheries were the blacksmith forges and the sail lofts, or upper floors of buildings where sails were cut and sewn.
2. *Central City*
   The central city was both the business district and a place where people lived. Typically, storekeepers and artisans used the ground floors of their houses for their workshops and sales, while they resided on the upper floor(s). Along the streets, then, were the shops of cobblers, tailors, bakers, hatters, candlestick makers, and glaziers (glass setters). Professionals such as lawyers and doctors also lived in the central city—the district that provided goods and services to the local inhabitants.
3. *Residential*
   The shops and offices of the central city gave way gradually to a more uniformly residential district. Here lived the workers who walked to jobs in the central city or on the waterfront. Houses were densely packed along often narrow and sometimes winding streets. As one went inland, the density of people decreased, and on higher land that might overlook the harbor could predictably be found the largest and least densely packed houses of the elite.

4. *Periphery*
   On the urban fringe was found a land-use zone where both space-consuming and noxious activities were located. "Ropewalks," or long buildings that took up an acre or more and were used for the manufacture of rope needed for sailing vessels; brickyards; and iron foundries were examples of space-consuming activities. Both the tanning of leather and slaughtering of animals were examples of noxious industry. If laid out wisely, the slaughterhouses and tanneries would be located downwind and removed from the view of the houses of the elite.

At some point, the buildings of these port cities gave way to farming and countryside. However, at waterfalls on streams flowing to the coast, small towns developed around gristmills, sawmills, woolen mills, and cotton mills. The products of these mill towns—flour, lumber, and textiles—joined those from the farms in their journey by wagon over dirt roads in the countryside that led to the markets lining the cobblestone streets of the port cities.[17]

Urban, or "mercantile," communities at the end of the colonial period were basically small and compact circles with high population densities. Transportation was by foot or by horse and wagon, and power was generated at waterfalls if not by sheer animal or human muscle. Much of this would change in the nineteenth century with the growth of "industrial" cities. Handicrafts manufactured in a single shop would yield to workshops and factories and a whole new category of industrial land use. Waterpower would be replaced by coal and steam engines, and the population densities of urban communities would decline as people moved out along the tracks of new railways and horsecars. The once small urban circles became more starlike in shape with the outward expansion of people following new transportation radii. A heavily rural United States in 1800 was on its way to becoming nearly half urban by 1900.

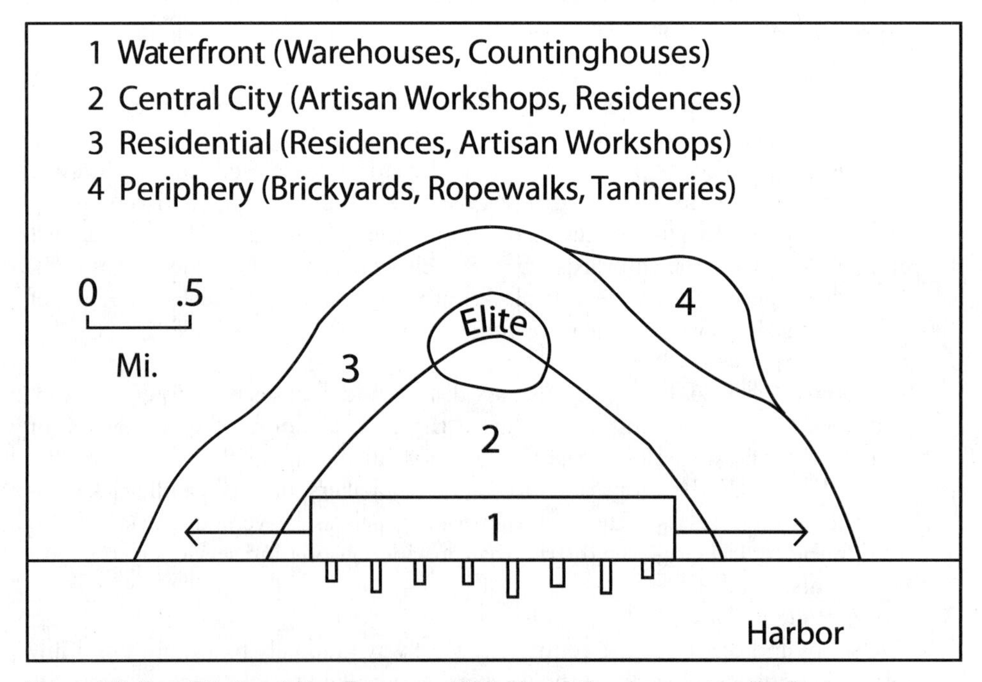

Figure 7.3 Port City Land-Use Zones, c. 1800. Mercantile urban centers like Boston and Baltimore had these zones.

Sources: McManis, *Colonial New England*, esp. 76; Muller, "Distinctive Downtown," 750.

Notes

[1] Mikesell, "Comparative Studies in Frontier History," 65, for frontiers of inclusion and exclusion; 66, for static and dynamic frontiers.

[2] Meinig, *The Shaping of America*, 1:247, 249, 348.

[3] Ibid., 1:417.

[4] Ibid. 1:52. And see Hudson, "North American Origins of Middlewestern Frontier Populations," esp. 411 (map).

[5] Hudson, "North American Origins of Middlewestern Frontier Populations," 403–4.

[6] Meinig, "The American Colonial Era," 22, for "springboard"; 21, for "mainstream." Meinig, "Midlands Extended," in *The Shaping of America*, 1:279–84, reinforces the Middle Atlantic column as mainstream.

[7] Kniffen, "Folk Housing," 549–77; Glassie, *Pattern in the Material Folk Culture of the Eastern United States*, 33, for folk- versus popular-culture distinctions.

[8] Kniffen, "Folk Housing," 553, for "I-house"; 555, for "most widely distributed."

[9] Everything on I-houses in this paragraph is from Glassie, *Pattern in the Material Folk Culture of the Eastern United States*, 49, from England; 107, most common; 49, Pennsylvania internal-end chimney; 184, Tidewater Chesapeake external-end chimney; 234, "bind" both Upland South and Lowland South; 124 and 184, New England central chimney; 129, west of New England.

[10] Ibid., 124–25, for "characteristically British."

[11] Kniffen, "Folk Housing," 559, charts house type popularity west from New England. And see Glassie, *Pattern in the Material Folk Culture of the Eastern United States*, 128, for "folk essence," and 129, for "descendent," in a discussion of Cape Cods.

[12] Kniffen, "Folk Housing," 558–59.

[13] Glassie, *Pattern in the Material Folk Culture of the Eastern United States*, 96, and 109, for "most usual home"; Kniffen, "Folk Housing," 549, 565, 570.

[14] Glassie, *Pattern in the Material Folk Culture of the Eastern United States*, 89, for Tennessee Valley circa 1825; 96, 98, 101, for other references to dogtrot houses; 218, 220, and 221, for shotgun houses.

[15] Holdsworth, "Revaluing the House," 95–109.

[16] Brown, *Mirror for Americans*, 35–36, for the fifteen seaports. For "connecting points," see Muller, "From Waterfront to Metropolitan Region," 109.

[17] McManis, *Colonial New England*, 72–84, esp. 76, for model, with reference to Boston; Muller and Groves, "The Emergence of Industrial Districts in Mid-Nineteenth Century Baltimore," 160–68, for Baltimore in 1833. In personal correspondence dated March 31 and April 28, 2011, Muller refined my port city diagram. He would not have made zone 2, the business area of the Central City, a "truly separate" land-use zone.

THE HUMID EAST II PART

The Upland South 8

Two items bind the Upland South as a region: its rugged terrain and its people's distinctive cultural attributes. Under "Appalachian Routes," we look first at the eastern segment of the region and the two major pathways that led west through Appalachia: the Wilderness Road and the forerunners of the National Road. Under "Upland South Attributes," we note how the region's settlers, notably people of Scotch-Irish, English, and German ancestry, left features diagnostic of the region's character—for example, structures built of log, dogtrot houses, and Shelbyville courthouse squares. We then turn to the creation of "The Public Domain and New States": In the vanguard of the westward movement were people who pushed through the middle of Appalachia to Kentucky and Tennessee, which became the first two western states. Finally, we discuss "Washington, DC," the new national capital, with its planned baroque grandeur positioned at the foot of the National Road.

Appalachian Routes

One of the significant facts about the geography of the United States, as noted in Chapter 1, is that its eastern half is humid and its western half, except for the west coast, is dry (see figure 8.1). What this meant for those moving west is that in the West supplemental water in the form of irrigation or measures to conserve soil moisture were necessary for growing crops, whereas in the East precipitation amounts were adequate. In colonial times settlers in the three Atlantic Seaboard regions were a rural people, and nearly all were farmers. When these people moved west—within the same approximate bands of latitude—the farmers among them had no difficulty growing their same grains and vegetables and orchard trees—until they reached that zone in the middle of the country where precipitation became scant and unreliable. There was no climatic "barrier" to agriculture moving west through the humid East. But inland from the Atlantic Seaboard, there was an environmental obstacle of a different kind—namely, Appalachia.

Appalachia is an upland. It stretches from northern Alabama and Georgia northeast to upstate New York and then clear to Newfoundland (see map 8.1). The "grain" of this three-hundred-mile-wide (480-kilometer-wide) physiographic province runs southwest to northeast and is composed of the Piedmont, the Ridge and Valley Province, and the Appalachian Plateau (known as the Cumberland Plateau in Kentucky and Tennessee—see Chapter 5). As an upland, with its hills, low mountains, and otherwise rugged terrain, Appalachia was an obstacle in the westward movement. Trails once used by buffalo and Indians identified the easiest grades, which wound through valleys and led to mountain gaps; following these paths would slow travel and challenge those on horseback or in wagons. But in the early twentieth century a generation of geographers overemphasized Appalachia's role as a landform province, calling the upland a "barrier" and attributing to it unwarranted examples of significance.

The champion of this physical-barrier thinking was geographer Ellen Churchill Semple (see spotlight). In her *American History and Its Geographic Conditions* (1903), Semple argued in Chapter 3, "The Influence of the Appalachian Barrier upon Colonial History," that the presence of a "mountain wall" gave colonists along the "narrow strip of coast" a "certain solidarity" in their Revolutionary War efforts. However, the colonists were not united against the British; an estimated third were

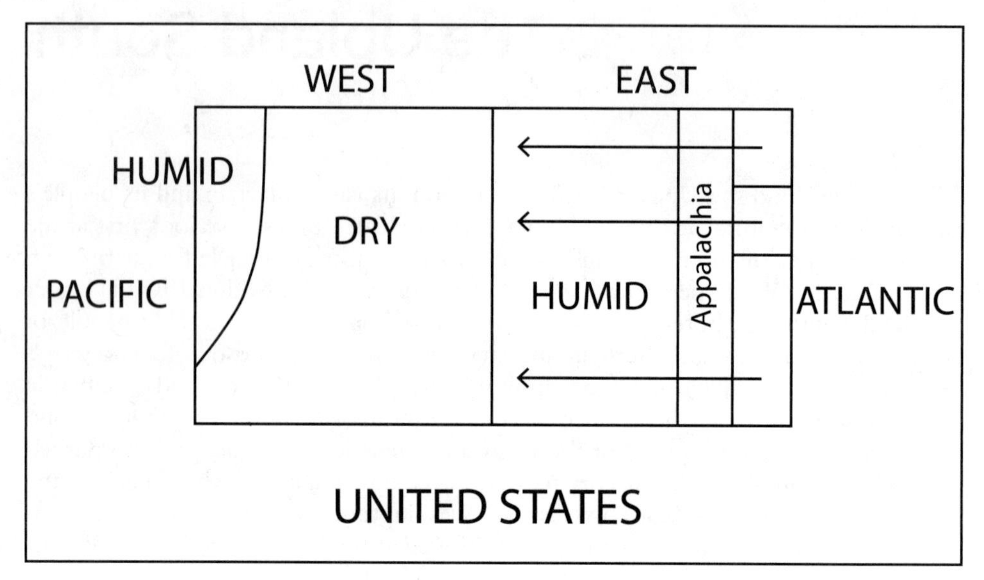

Figure 8.1 Humid East and Dry West. Crops grown in the three Seaboard regions spread west in latitudinal bands without problems—until they reached the dry West.

opposed to the revolt, about a third were uncommitted, and only a third were true rebels. Semple also argued that the barrier "protected" the thirteen colonies from the Indians, the assumption being that Appalachia separated the colonists on the coast from the Indians west of the mountains. In reality, Native Americans who were hostile to white people used the upland as a hideout while engaging in coastal raids and depredations. In addition, Semple argued, in what was a debatable contention, that the presence of Appalachia tended to confine colonists to the coastal plain, thus lengthening their maritime orientation and attachment to the Old World. This thinking, called environmental determinism, dominated American geography in the early twentieth century, but in the 1920s it came to be gradually replaced by possibilism, or the idea that the environment did not determine the actions of humans but rather gave them possibilities based on their objectives and technical skills.[1]

So Appalachia was not the physical barrier some believed. Indeed, if anything, where there were lowlands within or near the uplands, Appalachia was a "human" barrier. The easiest way to cross from east to west through Appalachia was to ascend the Hudson River to the Mohawk, and to follow the Mohawk Valley west to the Ontario Lake Plain. This so-called Hudson-Mohawk depression had a modest maximum elevation above sea level of only 445 feet (135 meters) between Albany and Syracuse. But at the western end of this depression lived the Iroquois Nations, a formidable alliance, who wanted no movement through their land. Likewise, where Appalachia ends in northern Georgia and Alabama, colonists were blocked from crossing on lowlands around the southern end of Appalachia by the powerful Cherokee and Creek nations. And a third possible lowland route through Appalachia, the St. Lawrence Valley, was blocked by the British, with whom the Americans had just been at war. And so, when pioneers began to move west, they pushed right through the middle of the upland itself.

A number of routes through Appalachia came to be followed by those seeking western lands, but the earliest and most important was the Wilderness Road and its famous Cumberland Gap. The Wilderness Road began at a number of points—Philadelphia, Baltimore, Richmond—but all feeders converged on the Shenandoah Valley. Colonists ascended the valley to its beginning point at a low divide near the headwaters of the James River south of Staunton. They then traversed another low

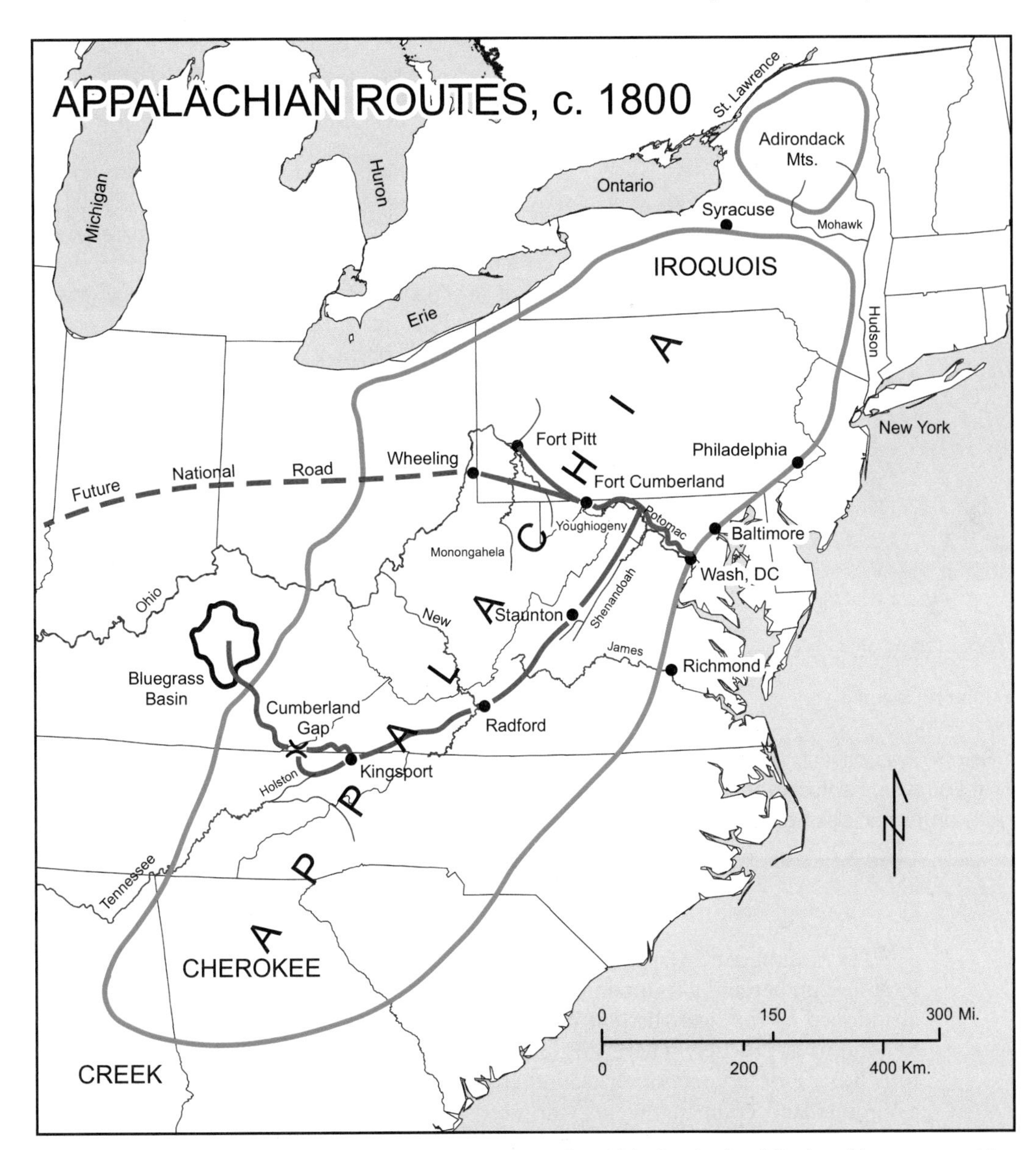

Map 8.1 Appalachian Routes, c. 1800. The Wilderness Road (via Cumberland Gap) and forerunners of the National Road (up the Potomac) led through Appalachia.

divide to the New River, which they crossed at Ingles Ferry (Radford, Virginia), and followed the middle fork of the Holston River to Long Island (Kingsport, Tennessee, shown on map 8.1). The branch cut out by Daniel Boone in 1775 wound for some forty miles through southern Virginia from Moccasin Gap in the Clinch Mountains to Cumberland Gap in Virginia's southeastern corner. After 1785 a somewhat easier alternate route followed the Holston Valley south into Tennessee before bending north to cross through Cumberland Gap from the south. Once through Cumberland Gap, the Wilderness Road led north across the rugged Cumberland Plateau to Kentucky's Bluegrass Basin, initially to Harrodsburg (1774) or to Boonesborough (1775, near Lexington). From 1775 until well past 1800, more people used the Wilderness Road than all other Appalachian routes combined—an estimated two hundred thousand in all.[2]

Spotlight

Ellen Churchill Semple (1863–1932)

Ellen Churchill Semple. Photograph courtesy of Geoffrey J. Martin.

Ellen Semple was brilliant. Born into a prominent family in Louisville, Kentucky, at nineteen she was valedictorian of her class at Vassar College. She taught in private schools in Louisville for several years, returned to Vassar to earn a master's degree in 1891, and in 1891–1892 and again in 1895, she studied under the famous German anthropogeographer Friedrich Ratzel, in Leipzig, Germany. Greatly inspired by Ratzel, Semple took Ratzel's concepts about how physical geography shaped human development to a new level, becoming geography's foremost advocate for what came to be known as environmental determinism. Her eloquent writing in three books and dozens of journal articles focused on the United States (and Appalachia) and, later, on the Mediterranean Sea. Between 1906 and 1924 she taught in roughly alternate years at the University of Chicago, and from 1921 to 1931 she held a permanent appointment at Clark University in Worcester, Massachusetts. These were geography's two leading university departments at the time. In 1921, Semple was elected president of the Association of American Geographers; she was the first woman to hold that office.

The Cumberland Gap was a low pass of some 1,600 feet (488 meters) above sea level in Cumberland Mountain. Technically a "wind" gap, because it had long been abandoned by the river that cut it, this pass is located in the far southwestern corner of Virginia at precisely where Virginia, Kentucky, and Tennessee come together (see map 8.2). First reports of the Cumberland Gap are attributed to Thomas Walker, a Virginia land company agent, who used the gap in 1750. In 1775 Daniel Boone and some thirty axmen were hired to cut a pack trail, the Wilderness Road, through the gap. After 1775, pioneers headed primarily for Kentucky's Bluegrass Basin filed through the gap, and when widened to a wagon road in about 1800, cattle and hogs were driven east, making the Wilderness Road two-way. Today, the Cumberland Gap is maintained and interpreted by the National Park Service as the Cumberland Gap National Historical Park, authorized by Congress in 1940. And—happily—the US 25 East highway, which once crossed through the gap, thus compromising its history, has been removed. Built between 1991 and 1996, the Cumberland Gap Tunnel, an engineering feat measuring 4,600 feet (1,400 meters) in length, diverts four lanes of US 25E traffic through twin tunnels southwest of the gap.

Meanwhile, to the north, the Potomac Valley became a major Appalachian gateway west. On the upper Potomac in the Maryland panhandle, Fort Cumberland, a trading post established in 1750, became the jumping-off point. To reach the Ohio River, westward travelers left the Potomac at Fort Cumberland and crossed against the rugged grain of Appalachia in southwestern Pennsylvania to the Youghiogheny River. One could descend the Youghiogheny Valley to the Monongahela Valley, which led

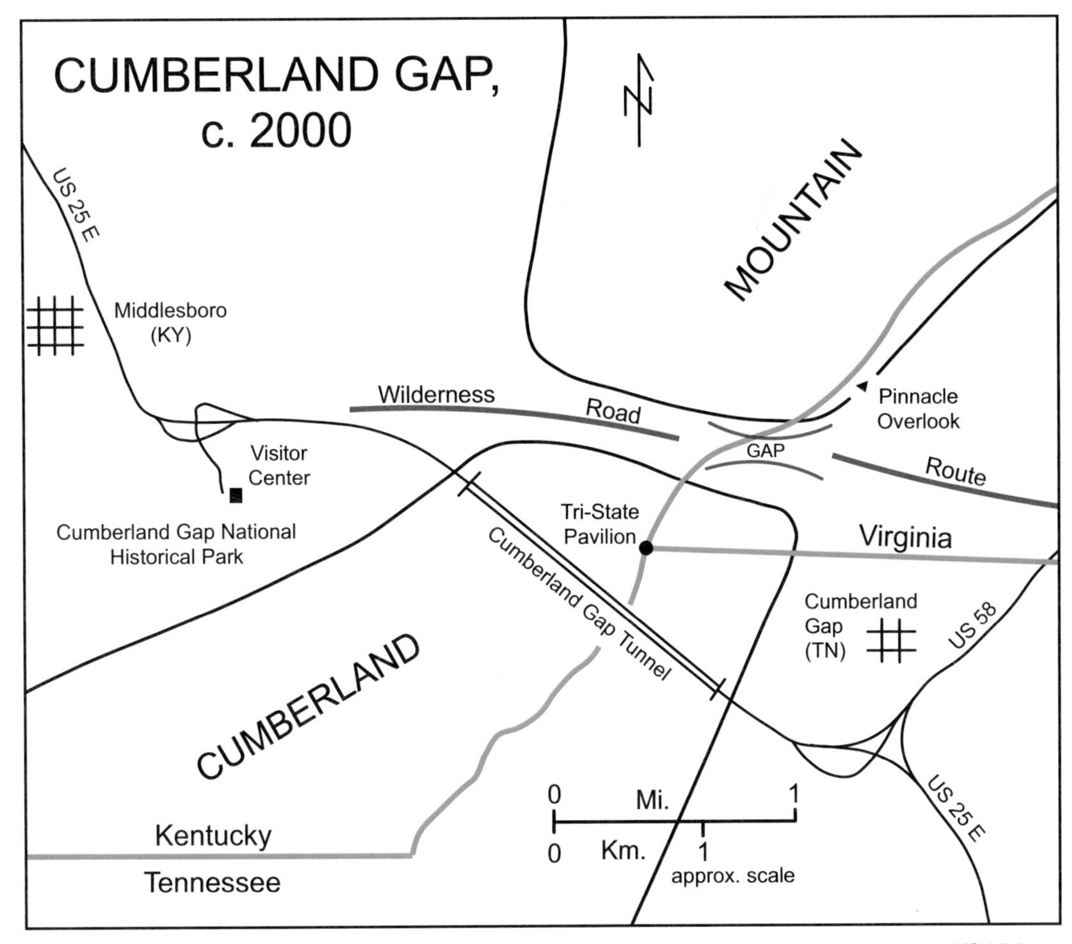

Map 8.2 Cumberland Gap, c. 2000. Completed in 1996, the Cumberland Gap Tunnel removed US Highway 25E from the Cumberland Gap.

downstream to the "Forks of the Ohio" and Fort Pitt (1754). One could also traverse the rugged Appalachians west from Fort Cumberland on the so-called Cumberland Road, whose name, by analogy with Cumberland Gap, has confused generations of geography students. The Cumberland Road was the forerunner of the National Road, appropriately named because it led west from Washington, DC, on the lower Potomac. Construction of the National Road, the country's first long-distance road, began at Fort Cumberland in 1808, crossed the Ohio River at Wheeling, West Virginia, and, albeit in piecemeal fashion, reached Vandalia, Illinois, by 1850.[3]

Upland South Attributes

The Upland South is pretty much just that—an upland. Appalachia itself makes up roughly half the region, and the Ozark Plateau and Quachita Mountains west of the Mississippi River constitute a part of the other half (see map 8.3). The region stretches from southeastern Pennsylvania to northeastern Texas. It overlaps with the Old Northwest north of the Ohio River because Upland Southerners settled there. Likewise, it overlaps with Indian territory in eastern Oklahoma. It is a bit of a stretch to include gently rolling eastern Kansas and the Blackland Prairie of Texas as upland, but they are also part of the region because they, too, were settled by Upland Southerners.

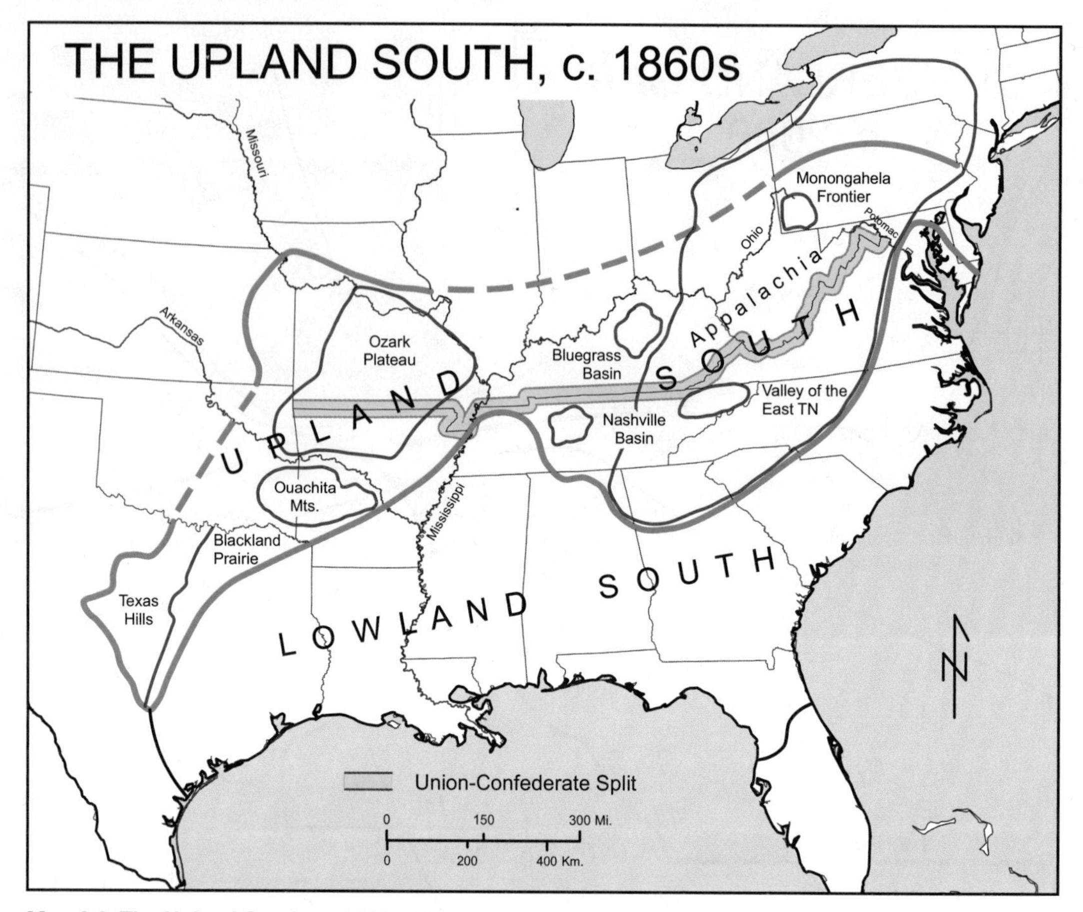

Map 8.3 The Upland South, c. 1860s. The Civil War split reflected the region's middle location.

Source: Modified from Jordan, "The Imprint of the Upper and Lower South on Mid-Nineteenth-Century Texas," 668 (map).

So who were the Upland Southerners? In *The Upland South*, Terry G. Jordan-Bychkov emphasizes that Upland Southerners are a "blending" of peoples originally from Pennsylvania, Tidewater Virginia-Maryland, and the Carolinas. The Scotch-Irish, perhaps the largest single contributor, Jordan-Bychkov notes, helped shape the regional culture with their Presbyterianism, country music, and whiskey making, but because of their intermixing with others, a Scotch-Irish ethnicity does not survive today. English people are a second main ingredient, and some were Quakers, like famous regional icon Daniel Boone, whose grandfather sought religious freedom in Pennsylvania in 1717, whose father moved his family from Pennsylvania to the Yadkin Valley of western North Carolina about 1752, and who himself moved his own family from North Carolina to Kentucky in 1775 and then to Missouri in 1799. Germans were a third, if lesser, constituent, and there also came Swedes and Finns from the Delaware Valley. Native Americans constituted a relatively significant proportion of the population blend that was Upland Southern, and there were also some blacks, who arrived initially as slaves.[4]

Jordan-Bychkov identifies five attributes that are diagnostic of the culture of this blend of Upland Southern people, and in his book he devotes a chapter to each. They include log constitution, especially half-dovetailed notching; the dogtrot folk house; the transverse-crib barn; the courthouse square, notably the Shelbyville plan; and Upland Southern cemeteries. He mentions but does not develop such additional cultural attributes as a unique nasal dialect, mournful country music, folk medicine, food and drink, crafts, clan-based feuding, and an emotional and a vigorous Calvinistic religion. Jordan-Bychkov emphasizes that middle Tennessee played a pivotal role in the formation of the culture of the Upland South. One manifestation is the Shelbyville courthouse square (see Chapter 5), which diffused from Bedford County in middle Tennessee notably to Texas, where fully 114 courthouse squares (62 percent of Texas counties with courthouse squares) came to be Shelbyvilles or variations thereof.[5]

Middle Tennessee attracted pioneers in a second surge in Upland South colonization. The first push destined for areas deep within Appalachia went to the Monongahela frontier and the Valley of the East Tennessee (map 8.3). In the 1750s the Monongahela Valley area was the destination for a small number who had traveled up the Potomac Valley. The area became safer after 1768, when the Iroquois ceded their local land claims in the Treaty of Fort Stanwix. In 1780, Pennsylvania emancipated its slaves, prompting some who owned slaves to leave southwestern Pennsylvania for Kentucky. Beginning in the late 1760s, the Valley of the East Tennessee, reached via forerunners of the Wilderness Road, attracted colonists. In the far northeastern corner of present-day Tennessee, rather cohesive communities developed in the Watauga and Holston valleys. People in these communities proposed a new state—Franklin—as we shall see.

In the 1770s settlers were pulled to the Bluegrass and Nashville basins of Kentucky and Tennessee (map 8.3). As large, basically circular areas with fertile brown silt and loam soils derived from underlying limestone, these two basins became islands of agricultural prosperity. Major crops were corn (distilled to make bourbon in Kentucky), tobacco, and hemp; livestock included hogs, high-quality cattle, and horses. By the time of the Civil War, Kentucky had become the leading tobacco-producing state. Kentucky's Inner Bluegrass around Lexington would eventually gain notoriety for its breeding and training of horses, and today tourists marvel at its landscape of elite people: roads winding through park-like pastures behind white board fences, and elegant mansions behind limestone walls. These two limestone basins illustrate nicely the axiom that the westward-moving frontier did not advance along a broad continuous front; rather, it advanced unevenly and in a highly selective fashion, with choice areas attracting colonists ahead of other areas.[6]

Upland Southerners carried their distinctive attributes west of the Mississippi River. Along the Mississippi they were preceded by French people who had arrived about 1700 and farmed on long lots in the "bottoms" (see Chapter 3). Beginning with Ste. Genevieve (1735), the French had also established mining communities at lead deposits west of the Mississippi. Upland Southerners, largely Kentuckians and Tennesseans, began to arrive in the 1790s, outnumbering the French by 1800, and would advance west the length of the Missouri Valley in Missouri by 1830. The Ozark Plateau of southern Missouri and northern Arkansas, perceived favorably by Upland Southerners despite its soils of marginal quality, attracted these people, who pushed into the broken country south from the Missouri Valley and west from the Mississippi Valley. By 1860 they had carved much of the highland into farms. Meanwhile, immigrants, primarily Germans, coming directly from Europe, arrived beginning about 1830. Geographer Russel Gerlach tells how they bought out the land of the Upland Southerners, created their own landscapes, and by 1860 constituted 10 percent of Missouri's Ozark population. At the same time, Upland Southerners were settling the Quachita Mountains, located south of the Arkansas River (map 8.3).[7]

In the 1840s and '50s, Upland Southerners, many from the Ozark-Quachita highlands, settled in northeastern Texas (map 8.3). Besides Arkansas and Missouri, the main contributing state was Tennessee. In the Blackland Prairie, the eastern section of Texas to which they moved, fertile soils produced surpluses of wheat, the principal cash crop, yet getting that surplus to market by ox wagon—railroads would not arrive until the 1870s—was a challenge. To the Texas hills just to the west of the Blackland Prairie went Upland Southerners who were more representative of southern Appalachia's "mountaineer" folk. Rural and less prosperous, the Texas hills came to be characterized by attributes of the larger region, including log construction, emphasis on livestock herding, corn patches for subsistence, low educational attainment, and feuding. In February 1861, when Texans voted to secede from the Union, the state's Lowland Southerners by the Gulf coast were solidly behind joining the Confederacy, but its Upland Southern population was, quite understandably, split in its vote.[8]

The reason for this split is that the Upland South as a region lay midway between the North and the Lowland South, and its blend of people drew from both. Indeed, many geographers prefer the label Border South (and its pair, Deep South) because "Border" suggests this middle position. When the Deep or Lowland South states seceded, beginning with South Carolina in December 1860, the Upland South quite literally split in half (map 8.3): Kentucky went with the Union and Tennessee went with the Confederacy. Likewise, Missouri aligned with the North while Arkansas went with the South. In Virginia the Appalachian people broke away from the Tidewater people as forty-eight western upland counties voted to form the new state of West Virginia, which joined the Union in June 1863.

The Public Domain and New States

All this business about a Union-Confederate split happened long after the creation of the public domain and the first western states. In colonial times six of the thirteen colonies had grants of land that read "from sea to sea." They were Massachusetts, Connecticut, Virginia, North Carolina, South Carolina, and Georgia (see map 8.4). Each of the six made claims to western land. Meanwhile, a seventh colony, New York, had negotiated treaties with the Iroquois, and on that basis, it too claimed western land. After the seven became states, each ceded its claims to the new federal government (or to one of the other original states) over a twenty-year period. New York was the first to do so, in 1782; Georgia was the last, in 1802. These land cessions, also the treaties that the federal government would negotiate with the various Indian confederacies, led to the creation of the public domain. Early in our national experience, then, the principle was established that the federal government controlled unappropriated public lands. Indeed, half of the American West and 89 percent of Alaska are still part of the public domain today (see Chapter 20).

In June 1784 North Carolina relinquished its claim to western land. This left the people of the Valley of the East Tennessee, already a cohesive community, without a government. These frontier residents had already grown impatient with North Carolina's political neglect, and they were also tired of the nearly 350-mile (560-kilometer) trip to New Bern, the capital (Raleigh, located one hundred miles [160 kilometers] closer, was made the capital in 1788). So in August 1784, these residents of present-day northeastern Tennessee petitioned the Continental Congress to create a state called Franklin, named after Benjamin Franklin (map 8.4). They established a capital at Greeneville; drew up a constitution; elected one of their leaders, John C. Sevier, governor; elected legislators; and formed a militia. But the Continental Congress declined to recognize the new state; in March it had enacted the Ordinance of 1784, requiring *all* states to cede their claims before a new state could be created. At that point, the momentum behind a state of Franklin—what could have been the fourteenth state—waned, and by 1788 it ended. The Valley of the East Tennessee would become part of the state of Tennessee, and Tennessee's first governor would be John C. Sevier.[9]

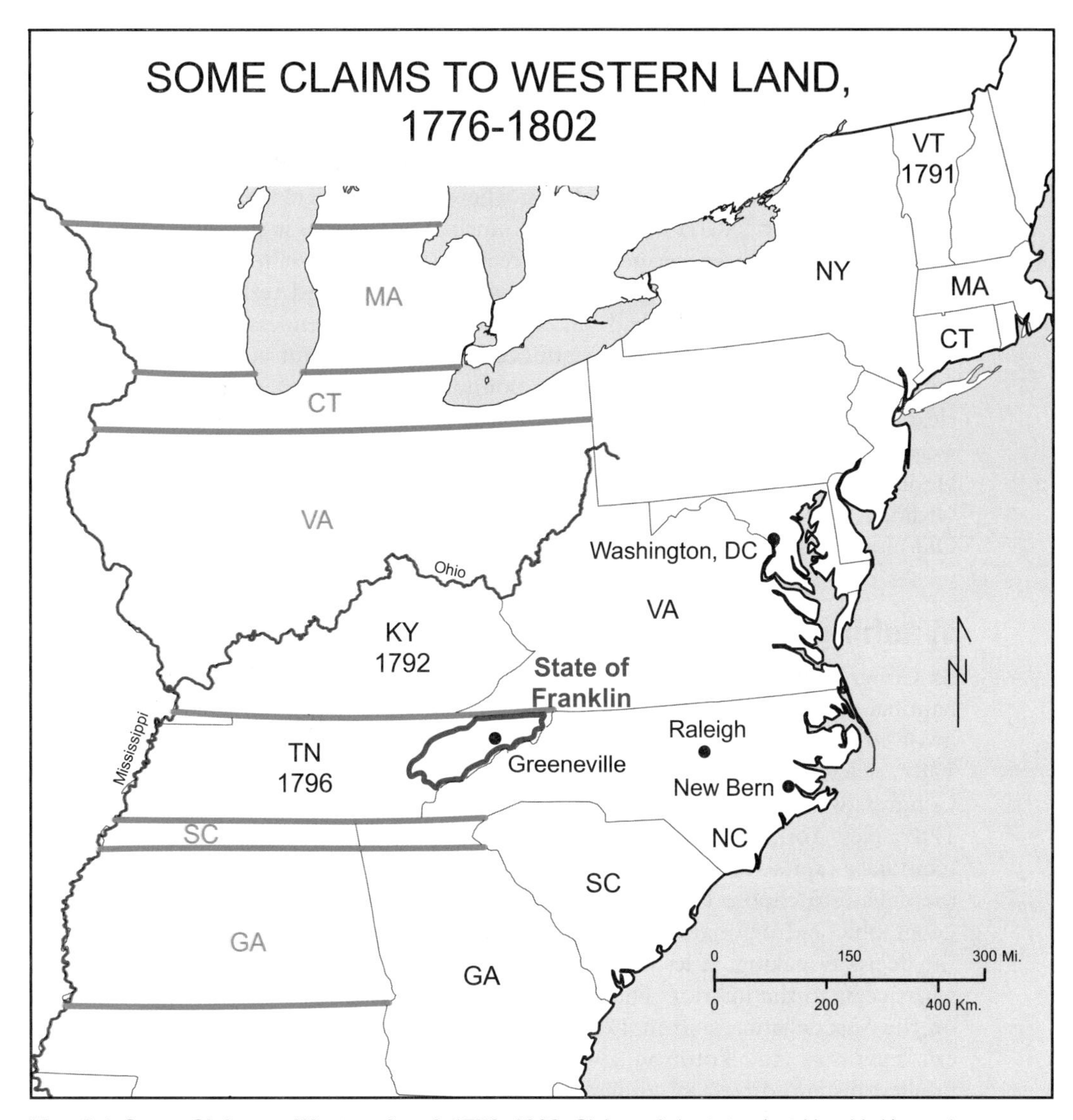

Map 8.4 Some Claims to Western Land, 1776–1802. Claims of six states (not New York) are shown; part of Virginia's claim became Kentucky, and North Carolina's claim became Tennessee.

Sources: The state of Franklin is from Paullin and Wright, *Atlas of the Historical Geography of the United States*, plate 41C; claims are in Meinig, *The Shaping of America*, 1:350.

The mechanism whereby new states could be created was spelled out in the so-called Northwest Ordinance of 1787. Directed at the "territory northwest of the Ohio River," this law said that in that area, not less than three nor more than five states were to be formed. It laid down a three-step process that would define a procedure for creating new states:

1. Congress would organize a territorial government and appoint a governor, a secretary, and judges.
2. On reaching five thousand free adult males, the new district could elect a territorial legislature.
3. When the territory had sixty thousand free inhabitants, it could petition for admission as a state.

These new states would be equals in every way with the original thirteen states.[10]

Vermont became the first state to be admitted beyond the original thirteen. Claimed by both New Hampshire and New York, lightly settled Vermont declared itself to be a state in 1777, petitioned the Continental Congress for admission in 1781, accepted the conditions for admission in 1782, but did not become a state (the fourteenth) until 1791. In 1792, Kentucky became the fifteenth—and first western—state. This happened in a plan whereby Virginia granted Kentucky its independence and the federal government immediately granted it statehood. Four years later (1796) Tennessee became the sixteenth state, the first to be admitted under the Northwest Ordinance; it qualified with sixty thousand residents but bypassed territorial status. Many who had settled in Kentucky and Tennessee were "squatters," meaning that they simply took up unoccupied land without authorization or title. Subsequent surveys made to establish boundaries found that claims overlapped; to clear up clouded titles meant expensive litigation, and some who had lost land and were out money, like Kentucky's Daniel Boone, left the area in dismay. As we shall see in Chapter 10, orderly surveys in advance of settlement, called for by the Land Ordinance of 1784, would not be implemented until the settlement of Ohio and the Old Northwest.[11]

Washington, DC

As Congress went about the business of forming new states, it also created a new national capital, and in this undertaking George Washington himself was very much involved. Wishing to avoid awarding any one state with the prize of a new capital, in 1787, delegates to the Constitutional Convention decided that for a future capital, Congress would create a federal district that would not exceed ten miles square. In 1789, New York City, where Washington was inaugurated president, was made the temporary capital, and during the 1790s Philadelphia had that honor. Meanwhile, in the debate to choose the location of a permanent capital, the issue of centrality—a geographic and demographic middle location between Maine and Georgia—drove the decision-making. A lesser consideration was a location on a navigable waterway with access to the interior. The choice came down to a location on the Potomac or on the Susquehanna, and in 1790, in a compromise between northern and southern legislators, the Potomac got the nod. In January 1791 George Washington fulfilled his authorized assignment of choosing a site on the Potomac. The federal government then acquired one hundred square miles (260 square kilometers), some sixty-eight on the Maryland side of the river and thirty-two on the Virginia side, and created a new "Territory of Columbia" (see map 8.5A). Two communities of consequence, Alexandria and Georgetown, were already there, as were two hamlets, Hamburg and Carrollsburg. For the "Federal City," George Washington chose a site on the Maryland side just south of Georgetown.[12]

In March 1791, to design the Federal City, Washington selected Pierre Charles L'Enfant, a French artist and military engineer who had served as an officer under Washington in the Revolutionary War. Working desperately hard for the next six months, L'Enfant finished his plan and presented it to Washington in late August (see map 8.5B). The plan called for streets to be laid out on a grid pattern within which would be fifteen squares named for each of the fifteen states (a square for Kentucky was apparently added in 1792). These squares were to be the nodes from which fifteen avenues (carrying the states' names) would be superimposed over the grid as diagonals to connect the squares. The two major features, the Capitol and the President's House (called the White House after the War of 1812) sat on slightly higher ground above the flat Coastal Plain, and both were positioned along a principal avenue (Pennsylvania Avenue) that L'Enfant envisioned would connect

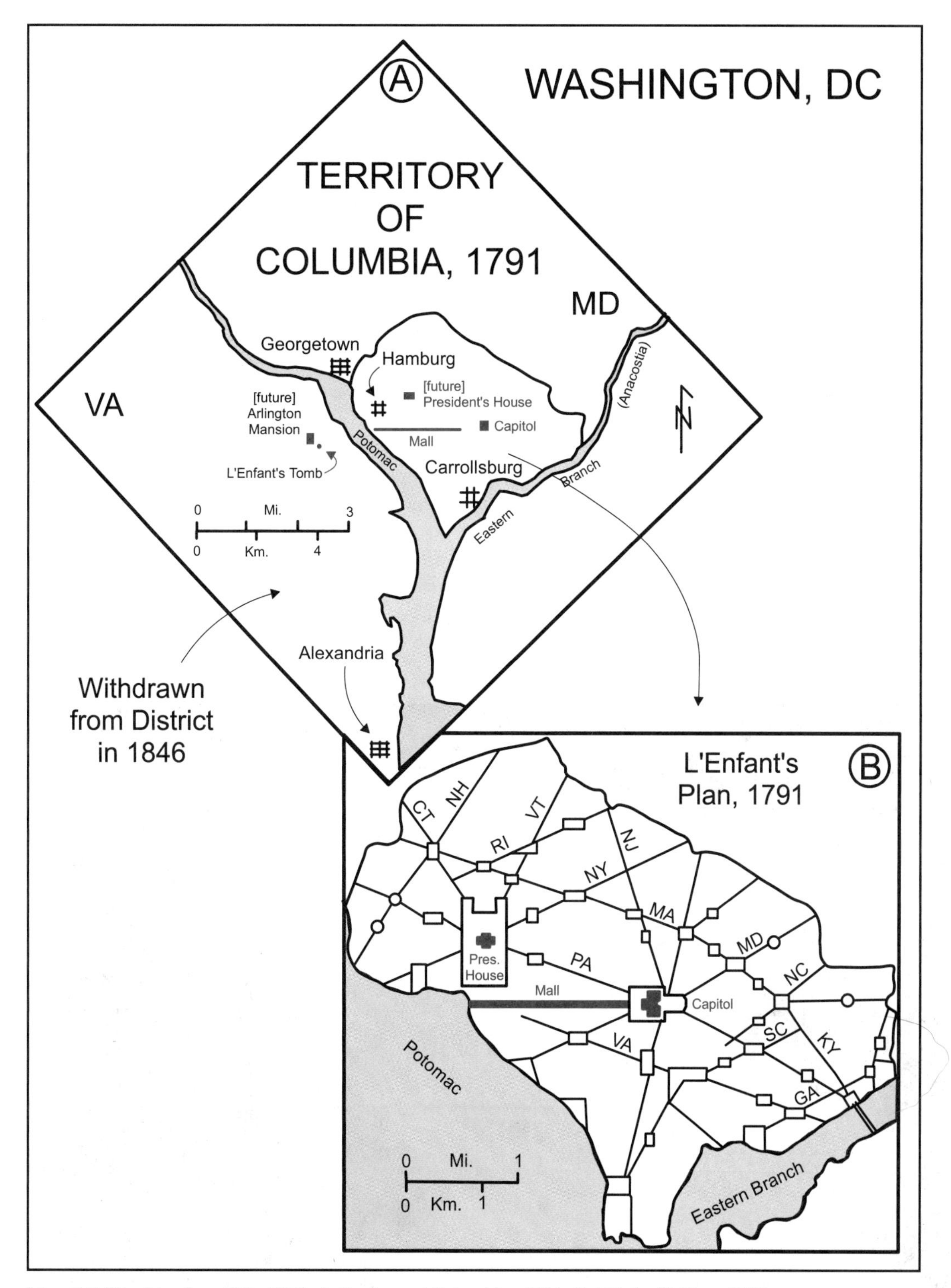

Map 8.5 Washington, DC, 1791. A: Territory of Columbia, 1791; B: L'Enfant's Plan, 1791.

Source: Redrawn (with some details omitted) from Harper and Ahnert, *Introduction to Metropolitan Washington*, maps on 31 and 32.

Georgetown with a bridge over the Eastern Branch (the Anacostia River). Where the mall is today, L'Enfant planned a "grand avenue" four hundred feet (122 meters) wide with gardens on either side. As recently as 1900, the mall ended in front of the White House at the Washington Monument; the western end of the mall lies on land since reclaimed from the Potomac.

L'Enfant's plan for the City of Washington, so named in October 1791, was brilliant. Drawing on his firsthand knowledge of Paris and its baroque architectural grandeur, L'Enfant had designed a city with breadth of scale and monumentality that would befit the new and promising nation-state. At the same time, ironically, L'Enfant's design of a city for emperors turned its back on the democratic values and egalitarian spirit that underlay the new republic. Nonetheless, his plan was implemented and, with many exceptions, endures today. (One exception is the one-time canal that followed Constitution Avenue to link the Potomac and the Anacostia; a second is that many of the fifteen [and other] squares became circles to facilitate traffic flow.) Helping to preserve L'Enfant's vision would be a law enacted by Congress in 1910 that would limit the height of new steel-frame buildings in order to retain the Capitol dome as the city's only skyscraper (see figure 8.2). Running counter to his vision would be the fact that in 1846, residents of the District of Columbia on the Virginia side, discovering that they had not shared in the city's prosperity, and with plans to build a railroad, would vote to withdraw their thirty-two square miles from the district. What happened to L'Enfant himself is most unfortunate: His demands and tirades after August 1791 forced George Washington to terminate his position

Figure 8.2 The US Capitol in Washington, DC, 1965. Thomas Jefferson was among those who selected William Thornton's design for the Capitol in 1793. Following 9/11 (September 11, 2001), security measures have severely restricted public access to the Capitol.

Photograph by the author, June 13, 1965.

in February 1792, and L'Enfant died in obscurity in Maryland at the age of seventy in 1825.[13]

Quite amazingly, L'Enfant received some recognition as events played out on the Virginia side of the Potomac. Directly across the Potomac from Washington was Arlington: the plantation, on the eve of the Civil War, of Robert E. Lee (1807–1870). Lee, a Virginian, had been graduated second in his class at West Point in 1829; two years later he married Mary Custis, a great-granddaughter of George Washington and Martha Custis Washington. Through marriage, Lee acquired Arlington, whose mansion house had a commanding view of the mall and the Capitol (map 8.5A). In April 1861, as the divide between North and South widened, Lee was offered the command of the Union Army. As a Virginian with loyalties to the South, he refused, promptly resigned his commission in the US Army, and soon took command of the Confederate Army of Northern Virginia. Lee and his family then abandoned Arlington, and they never returned. As the war progressed, the mall became a makeshift hospital for Union troops, and in early 1864 Lee's plantation became a cemetery—the beginning of today's Arlington National Cemetery. Then in 1909, L'Enfant's accomplishments were officially recognized. In a ceremony overlooking Washington in front of Lee's Arlington mansion house, L'Enfant's remains were reinterred from a grave in Maryland, and on his new marble tomb was engraved his plan for the great City of Washington.

Notes

1. Semple, *American History and Its Geographic Conditions*, 37, 38, 47, for "solidarity"; 42 and 51, for "protected."
2. Kincaid, *The Wilderness Road*, 13–14 (unnumbered), 77, for route and map.
3. The National Road became US 40 about 1926. See Raitz, *A Guide to the National Road*, 24.
4. Jordan-Bychkov, *The Upland South*, Chapter 1, esp. 8, for "blending." (Terry G. Jordan married Bella Bychkova on March 8, 1997, and several years later he adopted the masculine form of her family name as his nom de plume.)
5. Ibid., 69.
6. Raitz, *The Kentucky Bluegrass*, 8, for Kentucky as the leading tobacco state. Not a "broad sweep" but an "uneven advance" is from Meinig, *The Shaping of America*, 2:224.
7. Gerlach, *Immigrants in the Ozarks*, 57, for 10 percent German.
8. Terry G. Jordan (pre–name change) addresses Southerners in Texas in two articles: "The Imprint of the Upper and Lower South on Mid-Nineteenth-Century Texas" and "The Texan Appalachia," the latter of which develops the theme of "mountaineers" in the Texas hills.
9. Paullin and Wright, *Atlas of the Historical Geography of the United States*, text 22, plate 41C, for Franklin's boundaries.
10. See Meinig, *The Shaping of America*, 1:341–43, 391; 2:432–34.
11. For Vermont, see Paullin and Wright, *Atlas of the Historical Geography of the United States*, text 73; for Kentucky and Tennessee, see Meinig, *The Shaping of America*, 1:349–51.
12. For Washington's creation, see Reps, *The Making of Urban America*, 240–62.
13. Reps, *Monumental Washington*, 21, points out the "supreme irony" of L'Enfant's plan.

New England Extended 9

New England Extended refers to those areas west of New England where Yankees settled in large numbers. The label is attributed to geographer Peirce F. Lewis (see spotlight), whose writings have contributed much to our understanding of the region. We begin with the big picture, "Yankeeland," a discussion of where Yankees settled beyond New England and what they left as a landscape imprint. We then look in greater detail at a part of Yankeeland, Connecticut's "Western Reserve" in northeastern Ohio and its special Firelands. Next, we see how the gamble of digging an "Erie Canal" paid off in multiple ways, giving an economic boost to virtually all of Yankeeland. Finally, in the "Classical Belt," we examine a Greek- and Roman-influenced yet distinctively Yankee legacy that came to stretch in a fan-shaped zone well west of the Mississippi River.[1]

Spotlight

Peirce F. Lewis (1927–)

Peirce F. Lewis. Courtesy of Peirce and Felicia Lewis, October 15, 2011.

Peirce Lewis admonished his students at Pennsylvania State University, where he taught from 1958 to 1996, to put aside their books and observe the world around them. He followed this advice himself. Equipped with an understanding of geomorphology, plant ecology, history, and the human geography he taught, Lewis read and interpreted ordinary American landscapes, contributing journal articles and books written in titillating prose to the topics of old houses, small towns, preserving downtowns, and butchered townscapes. He undertook field reconnaissance in New England and its westward extension, which he delighted in comparing with his own Pennsylvania (see Chapter 4). In 2004 he won the J. B. Jackson Award for his *New Orleans: The Making of an Urban Landscape*. A Michigander by birth and education, Lewis earned his BA degree in philosophy and history at Albion College in 1950 and his MA and PhD in geography at the University of Michigan in 1953 and 1958, respectively. As president of the Association of American Geographers from 1983 to 1984, Lewis worked strenuously to improve public understanding of the nature and value of professional geography.

Yankeeland

In 1950 historian Stewart H. Holbrook published a book entitled *The Yankee Exodus*. Born in Vermont and a one-time resident of Vermont and New Hampshire, Holbrook saw his fellow Yankees leave the region and was moved to find out why they left, where they went, and what they contributed where they went. Holbrook was amazed to learn that by 1860 almost half of all New Englanders then alive had left the region. He also found the Yankee impact to be major: New England capital financed lumbering, mining, and real estate ventures, like the purchase of 1.5 million acres for the Ohio Company of Associates in 1787. Yankee "fanatical respect" for learning led New Englanders to dot the landscape with hundreds of schools, like Grinnell Academy (later College), founded in 1856 by Vermont-born Josiah Grinnell in Grinnell, Iowa. Yankee ingenuity overcame the problem of fencing grasslands when Joseph Glidden of New Hampshire invented a successful kind of barbed wire near DeKalb, Illinois, in 1873. And, as Yankees moved west, they founded hundreds of communities all the way to the Pacific Ocean—including, in 1853, Portland, Oregon, a settlement named after Portland, Maine, and the adopted home of Holbrook himself.[2]

Holbrook's study details the richness of Yankee contributions to American life, but it is imprecise on where Yankees went. Geographer John C. Hudson filled in this gap. Drawing on the 1880 manuscript census schedules, which he reinforced with accounts of Yankees in over two hundred contemporaneous county histories, Hudson carefully and resourcefully identified a "Yankeeland" column lying west of New England. Yankeeland lay between two other columns of westward-moving peoples; on the north was that of the St. Lawrence Valley French, and on the south was a Midland column composed especially of Germans and Scotch-Irish from Pennsylvania. Hudson's criterion for identifying Yankees in Yankeeland was that Yankee birthplaces had to be at least twice as common as birthplaces for those of the St. Lawrence or Midland columns, or from elsewhere. These Yankee birthplaces, Hudson pointed out, were sorted by generation in the westward flow from New England. Hudson's Yankeeland column and its generations are shown in map 9.1.[3]

In time and place, the Yankee exodus of people moving as individuals, as families, or in groups, went as follows: About 1790, relatively large numbers of New Englanders began to spread over upstate New York and into adjacent northern Pennsylvania. In the late 1790s Yankees, especially those born in Connecticut, also began to reach Connecticut's Western Reserve in northeastern Ohio. Hudson notes that for reasons not yet clear, northeastern Ohio contributed only modestly as a source area for subsequent western migrants. Those born in New York and in New England then became a second generation of migrants beyond New England as they moved to southern Michigan between 1825 and 1840, to southeastern Wisconsin between 1835 and 1850, and to southeastern Minnesota. And still a third generation of migrants beyond New England found significant numbers of the New York–born, Michigan-born, and Wisconsin-born moving to the grasslands of southwestern Minnesota and to east central South Dakota and North Dakota. Hudson's Yankeeland column, then, stretched from New York to southern Michigan to southeastern Wisconsin to southern Minnesota to the eastern and central Dakotas, and the time frame for this expansion was basically 1790 to the 1870s.[4]

The prospect of better lands was what pulled Yankees west into Yankeeland. And in their westward travels Yankees encountered in these better lands ecological features that they called "oak openings"—clearings in the hardwood forests that were mainly grass and only thinly timbered. Whether these oak openings were Indian Old Fields or were caused by some edaphic factor seems unclear. Geographer Bernard C. Peters points out that in the especially attractive "Genesee Country" of western New York,

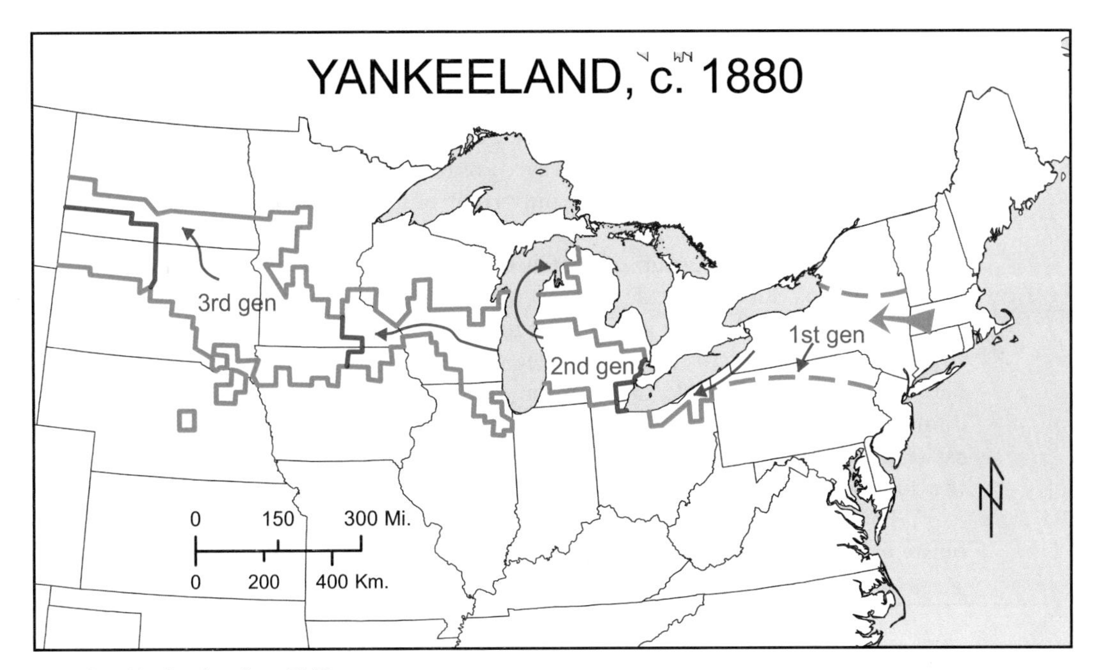

Map 9.1 Yankeeland, c. 1880.

After Hudson, "Yankeeland" (1986), 196. The labels 1st, 2nd, and 3rd generations are my own. See text for how Hudson defined Yankeeland and its Yankee generations beyond New England.

Yankees seemed to have equated the absence of trees in oak openings with infertile soils and, as a consequence, initially avoided these areas. As oak openings became more numerous in southern Michigan and Wisconsin, however, a second generation of Yankees found that these grassy areas produced excellent crops of wheat, and Yankees now began to select oak openings before forested lands. Wealthier New Englanders seem especially to have sought oak openings because they could afford the several yoke of oxen needed to plow the virgin grassland sod. At several levels, including how they came to exist and how they were perceived, oak openings remain something of an ecological enigma. And to complicate their understanding, southerners who encountered oak openings in Kentucky, Ohio, and Indiana, called these same features "barrens." Carl Sauer wrote that by "barrens," southerners in Kentucky meant areas barren of trees, but the term, according to Peters, may also have suggested poor soil to southerners.[5]

As Yankees moved west in Yankeeland, wheat became the leading commercial crop for each new generation. This is not surprising. Wheat was the preferred grain for virtually all colonists of European descent, and in newly plowed soils in the northerly latitudes of Yankeeland, wheat, a cool-climate grain, produced well. But in time soils became less productive, prompting the sons and daughters of the initial settlers to grow other grains, including corn, barley, and rye, and to focus on dairying for their incomes. Unlike Midlanders to the south, few Yankees seem to have fed their corn to cattle and hogs in order to market those livestock. The idea behind the Corn Belt—that one could grow corn and market it on the hoof—seems to have originated with Midlanders, not Yankees.

As Yankees moved west, land especially in upstate New York was made available in land tracts. Five land tracts in New York were notable (see map 9.2): In 1781 New York State set aside a Military Tract for Revolutionary War veterans, but few were attracted to its remote and uninviting location in the Adirondack Mountains. A year later the state created a New Military Tract out of 1.5 million acres in the eastern Finger Lakes district, and when it was opened for settlement in 1791, script that was issued to veterans passed quickly to speculators, who sold largely to New Englanders. Also in 1791 Sir William Pulteney of Scotland bought 2.25 million acres, which were known both as the Pulteney Purchase and as Genesee Country. Under company supervision lands were surveyed, roads were built, model farms were created, and New Englanders were the important buyers. And in 1791 William Morris bought four million acres in the western end of New York State, of which in 1792 he sold four-fifths to a group of Dutch bankers (thus the name Holland Purchase) who undertook surveys, made improvements, and sold mainly to Yankees. Finally, Alexander Macomb bought nearly all of four million acres, called the Macomb Tract, south of the St. Lawrence, a remote area that attracted settlers only slowly. In short, New York State had made the decision to raise revenue by selling its own public domain to war veterans or other settlers through the creation of several huge real estate tracts, and New Englanders who funneled through the Mohawk Valley were the dominant buyers.[6]

What can be said of the landscape impress of these predominantly Yankee colonists in upstate New York and in Yankeeland beyond? Recall that in about 1800,

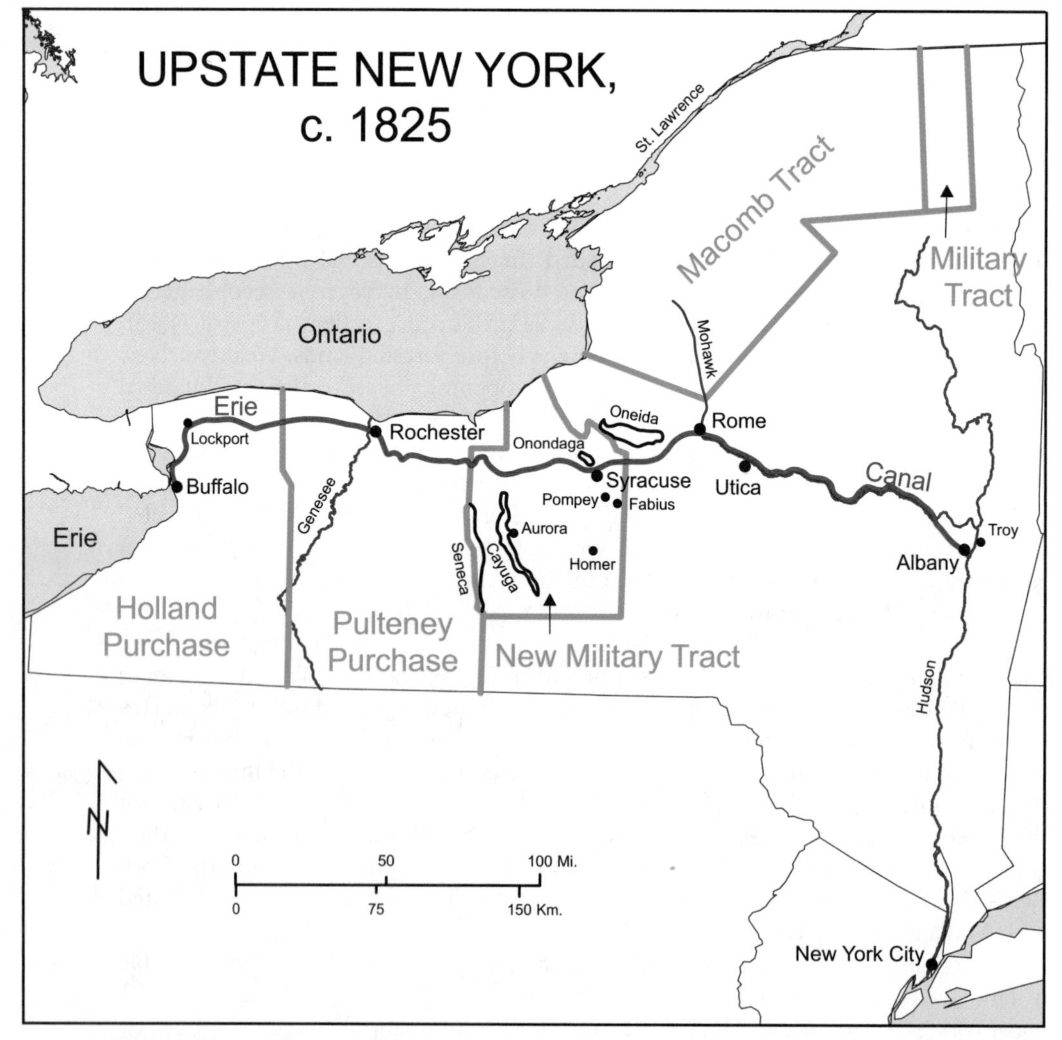

Map 9.2 Upstate New York, c. 1825. Shown are land tracts and the Erie Canal. The four villages in the New Military Tract exemplify those with New England qualities.

villages in New England itself were evolving as new, two-story houses and businesses filled in around a meetinghouse village focus. These larger, filled-in villages were apparently what Yankees re-created in the west. In the New Military Tract, for example, early New Englanders founded Pompey, Aurora, Fabius, and Homer, each with its characteristic Yankee attributes of a common or green, a Congregational meetinghouse, a school, and wooden houses painted white and set in large lots. For purposes of local governance, Yankees also created townships within the larger county political units. But the point about the Yankee landscape impress in New England Extended is that Yankees were not alone in Yankeeland. Dutch settlers from the Hudson Valley, a variety of peoples from Pennsylvania, and French settlers from Canada, among others, colonized Yankeeland, and the landscape impress reflected all streams. New England Extended was much more than Yankee villages with Yankee place names set within townships. As Meinig put it, "Although the marks of New England were readily apparent, they did not quite add up to a replica of New England."[7]

The Western Reserve

Connecticut left its imprint in a unique political way—in its so-called Western Reserve. Like five other colonies, Connecticut had a charter that said it stretched from "sea to sea." Connecticut's claim to western lands lay between latitude 41° and 42° (actually 42°2') north (see map 9.3A). In 1782 Connecticut relinquished to Pennsylvania the title to that band of latitude in Pennsylvania, and in 1786 it ceded to the federal government its claim to land within the same strip west of present-day central Ohio. But at the time, Connecticut kept for itself a Western Reserve in what would become northeastern Ohio, an area that extended west from the Pennsylvania border for 120 miles and lay between latitude 41° and the Lake Erie shoreline. The total area of this Western Reserve amounted to some three million acres. In 1800 Connecticut did cede to the United States its Western Reserve, but by then several developments had taken place.

In 1795 Connecticut sold most of its Western Reserve—some 2.5 million acres—to the Connecticut Land Company (see map 9.3B). Thirty-five men had formed the company, putting up $1.2 million in order to buy the land. A year later one of the landowners, Moses Cleaveland, also the company's general agent, led a party of some fifty men west to survey the purchase. Once on the land, most of the party broke into four groups, each led by a surveyor with accompanying axmen, chainmen, and rodmen. The survey process began with identifying the "Ellicott Line," by 1796 an overgrown twenty- to thirty-foot-wide swath that had been cut in 1786 by Andrew Ellicott to mark Pennsylvania's western border. The surveyors worked west from that line as they marked east-west range lines before they turned to the survey of north-south township lines. The objective was to create a grid with twenty-five-square-mile units called townships. The job stretched to a second survey season in 1797; inaccurate compasses (although the best available) and human error left behind a grid with many flaws.[8]

Meanwhile, in 1792, Connecticut did something special with the roughly 0.5 million acres lying at the far western end of its Western Reserve (map 9.3B). Out of that final twenty-five-mile-wide stretch, it created the Firelands (modern Erie and Huron counties), which were set aside as compensation for deserving Connecticut inhabitants. During the Revolutionary War, towns located primarily along Connecticut's Long Island Sound had suffered damages at the hands of the British, who burned homes, barns, churches, stores, warehouses, wharves, and even a courthouse (thence *Fire*lands). Claims to the Firelands, as they were known

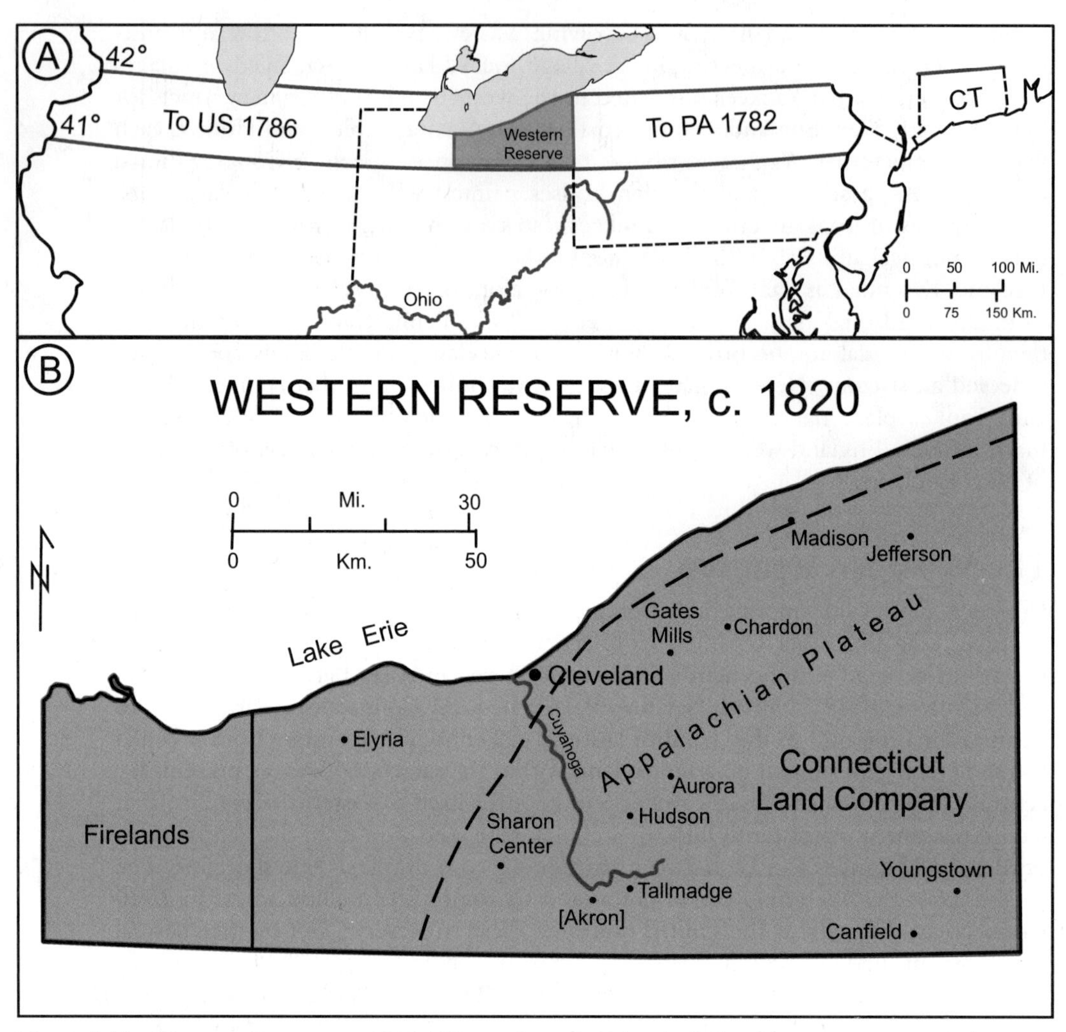

Map 9.3 The Western Reserve, c. 1820. Villages shown in B (not including Akron and Youngstown) have obvious New England qualities.

Source: Paullin and Wright, *Atlas of the Historical Geography of the United States*, plate 47B, for cessions shown in A.

in Ohio, or Sufferers Lands, as they were called back in Connecticut, were forthcoming from at least eight communities: from east to west, the coastal towns of Groton, New London, New Haven (also East Haven and West Haven), Fairfield, Norwalk, and Greenwich, and inland from Norwalk, Ridgefield, and Danbury. Surveying of the Firelands, again into twenty-five-square-mile townships, took place in 1806 and 1807; Indian title to the area had been extinguished only in 1805.[9]

Settlement of the Connecticut Land Company portion of the Western Reserve happened only gradually over the two decades following Cleaveland's surveys of 1796 and 1797. The colonists, mainly Yankees from Connecticut, had English surnames and, in keeping with the custom of the time, usually given names that were biblical, such as Ephraim and Ebenezer for men and Abigail and Prudence for women. About 1799, families and small groups of men who would later return for their families trickled in to identify the townships and the acreage they had purchased in drawings. Travelers in the Western Reserve reported these families to be living in isolated log cabins set in forest clearings sown to wheat. Sickness and hardship were, quite tragically, commonplace. Two routes of about equal length were used to reach the company lands: One crossed New York State to Buffalo and then proceeded overland or by boat to the Lake Erie shoreline, whence rivers were followed inland; the other led over rugged mountains to Pittsburgh and from there by pack trail to enter the Western Reserve in its southeastern corner near Youngstown (map 9.3B). The Firelands were not opened for settlement until 1808, and when opened, land was parceled out in a most complicated way—and thirty years too late to help many of the actual wartime sufferers. Not until the summer of 1817 did a major flow of Yankees begin to reach all of the Western Reserve—a flow that came after the War of 1812 had ended and after New England's disastrously cold summer of 1816.[10]

The Connecticut Land Company had predetermined that it would establish a major community on Lake Erie near the center of the Western Reserve. As the surveys of the range and township lines proceeded in 1796, Moses Cleaveland and a small party explored the lakeshore and chose as the site for this community the mouth of the Cuyahoga River (map 9.3B). Cuyahoga was to be the community's name, but Cleaveland was prevailed upon by his surveyors for permission to use his own name. What went wrong was that on a plat (dated October 1, 1796) that showed the new community, someone spelled "Cleaveland" as "Cleveland," and the misspelling stuck. Although Cleveland grew slowly (only fifty-seven people by 1810), it did become the region's projected major center, complete with a future Case Western Reserve University. It also had a ten-acre central public square that Moses Cleaveland had requested "after the New England custom."[11]

Even though settled early by Yankees mainly from Connecticut, the Western Reserve falls short of being a bit of New England reproduced to the west. Certain villages that Yankees founded do display clear Yankee qualities, including greens (in the shapes of squares), white-painted Congregational churches, and white-painted, New England–style, wooden houses. Modern-day examples of such villages can be seen particularly in the eastern Appalachian Plateau part of the Western Reserve, as at Jefferson, Madison, Chardon, Gates Mills, Aurora, Hudson, Tallmadge, Canfield, Sharon Center, and Elyria (map 9.3B). New England place names are also plentiful in the Western Reserve—examples are Amherst, Andover, Boston, Dorset, Greenwich, New Haven, Norwalk, Sheffield, and Windsor. But for the most part the Yankee impress is now faint and obscure and must be carefully sought out. Immigrants, especially those from southern and eastern Europe, lured by available jobs to various industries in Cleveland, to iron processing and steelmaking in Youngstown, and to rubber manufacturing in Akron, helped to obscure what was Yankee.[12]

The Erie Canal

Connecticut's Western Reserve, and virtually all of Yankeeland west of New England, benefitted economically from the completion of the Erie Canal. That a lowland corridor that followed the Mohawk Valley (to Rome) and the Ontario Lake Plain provided travelers the easiest way west from the Atlantic Seaboard was long recognized (map 9.2). Along this corridor, the high point located west of Rome was less than five hundred feet (1510 meters) above sea level. Recognizing their geographical good fortune, officials in New York State as early as 1785 proposed that a canal be dug across the corridor to connect the Hudson River with Lake Erie. DeWitt Clinton, a New York State commissioner and governor, became the project's leading advocate. In 1817 New York's legislature took the initiative and approved funds for the canal. The canal's construction over the next eight years brought into existence entirely new communities like Syracuse and rapid growth to places like Rochester, a mill site on the Genesee River. When officials opened the canal on October 26, 1825, Clinton and others celebrated the engineering triumph by riding on a slow-moving flotilla of boats from Buffalo to Albany, stopping for festivities in every town.[13]

Figure 9.1 Erie Canal and Mohawk River. The canal followed alongside the Mohawk, as shown here with two canal boats approaching a lock (unidentified location, n.d.).

Courtesy of the Erie Canal Museum, Syracuse, NY, http://www.eriecanalmuseum.org, cat. no. 1208a (Morganstein, Cregg, and Erie Canal Museum, *Erie Canal*, 38 [bottom]).

The route chosen for the canal followed in part an ancient beach line created by Lake Ontario before the lake shrank to its present size (map 9.2). Lake Ontario was purposefully avoided because some feared that goods would be diverted down the St. Lawrence River to Montreal. The route also avoided the Mohawk River itself because snags in the river and fluctuations in water levels posed greater problems than did digging a canal parallel with the river (see figure 9.1). The canal measured forty feet wide at the top and twenty-eight feet wide at the bottom, and had a uniform water depth of four feet. To cross the Genesee River at Rochester, an aqueduct had to be built, and to lift the barges up inclines, as at Lockport, locks that acted like elevators required constructing. In all, eighty-three locks were needed. Barges were pulled by horses and mules along a towpath on the canal's north side. Traveling at four miles per hour, barges could complete eighty miles (130 kilometers) in a day, and at that rate it took 4.5 days to cross the 363 miles (nearly 600 kilometers) from Buffalo to Albany.

Completion of the Erie Canal had major consequences. It immediately lowered the cost of transporting goods from Buffalo to Albany from about eighty to eight dollars per ton. Before the canal's completion, only valuable items like salt-cured pork, whiskey, and milled flour could bear the cost of transportation across New York State to markets. But through the canal, bulky and relatively low-priced corn, wheat, and lumber could also be transported economically, and items manufactured in the East could now flow west. The success the Erie Canal had in lowering transportation costs brought about a gradual reorientation of traffic: Goods that once flowed down the Ohio and Mississippi rivers on flatboats to New Orleans, and from New Orleans by coastal packets to Philadelphia and New York, now flowed east by canal boats to Albany, and then by steamboat down the Hudson River to New York City and on to Philadelphia. New York City had surpassed Philadelphia to become America's largest city soon after 1800, and the Erie Canal guaranteed that supremacy.

The success of the Erie Canal gave impetus to a canal-building era that lasted from the latter 1820s to the late 1840s. Many of these canals were constructed to breach the divide between the Great Lakes and the Ohio and Mississippi valleys. Details about the names of the canals and the old portage routes they followed between river headwaters are recorded and mapped in Ralph Brown's *Historical Geography of the United States.* In Ohio, Cleveland, on Lake Erie, was connected with Portsmouth, on the Ohio River; Toledo, on Lake Erie, was connected with Cincinnati, on the Ohio River. In Indiana a canal begun in Ohio along the Maumee River connected Lake Erie with the Wabash River, a tributary to the Ohio. And in Illinois a canal connected the short divide between the Chicago River and the Illinois River. None of these canals had the success of the Erie Canal. Then, in the 1840s and '50s, all canals, including the Erie, rather quickly lost out to the

Figure 9.2 Weighlock at Syracuse, New York, c. 1870. Once in the weighlock chamber, canal boats were weighed when the lock gates closed, the water was released, and the boat came to rest on a wooden cradle attached to a scale. Tolls were determined by the weight of cargo and its destination. Built in 1850, this Syracuse building is the only remaining canal-era weighlock.

Courtesy of the Erie Canal Museum, Syracuse, NY, www.eriecanalmuseum.org, cat. no. 79.52.784 22.01 (Morganstein, Cregg, and Erie Canal Museum, *Erie Canal*, 22 [bottom]).

railroads, which could transport goods more quickly and over routes that were much less rigid.[14]

On a personal note, I spent a geographer's dream day on Erie Boulevard in Syracuse early in my career. I came to a strange-looking building whose second story hung out several feet over the road (see figure 9.2). I was amazed to learn that the building, which housed the Erie Canal Museum, was the old Weighlock Building, under whose overhang had been scales for weighing canal boats and assessing tolls. And I was just as amazed to be told that Erie Boulevard itself was the filled-in trench of the Erie Canal. (I was also to learn why an early village near Syracuse was named Liverpool. The Erie Canal had, by design, been routed around the south end of small Lake Onondaga to tap the local salt deposits. Promoters of one lakeside community—now part of greater Syracuse—named the place Liverpool after a quality salt-producing community in England. The name enabled these promoters to label their casks "Liverpool Salt"—surely not the first example of fraudulent advertising in America.)

The Classical Belt

What prompted New Englanders in upstate New York to give their new communities names like Syracuse, Rome, Pompey, and Troy? After the Revolutionary War, Americans, quite understandably, rejected that which was British. In their search for a new identity, they revived the virtues and ideals of Athens and Rome. This "Classical Revival," as it is known, was manifested in various ways: in place names, domestic architecture, and the use of Greek letters and ideals for fraternities and sororities. Additional examples include city planning, as at the capital city, Washington, DC,

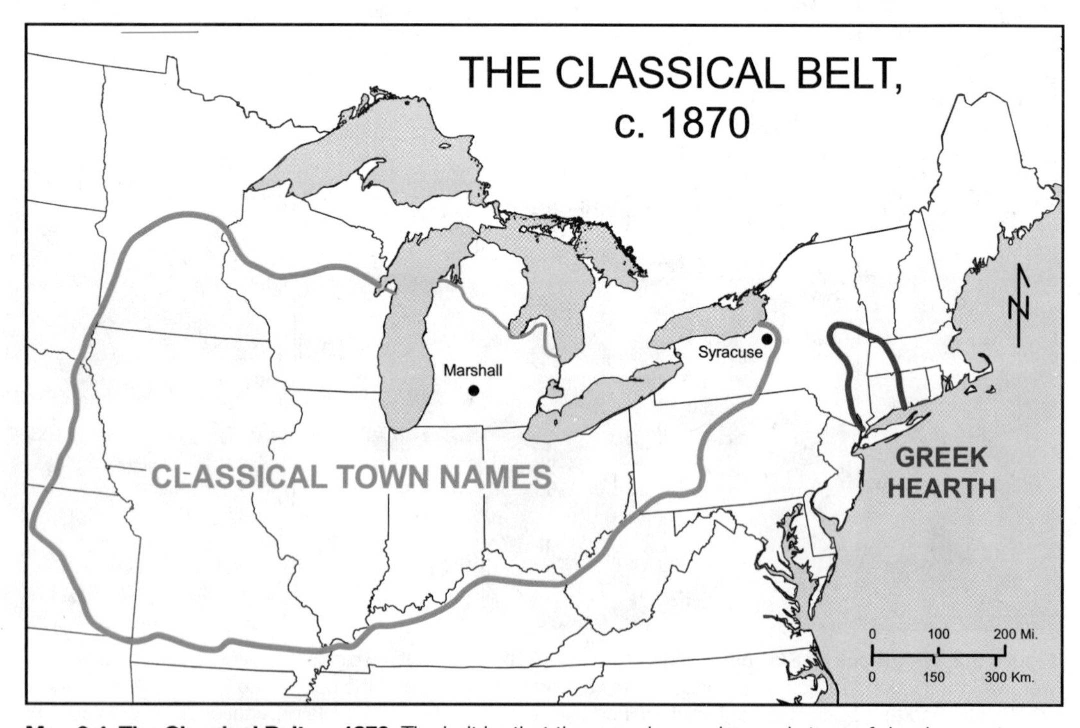

Map 9.4 The Classical Belt, c. 1870. The belt by that time was in an advanced stage of development.

Sources: Zelinsky, "Classical Town Names in the United States," 490; Torbenson, "College Fraternities and Sororities," 50, for the Greek fraternity/sorority culture hearth.

with its grid pattern and radials; personal names like Ulysses and Horace; and military terminology. New Englanders, the most literate of the new nation's peoples, were the strongest proponents of this classical rebirth. And west of New England this Classical Revival took geographical expression in the "Classical Belt."

How the term "Classical Belt" originated is not clear. It was used, although without a stated areal context, in 1945 by English professor George R. Stewart in a chapter about classical place names in his *Names on the Land*. As we have seen, geographers make it their business to try to define areal concepts, and in 1967 Wilbur Zelinsky undertook the job for classical town names. Zelinsky's series of temporal maps show that in the 1790s and early 1800s, dozens of new towns were given classical names in west central New York. The "swarm" of names like Manlius and Marcellus was "thickest" in the New Military Tract, which Zelinsky found to be the "culture hearth" for the practice of applying classical toponyms in a new place. After 1810 Zelinsky's maps show that the adoption of classical town names diffuse west with strength to about the time of the Civil War. The spread then weakens but continues until about 1920. Where classical names diffused with strength amounts, of course, to a toponymic definition of the Classical Belt. Zelinsky's map of classical town names in 1870 delimits a final stage in the Classical Belt's evolution. By 1890 the western edge of this roughly isosceles triangle (see map 9.4) would push into central Nebraska and the Dakotas, but its apex would remain anchored slightly east of Syracuse. Interestingly, in New England itself, although opportunities for naming new communities were limited, the modest number of new towns with classical names was greatest in central Maine.[15]

Figure 9.3 Greek Revival House in Hatfield, Massachusetts, 1973. A sense of pride finds some New Englanders displaying the dates their homes were constructed (here, 1792). The wing to the right was added in 1837. Greek Revival style has the gable end facing forward. Orthodoxy in style is greatest when the triangular pediment above the columns is a lower-angle triangle than is the case here; when the columns (fluted if Greek; smooth if Roman), for symmetry, are even in number, as is also true here; and when the doorway behind the columns is central, which is not the case here.

Photograph by the author, July 12, 1973.

A second Classical Revival manifestation was architectural. The use of Greek Revival architecture—triangular pediments supported by columns in templelike structures—was becoming common in America about 1790 for *public buildings* like state capitols, town halls, and county courthouses. With the Classical Revival, especially New Englanders now began to use Greek Revival architecture for their *private houses.* As with classical town names, people outside New England were participants. For example, America's premier innovator, Thomas Jefferson, who was always ahead of the curve, in 1768 built his Monticello in classical style. About 1790 we find in New England itself that Yankees who could afford to do so were beginning to build their houses in Greek Revival style (see figure 9.3). The practice then flourished among wealthier Yankees in upstate New York (see figure 9.4). And building houses in classical style was carried west in Yankeeland to places like Marshall, Michigan, where streets today are lined with gorgeous Greek Revival mansions built out of wood and painted white. Geography

Figure 9.4 Greek Revival House in Marcellus, New York, 1965. This example was built in 1830. Yankee temple-form houses often had one or two wings on the sides.

Photograph by the author, May 1965.

has yet to produce a Zelinsky clone to step up and map the correlation between New England Extended and classical architecture used domestically.

New Englanders also played a key role in the creation of Greek-letter fraternities and sororities. At the University of Oklahoma in 1992, Craig L. Torbenson, unbiased because he had never belonged to a fraternity and eager to test his computer mapping skills on a huge Greek system database, wrote a dissertation on the origin and diffusion of "College Fraternities and Sororities." Torbenson learned that Phi Beta Kappa, founded as a social fraternity in 1776 at William and Mary College in Williamsburg, Virginia, was the first fraternity. Responding to an antisecrecy movement in the 1820s, Phi Beta Kappa disclosed its secrets and became the scholastic honorary society it is today. Still, largely secretive fraternities were founded at Union College in Schenectady, New York (known as the Union Triad, one in 1825 and two in 1827), at Williams College in Williamstown, Massachusetts (one each in 1833 and 1834), and at Yale College in New Haven, Connecticut (one in 1836). These three colleges lay within Torbenson's Hudson River–straddling Greek system culture hearth (map 9.4). Thereafter, the number of fraternities and sororities grew, some being founded in well-touted "duos" and "triads." Torbenson's many maps show much of the founding of fraternities and sororities to have occurred within the Classical Belt, but also some beyond it. He points out that the founding of fraternities and sororities had two peaks, roughly 1840 to 1880 and 1900 to 1930; those founded in the nineteenth century have mainly lasted, while those founded after 1900 have mainly disappeared. The Greek system is significant if only because by 1990, when Torbenson wrote his dissertation, some two hundred national fraternities and sororities with over fourteen thousand chapters existed at nearly one thousand colleges and universities. The system's origins are complicated, and sorting out the obviously important role played by Yankees awaits a future scholar.[16]

Notes

[1] Wilbur Zelinsky credits Lewis with the term "New England Extended" in "Classical Town Names in the United States," 466, note 11.

[2] Holbrook, *The Yankee Exodus*, 4, for departure of almost half by 1860; 21, for Ohio Company; 9 and 319, for Glidden; viii, for the reference to "fanatical respect"; 138–39, for Grinnell; 227–28, for Portland, Oregon.

[3] Hudson, "Yankeeland in the Middle West," 196, for a map and a definition of Yankee birthplace dominance. Hudson is specific about his east-west sorting of generations: "The original settlers of southern Michigan and Wisconsin and southeastern Minnesota were born in New York or New England; the first settlers of southwestern Minnesota and eastern Dakota were about evenly divided between New York and either southern Michigan or southern Wisconsin birthplaces; a third generation, born in Wisconsin and southeastern Minnesota, settled west of the Missouri River."

[4] Ibid., esp. 196 and 198, for the Western Reserve.

[5] See Peters, "Changing Ideas about the Use of Vegetation as an Indicator of Soil Quality," 18–28, esp. 19, for barrens as poor soil. See also Peters, "Oak Openings or Barrens," 84–86; and Sauer, "The Barrens of Kentucky," 23–31.

[6] Brown, *Historical Geography of the United States*, 178–83 for land tracts. Meinig, *The Shaping of America*, 2:240–41, for New York State controlling its public domain. See Schein, "Unofficial Proprietors in Post-Revolutionary Central New York" and "Urban Origin and Form in Central New York," for facts about the New Military Tract. Wyckoff, *The Developer's Frontier*, 109, for the importance of Yankees in the Holland Purchase.

[7] Schein, "Urban Origin and Form in Central New York," 52, 58, 65, 67, for Yankee villages in the New Military Tract. See Meinig, *The Shaping of America*, 2:272, for the quotation.

[8] See Hatcher, *The Western Reserve*, esp. 1–61, and 20, for the Ellicott Line.

[9] Ibid., 40–48, for the Firelands.

[10] Ibid., 45, for Old Testament "baptismal" names; 59, for the two routes used. See Hoyt, "The Cold Summer of 1816," for crop loss and emigration.

[11] "After the New England custom" are Hatcher's own words for explaining Cleaveland's desire for a town square, in *The Western Reserve*, 26. See ibid., 27, for Cleveland's spelling.

[12] Ibid., 3–5, for Yankee community examples and place names. See Lindsey, "Place Names in Ohio's Western Reserve," for many Yankee-origin examples, and Reps, *The Making of Urban America*, 227–39, for Yankee towns in the Western Reserve.

[13] Meinig, *The Shaping of America*, 2:270, 318–23, 335, 353.

[14] Brown, *Historical Geography of the United States*, 264–68, map on 267.

[15] Stewart, *Names on the Land*, 185, for "classical belt"; Zelinsky, "Classical Town Names," 478, for "swarm" and "thick"; 486, for "culture hearth" in the New Military Tract; 478, note 28, for the Classical Belt in 1870; 479, for the Classical Belt as an isosceles triangle; 489, for classical names in central Maine.

[16] Torbenson, "College Fraternities and Sororities," 50 and 139, for maps showing a Greek system "cultural hearth"; 47, for a table showing peaks in fraternity and sorority founding by decade.

The Old Northwest 10

"Old Northwest" is a textbook term that refers to the area north of the Ohio River and east of the Mississippi River (see map 10.1). When created in the Northwest Ordinance of 1787, the Old Northwest was called the territory northwest of the Ohio River, or simply the Northwest Territory. But in the wake of the westward movement, there came to exist a "New Northwest" and then a "Pacific Northwest," both popular terms used regionally today (see map 10.1 inset). So, to differentiate the original northwest, historians have dubbed it the Old Northwest. We look first at the Old Northwest environmentally, especially in two problem areas, "The Black Swamp and the Grand Prairie." We then turn to the supremely important "Township and Range System" that was first implemented in the Old Northwest. People following down the Ohio River founded communities, the focus of "Ohio Valley Urban Growth." Finally, we look at America's agricultural breadbasket, "The Corn Belt," which is heavily anchored in the Old Northwest.

The Black Swamp and the Grand Prairie

As one crosses the Old Northwest from eastern Ohio to western Illinois, two notable environmental transitions occur. The first concerns landforms. In eastern Ohio's Appalachian section, the land is hilly to mountainous; by central Ohio it becomes more rolling; and by central Indiana and on through Illinois, it is mainly flat. Deep and fertile soils cover the flatter areas, Canada's much-appreciated contribution via glaciations. The second transition concerns vegetation. A rather dense broadleaf, deciduous forest covered eastern Ohio; by western Ohio and Indiana patches of grassland appeared in the forest (the oak openings or barrens described in Chapter 9); and by Illinois one reached open prairie. The forest of oak, maple, and elm trees in the East had given way to tallgrasses, known as prairie, in the West.

People who planned to farm had to clear the forest. Geographer Michael Williams has written extensively on this topic. Williams points out that Americans cleared their forests for three reasons: to create farmland; to provide fuel for houses, industry, and transport; and to supply lumber for construction. He makes the unexpected point that "the cutting and gathering of wood for fuel has probably consumed more wood than any other use . . . lumber included." As we shall see in Chapter 13, in the more northerly Great Lakes part of the Old Northwest, a white-pine commercial lumbering frontier swept through the region in the nineteenth century to clear its forests. Meanwhile, in Ohio, Indiana, and Illinois, pioneer families were the agents of deforestation. Williams reports that on average in the forested eastern United States, it took thirty-two man-days of labor to clear one acre of forest, and that one pioneer family would clear on average thirty to forty acres (twelve to sixteen hectares) in about ten years. The process, thus, was slowgoing, and pioneers in the more southerly Old Northwest were also confronted with large areas of swamp and ill-drained land. The first of these large wetland obstacles was the notorious Black Swamp.[1]

The Black Swamp was nearly impenetrable. It had heavy, wet, black soils (which apparently gave rise to its name) and a dense broadleaf, deciduous swamp forest cover made up of oak, maple, elm, ash, hickory, linden, and cottonwoods. Located in northwestern Ohio, the Black Swamp covered a reported 1,500 square miles (4,000 square kilometers) (map 10.1). The feature was formed in glacial times. Lake Erie, the shallowest Great Lake, was once a larger Glacial Lake Maumee, also very shallow,

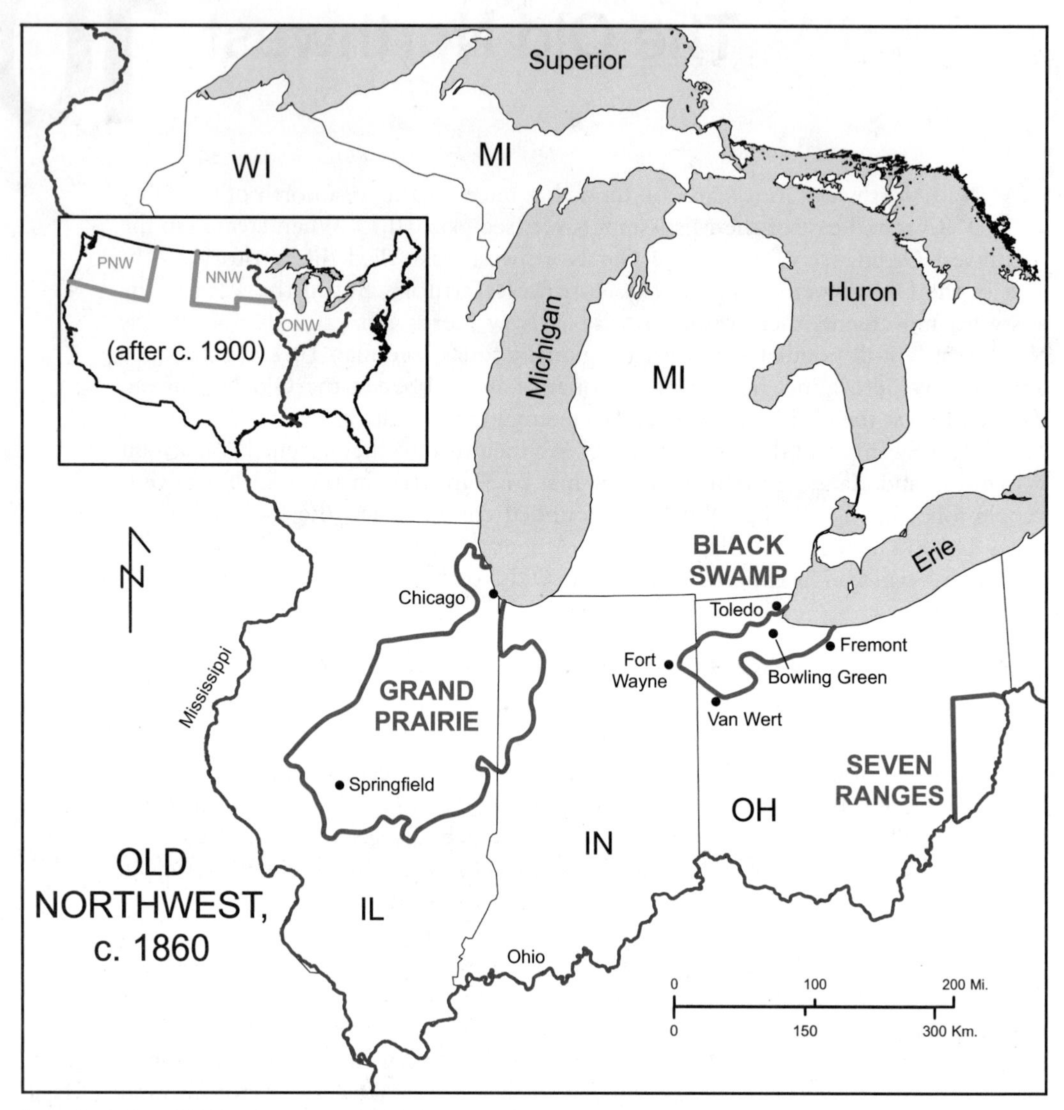

Map 10.1 The Old Northwest, c. 1860. The first of three northwests (see inset), the Old Northwest lay north of the Ohio and east of the Mississippi.

Sources: Kaatz, "The Black Swamp," 2; Urban, "An Uninhabited Waste," 655, for the Grand Prairie in Illinois; Hudson, *Making the Corn Belt*, 100, for the Grand Prairie in Indiana.

and as Glacial Lake Maumee shrank to become Lake Erie, it left a flat glaciolacustrine plain. This plain sloped in the direction of Lake Erie at an average rate of only four feet per mile. And because the glacial lake shrank in stages, it left wave-deposited beach ridges that stood some ten feet above the plain (see figure 10.1A). Also forested, these sandy ridges were used as pathways through the thirty-mile (fifty-kilometer) wide swamp, and because they were better drained, they were the first areas to be cleared and settled.[2]

Geographer Martin R. Kaatz tells us that problems of soil drainage delayed settlement of the Black Swamp until about 1850, after which a slow filling-in of people continued until about 1900. Once the trees were felled, early efforts at drainage included the following: "dead furrows" to catch excess water; placing stones and saplings in subsurface trenches for drainage; and laying inverted V-shaped wooden

Figure 10.1 The Black Swamp, 1985. Features near Bowling Green, Ohio, include: (A) a beach ridge above the plain; (B) an open ditch paralleling an east-west road; (C) a tile opening exposed in the ditch bank.

Photographs by the author, June 6, 1985.

underdrains within fields. In time, however, deep public ditches were dug alongside roads that ran east to Lake Erie, and by the 1860s and '70s, clay tile was available at reasonable prices for use within fields. Progress was slow: As late as 1870 more than half the Black Swamp was still in its natural state, but today the Black Swamp is a thing of the past, and the area is part of the Corn Belt that surrounds it. However, telltale clues do exist as testimony to its one-time presence: One is the presence of open, deep ditches next to east-west roads; exposed in the banks of these ditches are the ends of buried clay tiles several feet below the surface (see figures 10.1B and C). A second is a general absence of fences, apparently because farmers thought they could make better use of their fertile black soil by growing just corn and foregoing the raising of livestock in fenced fields. And because it was settled late, a third clue is the relative absence of large service communities. Urban centers like Fremont, Van Wert, Fort Wayne, and Toledo came to exist around the outside edges of the Black Swamp. Bowling Green was the only community of size to emerge within the Black Swamp.[3]

The settlers avoided the second area under discussion, the Grand Prairie, because it was swampy. Located in east central Illinois and adjacent western Indiana, this huge easternmost tip of a feature known as the Prairie Peninsula contained some 19,500 square miles (50,600 square kilometers; map 10.1). Unlike the Black Swamp, however, the Grand Prairie had only pockets of swamp and wet prairie within what was mostly dry prairie. Studies of the Grand Prairie by geographers Roger A. Winsor and Michael A. Urban confirm that the pockets of wetlands, because they were so difficult to cross and so unhealthy (malaria was the dominant disease in Illinois between 1780 and the 1850s), came to be overgeneralized as to their areal extent in the accounts of travelers. The image of swamps and bogs, while they actually constituted a rather small percentage of the total area, came to dominate public perception of the larger Grand Prairie. So what the Grand Prairie was thought to be—not what most of it really was—delayed its settlement until the 1850s. Ditches and clay tiling eventually drained the pockets of wet prairie, much as they had the Black Swamp. The Grand Prairie was the last large area to be settled in Illinois, but once occupied, like the Black Swamp, it became a productive part of the Corn Belt.[4]

The Township and Range System

Settlers taking up land in the Old Northwest had their properties described and recorded in a new, orderly township and range system. In many ways, this new system improved upon the metes and bounds way of describing and recording properties used in colonial times. The metes and bounds system, which came out of English common law, used natural and cultural features such as streams, large trees, roads, and stone fences for describing and recording land parcels. Properties were irregular in shape, making it difficult to determine precise acreage for accurate assessment of taxes. Moreover, property boundaries became issues between neighbors when streams changed course or large trees died or stone fences crumbled. During colonial times, lawyers were frequently occupied with property litigation. Recognizing these problems, early leaders—Thomas Jefferson at the forefront—sought to improve on metes and bounds, and their creation of the township and range system is carefully documented by geographer William D. Pattison.[5]

The township and range system became law in the Land Ordinance of 1785. This law mandated that surveys of the public domain would have two major coordinates: a north-south Principal Meridian and an east-west Base Line (lines of longitude and latitude, respectively; see figure 10.2). Township lines were to be surveyed every six miles parallel with the Base Line; likewise, range lines were to be marked every six miles parallel with the Principal Meridian. Created, then, was a rectangular

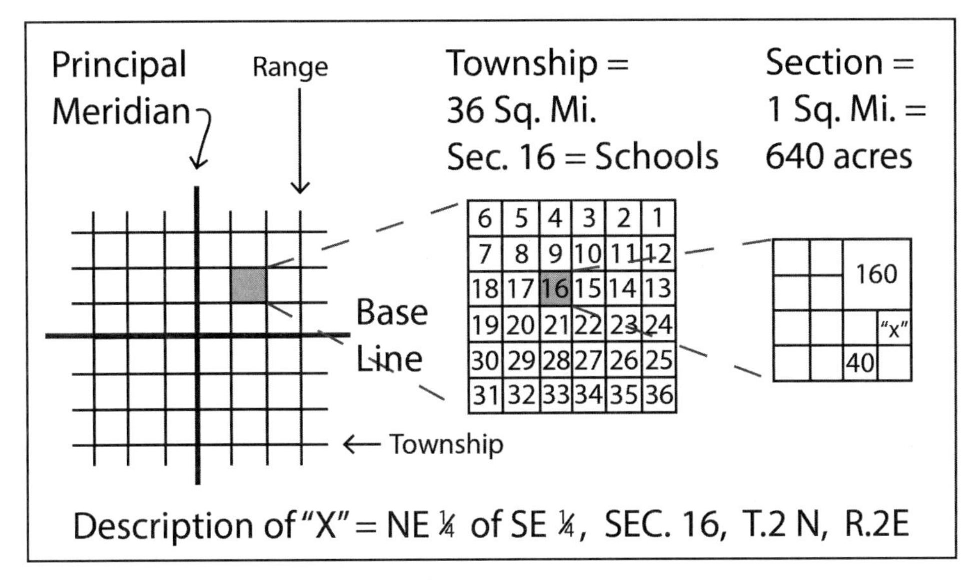

Figure 10.2 The Township and Range System.

grid with cells, called townships, which measured six miles on a side for a total of thirty-six square miles. The individual square miles, called sections, were numbered 1 through 36: A new numbering system created in the Land Act of 1796 is the one shown in figure 10.2. Section 16 in each township was to be reserved for the support of schools; observing the principle of separation of church and state, Congress rejected a proposal that a section be reserved for the support of religion. The creation of this new checkerboard meant that one's property could be rather precisely described and recorded by sections within townships. And with the Land Ordinance of 1785, the principle was established that the public domain was to be surveyed in advance of settlement.

The township and range system was first implemented in the Seven Ranges of southeastern Ohio (map 10.1). Between 1785 and 1787, seven north-south range lines (thus "Seven Ranges") were surveyed in a process that took three times longer and cost three times more than was planned. In all, thirty-three "great surveys," each with Principal Meridian and Base Line coordinates, were carried out in twenty-nine of the forty-eight contiguous states (see map 10.2). This means that nineteen of the lower forty-eight states lack the township and range pattern. In the classroom, it is always instructional to have students name the nineteen metes and bounds states. Besides the thirteen original states, they are Vermont, Kentucky, Tennessee, Maine, and West Virginia, and, for unique reasons to be explored in Chapter 12, also Texas. Within the twenty-nine township and range states, there are exceptions to the presence of a uniform grid. For example, Spanish metes and bounds surveys occur in New Mexico and California, and French long lots are found especially in Louisiana. Because it became an experimental ground for new surveys, Ohio, as geographer Norman J. G. Thrower has carefully shown, is fragmented by no less than seven survey patterns. But, as is made clear in map 10.2, the township and range system came to cover most of the Lower 48. It now also covers Alaska.[6]

The new survey system had advantages and disadvantages. On the plus side, the township and range system was rational in that a grid pattern facilitated accurate assessment of taxes. Land titles were also more secure, which meant fewer lawsuits. And surveying the straight lines called for was relatively easy to effect. On the minus side, the new survey was not always ecological. A "straight line," as geographer Hildegard Binder Johnson noted, is "rare in nature." It is easily observed that when

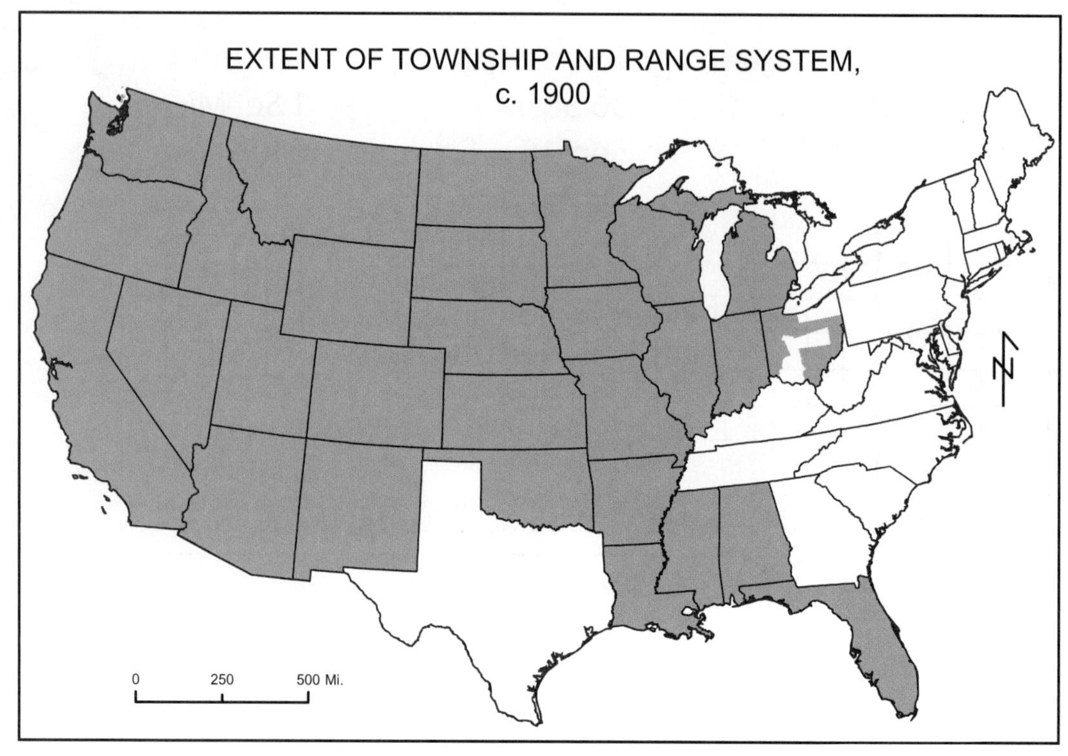

Map 10.2 Extent of the Township and Range System, c. 1900. Twenty-nine of the forty-eight contiguous states have the rectangular grid including Ohio which is fragmented.

Source: Pattison, *Beginnings of the American Rectangular Land Survey System, 1784–1800*, frontispiece.

a section line road goes straight over a steep hill instead of following its contour, as in New England, topsoil erodes in furrows next to that road. Moreover, on the minus side, a rigid checkerboard, many would agree, is flat-out monotonous, compared with the charm and grace of irregular surveys. And, for good or more likely for bad, a survey that left behind square properties insured that America's rural inhabitants would live in a dispersed pattern. The long lots of the French (and the Spanish in New Mexico) allowed greater proximity of farmhouses, and presumably greater neighborliness, than was possible for Americans living on squares. But once a survey is on the ground, it will be lasting, for good and for bad.[7]

As the township and range system was being implemented, how the public domain should be "alienated," or put into the hands of people, was being decided. After about 1790 the federal government ended its policy of selling large tracts to land companies and shifted its objectives to raising revenue by selling land to individuals. Beginning with the Land Act of 1796, individuals could alienate from the public domain a minimum of one section (640 acres) for two dollars per acre. However, subsequent land acts recognized that 640 acres was far more than a family could use, and that families working the land would raise more revenue in taxes than would a high initial price per acre. Two trends, then, characterized subsequent land legislation: The size of a parcel that people could acquire decreased from 640 acres in 1796 to 320 acres in 1800, 160 acres in 1805, 80 acres in 1820, and finally 40 acres in 1832—this "quarter-quarter" was as small as parcels would get. Meanwhile, the price charged was reduced from $2 per acre in 1796 to $1.25 per acre in 1820 to virtually free in the Homestead Act of 1862. Getting the public domain settled drove the decision-making.[8]

Ohio Valley Urban Growth

In the early 1800s the Ohio River came alive with boats. The reason was obvious: Here was a major river flowing in the direction people wanted to go—west (see map 10.3). Flatboats, built by strapping logs together, could float with the current downstream, carrying people to destinations where the logs could be used for building a log cabin or could be sold. Keelboats, whose pointed bows made them look like very large rowboats, were used for going upstream. Battling the current took great effort, for besides rowing, it required poling from running boards, and sometimes the use of cables attached to trees, to pull the craft ahead. In 1811 the first steamboat on the Ohio descended the river from its place of construction, Pittsburgh, and by the 1820s, steamboats carried much of the traffic along the Ohio's six-hundred-plus–mile (970-plus–kilometer) course. The Ohio, then, became the great artery leading west, and the settlement frontier advanced along its banks.

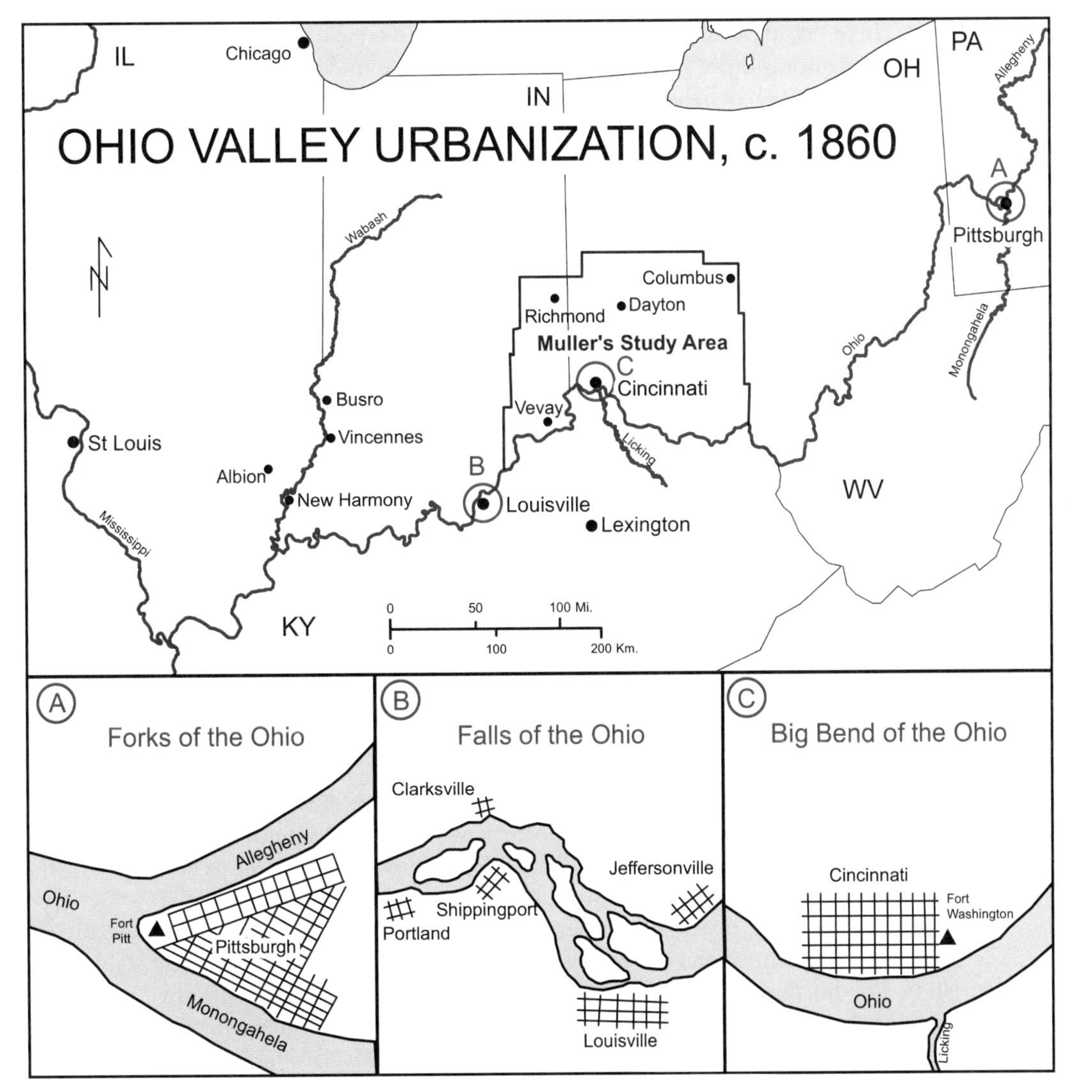

Map 10.3 Ohio Valley Urbanization, c. 1860. The Ohio River facilitated people in their westward movement. Towns developed quickly, as did intermediate trade centers and several colonies settled by emigrants from Europe.

Three places along the Ohio had natural advantages for town founding, and at those places Pittsburgh, Louisville, and Cincinnati came to exist. For Pittsburgh, the natural advantage lay in the so-called forks of the Ohio (map 10.3A). A strategically located point of land existed between the Allegheny and Monongahela rivers near where they converged to form the Ohio. On this point in 1754, British soldiers built a log fort that the French quickly took control of (Fort Duquesne) before it again became British Fort Pitt in 1758. The fort became the nucleus around which Pittsburgh grew—early on, a community of log houses built along streets laid out on a grid. Reached by roads leading west from Philadelphia and Baltimore, Pittsburgh developed quickly as the place of departure for colonists traveling by flatboat or raft down the Ohio. From the Monongahela River's west-side high-bluff overlook, visitors today have a magnificent view of Pittsburgh's downtown set within its "golden triangle."

For Louisville, the natural advantage consisted of the so-called falls of the Ohio (map 10.3B). At Louisville, a two-mile-long limestone ledge passed under the mile-wide river, creating islands and cascading water and a virtual break in river transportation. Three "chutes" led through this whitewater obstacle, one on the Kentucky side of the river, one on the Indiana side, and one in the middle. Louisville came to exist in 1779 at the head of the falls on the Kentucky side, as the place where boats could stop, unload or partially unload, and hire a pilot to guide the craft through one of the chutes. Eventually three additional towns came to compete in the business of guiding boats: on the Indiana side, Clarksville (1783) and Jeffersonville (1802), at the foot and the head of the falls, respectively, and on the Kentucky side, Shippingport (1803), at the foot of the falls. In the urban economic rivalry, low riverbanks gave the Kentucky side the advantage. Indeed, flat land on the Kentucky side prompted the digging of a canal (completed in 1830), and in its anticipation, the founding of still a fifth town at Portland (1814). Even after the canal, however, the falls continued to be a break in river transportation, and steamboats would work either the upper basin (Pittsburgh to Louisville) or the lower basin (Louisville and below). In the twentieth century, due to the construction of dams, canals, and locks, the depth of the Ohio is controlled, and the limestone obstacle at Louisville is hidden below some nine feet of water.[9]

Cincinnati had as its natural advantage a location on the north-looping "Big Bend" of the Ohio (map 10.3C). On a high bluff on the Ohio side of the river across from the mouth of Kentucky's Licking River, Losantiville came to exist late in 1788. In 1789 the place was renamed Cincinnati, and soon an army post, Fort Washington, was built at the site. Cincinnati grew rapidly as the jumping-off point for colonists who left the river and continued their journey by wagon overland into Indiana and Illinois. By 1820, Cincinnati, with 9,642 residents, had surpassed both Pittsburgh (7,248) and Louisville (4,012) in population, and at mid-century it was the largest urban center in the Old Northwest. Cincinnati became a place for the production of such items as saddles, glass, pottery, whiskey, and steamboats. It had already assembled sixty steamboats by 1828. And Cincinnati's major meatpacking industry by 1840 would earn it the rather unfortunate nickname, "Porkopolis."[10]

As these three river ports emerged to be the Ohio Valley's regional giants, a host of small "central places" or towns was founded in the river's agricultural hinterlands. For his dissertation, geographer Edward K. Muller (see spotlight) asked why some of these small, established central places grew to become regional centers while others did not. For the period from 1800 to 1860, Muller studied central places in the forty-one counties of Ohio and Indiana that lay north of the Ohio River behind Cincinnati, which he soon excluded from his study area when the city grew quickly to be in a class of its own (map 10.3). Muller found that for his study area, two major reasons explained why certain towns grew: First, they had good transportation nodality, meaning that they were connected with their local hinterlands and communities beyond by roads, turnpikes, canals, and after 1850, railroads. And second, they developed manufacturing for export, which early on meant processing of crops and livestock (flour milling, meatpacking, brewing, distilling), and later meant secondary or nonprocessing manufacturing (hardware, leather, paper, agricultural implements). Thus, places like Columbus, Dayton, and Richmond emerged as regional trade centers while most other towns did not. After 1850, railroads had a major role in determining which centers would be the more competitive.[11]

The historian Richard C. Wade observed that the three Ohio river ports—Pittsburgh, Louisville, and Cincinnati—and Lexington and St. Louis came to exist "before the surrounding areas had fallen to the plow." In his *The Urban Frontier*, Wade made the case that the trans-Allegheny frontier was, from the very beginning, both urban and rural. Geographer Howard J. Nelson analyzed the question Wade posed about the frontier role of urban centers and determined that Wade's five cities were, for the most part, exceptions. Nelson examined 168 city foundings over the seventy years between 1790 and 1860. He noted that town founding was most active in those areas already settled by pioneer farmers. The census frontier line of two or more people per square mile had already been surpassed in the previous decade before urban places came to exist, reported Nelson. Wade's five cities were, indeed, frontier spearheads, but Nelson suggests that they were exceptions and not the rule.

Spotlight

Edward K. Muller (1943–)

Edward K. Muller, Homestead, Pennsylvania, May 25, 1993. Photograph by the author.

Edward (Ted) Muller specializes in the historical geography of nineteenth- and early-twentieth-century American cities. We draw here on his early work about urbanization in the middle Ohio Valley. While teaching at the University of Maryland (1970–1977), Ted published articles on the changing industrial geography of Baltimore. And after moving to the University of Pittsburgh in 1977, he wrote extensively on regional structural change and planning in Pittsburgh. In 1993 Ted gave several of us a memorable tour of change in the steel industry along the Monongahela River. Muller earned his BA at Dartmouth College (1965) and his MA (1968) and PhD (1972) degrees at the University of Wisconsin–Madison, all in geography. At the University of Pittsburgh, he has been a professor of history. By 2011 he had coauthored, edited, or coedited six books and was the author or coauthor of forty-five articles and book chapters. He was a Fulbright Research Scholar in 1992, and in 2011 he received the Pennsylvania Geographical Society's Distinguished Geographer Award.

Recognizing that even very small (three to five hundred people) settlements furnished important urban functions, scholars, then, should be cautious when they refer to America's westward movement as an "urban frontier."[12]

One of America's epic stories is how Chicago, in its meteoric rise, outstripped in size the river ports of, notably, Cincinnati and St. Louis. Chicago's rise equates to the story of how railroads surpassed inland waterways in the transport of people and commerce, and how midwestern products moved east to markets by rail, not through canals to the Hudson River and New York or down the Ohio and Mississippi to New Orleans and the Atlantic coast beyond. Trains first reached Chicago in 1851. Thereafter, as commerce began to flow overwhelmingly eastward by rail, Chicago added railroad facilities more quickly than did its competitors. By 1880, Chicago had surpassed St. Louis to become the largest city in the Middle West, and its continued growth would make it second only to New York in size. The disruption of river traffic on the Ohio and Mississippi rivers during the Civil War (1861–1865) was also a factor. But transportation by rail was simply faster and more flexible than by water. The trend was captured by economic historian Margaret Walsh in her study of meatpacking in the Middle West between 1835 and 1875: About 1850, river-port packing plants, reported Walsh, dominated the pork industry, with some twenty-five percent of the "packs" occurring in the "Queen City" of Cincinnati. By 1870, railroad-located packers dominated, notably at Chicago, which advertised itself as the "Hogbutcher to the World."[13]

In the urbanization of the Ohio Valley, especially its downriver stretch, colonies of European-born national groups added important ethnic variety. In 1804 a group of German- and French-speaking Swiss vintners, after locating initially in Kentucky, moved to Vevay, Indiana, and a hilly area just north of the Ohio River to cultivate wine grapes (map 10.3). Disappointing yields found these Swiss turning to other pursuits, yet their village of Vevay, located in "Switzerland" County, endures. In 1814 German Lutheran "Rappites," or followers of George Rapp, arrived via Pennsylvania to establish the bustling community of Harmony, located in Indiana some forty miles (sixty kilometers) up the Wabash River. At Harmony men and women lived in separate quarters and marriage was discouraged; lacking children, the Rappites recruited more Germans and adopted children. They prospered, and for a decade their agricultural surpluses and manufactured goods helped neighbors to survive—until 1824, when they sold all their holdings to Robert Owen, a wealthy Scot, and moved away. Under Owen, Harmony became New Harmony. Because too many of the Scottish-born "Owenites" were professionals and too few were artisans and skilled workers, productivity at New Harmony declined. Yet the restored Rappite-Owenite buildings and facilities attract many tourists to the village today. Meanwhile, in 1818, emigrants from England settled on what is generally known as the "English Prairie," located in Illinois across the Wabash some twenty miles northwest of New Harmony. What remains of the English colony is Albion, the community of colony-promoter George Flowers. In addition to these colonies, several "Shakertowns," founded by American-born members of the Shaker religion, added to the Ohio Valley's urban variety. Busro (1808), located sixteen miles north of Vincennes on the Wabash in Indiana, was the westernmost of these Shakertowns.[14]

The Corn Belt

The Corn Belt is an old and geographically stable agricultural region, and much of it lies in the Old Northwest. No one should be surprised that corn is its major crop. What may be surprising, however, is that relatively little of this corn is consumed by humans. Corn grown in the Corn Belt is not the sweet corn people eat but a coarse-and-starchy-kernel variety that is fed to livestock, primarily beef cattle and hogs. Thus corn goes to market mainly as beef and pork, which suggests that the region should also be recognized as the Meat Belt. The term *Corn Belt* apparently first appeared in print in 1882, and the region it describes reached its present approximate areal dimensions about 1900. All aspects of the Corn Belt have been analyzed, but what is still debated is how it originated.

The conventional view of its origin, as expressed for example by geographers Joseph E. Spencer and Ronald J. Horvath in "How Does an Agricultural Region Originate?" goes something like this: Farmers from three regions, New England, the Middle Colonies, and the Upland South, converged in the Old Northwest in the early 1800s. Each group brought its crops, notably wheat and corn, and its livestock, notably cattle and hogs. Wheat, the preferred grain, sometimes failed, but in the Old Northwest, with its hot and rainy summers, corn seldom failed. The problem with corn was that it was a bulky, low-value grain that could not bear the cost of transportation to market. Farmers found, however, that if they fed their corn to beef cattle and hogs, they could "walk" their product to market. So a new practice developed whereby corn was grown to be fed to livestock, and the livestock would go to market. Because a mixed grain-livestock agriculture existed in Southeastern Pennsylvania, some scholars credited Pennsylvanians with the antecedents of this innovation. They also speculated that feeding corn to livestock and marketing the livestock happened first in southern Ohio's Scioto Valley about 1805—the place and time from which the first reported corn-fed cattle were driven east to market.[15]

In his thoughtful, revisionist book, *Making the Corn Belt*, geographer John C. Hudson substantially revises our understanding of how the Corn Belt originated. Hudson credits Upland Southerners, not Pennsylvanians, as being the major innovators, and he asserts that marketing corn-fed livestock happened first in five "islands" of good agricultural land, three of them (the Bluegrass Basin, the Nashville Basin, and the Pennyroyal Plateau of south central Kentucky) south of the Ohio River, and two of them (the Scioto Valley and the "Miami" Valley—the area of the Little Miami and Great Miami rivers) north of the Ohio (see map 10.4). Of these islands, said Hudson, the Bluegrass Basin was the most important. Upland Southerners, noted Hudson, in the early 1800s grew southern dent corn, which they fed to livestock in feeder barns or in open feed lots. In 1796 Upland Southerners from Berkeley County in eastern (West) Virginia settled in the Scioto Valley, and by 1830 many Upland Southerners from Kentucky and Tennessee lived north of the Ohio River in Ohio and Indiana. They and their Midland and Yankee farmer neighbors then moved west, resulting in the diffusion of this distinctive crop and livestock assemblage and the creation of the Corn Belt.[16]

Hudson chronicles this westward diffusion. By 1850 the Corn Belt was "unbroken" from the Scioto Valley to the Wabash Valley and on into Illinois. Migrants from all five islands then spread onto new lands in the valleys of the Illinois, the Mississippi, and the lower Missouri rivers. By 1860 the Corn Belt covered parts of the seven states of Ohio, Indiana, Illinois, Iowa, Missouri, Kentucky, and Tennessee (map 10.4). Corn production in Kentucky and Tennessee declined after 1880. But it grew in Kansas and Nebraska as farmers pushed west, and it grew in South Dakota and Minnesota to the north. With exceptions, the Corn Belt reached its approximate present size about 1900; it is delimited as a region in the 1920s in map 10.4. In the twentieth century, use of groundwater in center-pivot irrigators "erased" the region's western limit, and Hudson notes that gradual movement north into areas of dairy farming is ongoing.[17]

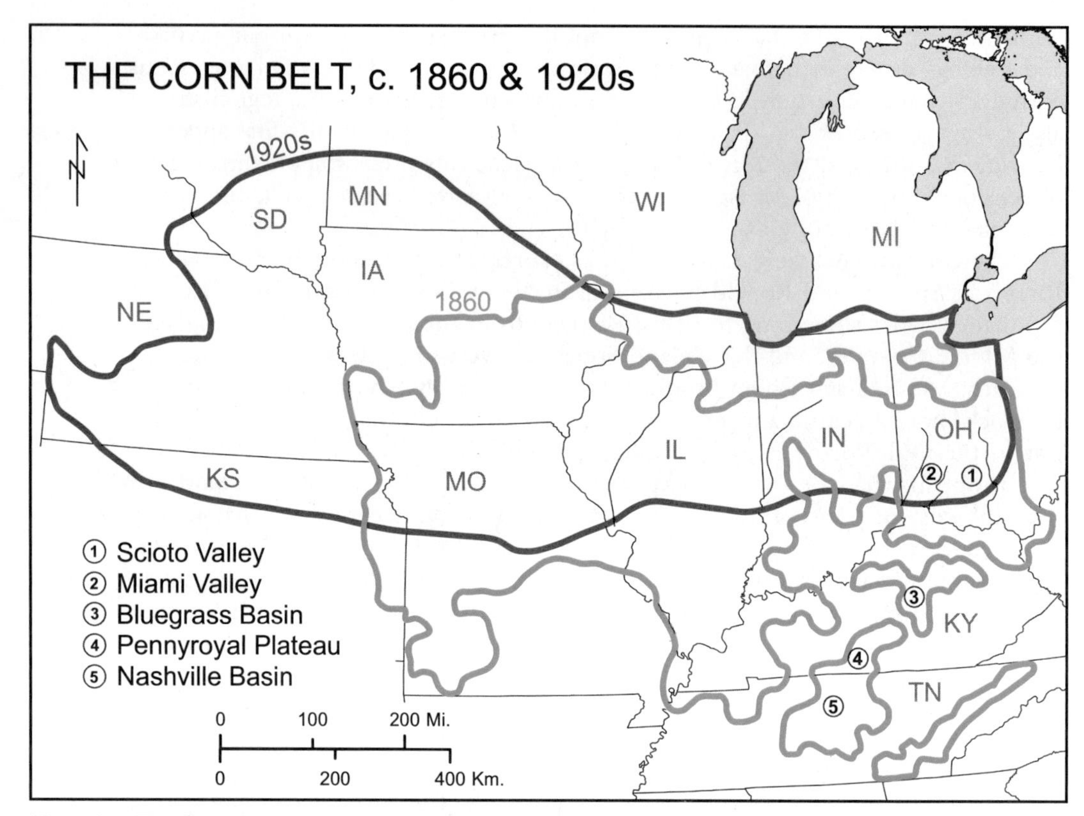

Map 10.4 The Corn Belt, c. 1860 and 1920s.

Source: Hudson, *Making the Corn Belt*, 11, for the 1860 delineation; 2, for O. E. Baker's 1920s delineation.

In the twentieth century, the Corn Belt changed in significant ways, and relic features are a consequence. John Fraser Hart understands, probably better than does any other geographer, what happened: Yields of corn increased dramatically with the introduction (in 1933) of new hybrid seed. The old three-year crop-rotation pattern (corn–small grains–legumes), with the introduction of chemical fertilizers, went to a two-year rotation cycle (corn-soybeans), and now corn often follows corn with no rotation. With mechanization, one person in a combine can harvest (cob and kernels shelled) forty acres in one day; as recently as 1940, it took one person with a horse and wagon one day to harvest just one acre (cob and cover only). Wooden corncribs with slats opened for air circulation so that ears of corn could dry naturally and be stored were replaced by steel cylindrical bins for storage after the shelled corn has been dried artificially. High-powered tractors replaced horses as the source of power for plowing and other needs. These changes made relics of facilities like the horse barn and the wooden corncrib that were ubiquitous in the nineteenth century (figure 10.3, A and B). And as hog production is increasingly concentrated in vertically integrated plants, the barbed-wire fences that once enclosed fields where pigs would "hog down" on ears of corn left behind after a harvest are rapidly disappearing to make more acreage for tilling.[18]

Figure 10.3 Corn Belt Relic Features, 1993 and 2011. In southwestern Tippecanoe County, Indiana: (A) a horse barn (on the Levering-Nostrand farm) sits unused; (B) a corncrib (two photographs) on the Raub farm is totally empty.

Photographs by the author, May 3, 2011 (horse barn); October 30, 1993 (corncrib).

Notes

[1] Williams, "Clearing the Forests," 180 and 182, for quotes concerning the use of wood for fuel; 166, for thirty-two man-days of labor to clear an acre of forest; 164, for the fact that in ten years about thirty to forty acres would be cleared by a family.

[2] Kaatz, "The Black Swamp," 9, for the name probably derived from the black soil; 14 and 18, for the species of trees (no conifers); 1, for the size—1,500 square miles, and thirty miles wide; 17, for the four-foot-per-mile average slope; and 18, for the ten-foot ridges.

[3] Ibid., 29, for settlement 1850 to 1900; 23–24, for efforts to drain; 26, for half in natural state until 1870; 33–34, for the extremely high acreage that is planted to crops; 34, for a list of ten cities on the periphery of the Black Swamp.

[4] Winsor, "Environmental Imagery of the Wet Prairie of East Central Illinois, 1820–1920," 375, for the area of the Grant Prairie; 388, for malaria as the "dominant disease." In "An Uninhabited Waste," 654, Urban states that mosquitoes came to be recognized as the biological vector for transmitting malaria in 1898.

[5] See Pattison, *Beginnings of the American Rectangular Land Survey System, 1784–1800.*

[6] Ibid., part 2, for the Seven Ranges. Also see Thrower, *Original Survey and Land Subdivision*, 129.

[7] Johnson, "Gridding a National Landscape," 142, for the quote about a straight line.

[8] Ibid., 148, lists the decreases in acreage by year.

[9] For the falls of the Ohio, see Brown, *Historical Geography of the United States*, 240–44, 263. Also see Raitz, *The Kentucky Bluegrass*, 30–36.

[10] Brown, *Historical Geography of the United States*, 197–98, 231–35, for Cincinnati; 233, for sixty steamboats in 1828. Population figures for 1820 are from Muller, "Early Urbanization in the Ohio Valley," 22.

[11] Muller, "Selective Urban Growth in the Middle Ohio Valley, 1800–1860," for a report of his dissertation findings. See also Muller, "Early Urbanization in the Ohio Valley," for the bigger picture of Ohio Valley urbanization; and Muller, "Regional Urbanization and the Selective Growth of Towns in North American Regions," for a model pulled from this research.

[12] Wade, *The Urban Frontier*, foreword, for the quotation. Nelson, "Town Founding and the American Frontier," 7–23, esp. 23.

[13] For Chicago, Conzen, "A Transport Interpretation of the Growth of Urban Regions," 365, for the first train in 1851; 370, for more trains than the competitors had. See Walsh, "The Spatial Evolution of the Mid-Western Pork Industry, 1835–75," 1–22.

[14] For these "Down-River Country" colonies, see Brown, *Historical Geography of the United States*, 236–54. Much has been written on the Shakers, including a Ken Burns 1984 documentary for American public television. See Andrews, *The People Called Shakers*, and Newman, "The Shakers' Brief Eternity," 310, which includes a map showing twenty-four Shaker communities.

[15] Spencer and Horvath, "How Does an Agricultural Region Originate?" 75–82, esp. 78; 81, for 1882 as earliest date for *Corn Belt*; 82, for "about 1900," when the Corn Belt reached its present form.

[16] Hudson, *Making the Corn Belt*, 10 and 92, for Upland Southerners; 63, for not Pennsylvanians; 6 and 103, for five islands; 94, for Bluegrass Basin importance; 68, 102, 108, for southern dent corn, open feed lots, and feeder barns; 64, 96, 102, for migration to Scioto Valley in 1796 and many settlements before 1830.

[17] Ibid., 96, for "unbroken" by 1850; 101 and 103, for migrants from all five islands spread along rivers; 109, for coverage in seven states by 1860; 12, for decline in Kentucky and Tennessee after 1880, and spread west "erased" but continuation northward.

[18] Hart, "Change in the Corn Belt," 51–72, esp. 52, for hybrid seed introduced in 1933.

The Lowland South 11

The Lowland South is, arguably, America's most distinctive region. In this chapter we focus on four reasons for this distinctiveness: a subtropical lowland covered with "The Piney Woods"; a strong emphasis on one commercial crop, "King Cotton," in the antebellum period; in that same pre–Civil War period, a reliance on slaves in "The Plantation"; and finally, the consequences of trying to preserve a way of life under "The Confederate States of America." Other reasons why the Lowland South is distinctive are only touched on: a Southern dialect ("y'all," "fixin' to") derived apparently from the sizable black minority; an absence of foreign-born people, since emigrants from Europe went elsewhere rather than compete with cheap black labor; a strong preference for pork, corn (pone, grits, cornbread), and turnip greens in foodways; and an Evangelical (largely Baptist) religion that makes the Lowland South the heart of the Bible Belt.

The Piney Woods

The Lowland South stretches generously from the Atlantic coast to East Texas (see map 11.1). The Atlantic Coastal Plain and the Gulf Coastal Plain underlie most of the region. This large flat area lies in the subtropics, explaining why summers are hot and winters are mild in temperatures. Lying within the humid East, the Lowland South has ample annual precipitation (forty to sixty inches), with the heaviest amounts coming in the summer months to make its sensible temperatures in that high-sun season absolutely oppressive. Rivers rising in the Upland South to the north and west cross the flat lower region to flow into the Atlantic Ocean and the Gulf of Mexico. Their valleys have heavy black soils while the broad interfluves between them have light sandy soils.

That the Lowland South is covered with a pinewood is something of an environmental anomaly. Given the region's subtropical climate, one would expect the Lowland South to be covered with broadleaf deciduous trees much like those that grow to the north. But pine trees outcompete broadleaf trees in sandy soils: The root systems of pine trees penetrate more deeply to tap subsurface moisture. So the explanation for the presence of some twelve species of yellow pine that are commonly called the piney woods seems to be edaphic. In the fertile river-bottom soils grow cypress, tupelo, and gum trees. Because the bottoms were harder to clear and subject to flooding, they were generally settled later than were the more expansive sandy interfluves.

Two rather large areas in the Lowland South have superior soils; consequently they became major cotton-growing districts and the locations of many of the region's large cotton plantations. The first was the Black Belt, so named because of its black soils—which were black because they were formed under prairie grass. The Black Belt is a twenty- to thirty-mile-wide, crescent-shaped area that stretches for three hundred miles (five hundred kilometers) through central Alabama and into northeastern Mississippi (map 11.1). While the area has rich prairie-derived soils, geographer Erhard Rostlund asked whether the Black Belt was actually the "natural prairie" it was labeled to be on many maps. Rostlund carefully reviewed the reports about vegetation made by numerous travelers through the Lowland South, beginning with DeSoto, who explored the area from 1539 to 1542, to determine whether, in historic times, travelers had noted the Black Belt to have been an open prairie. He found

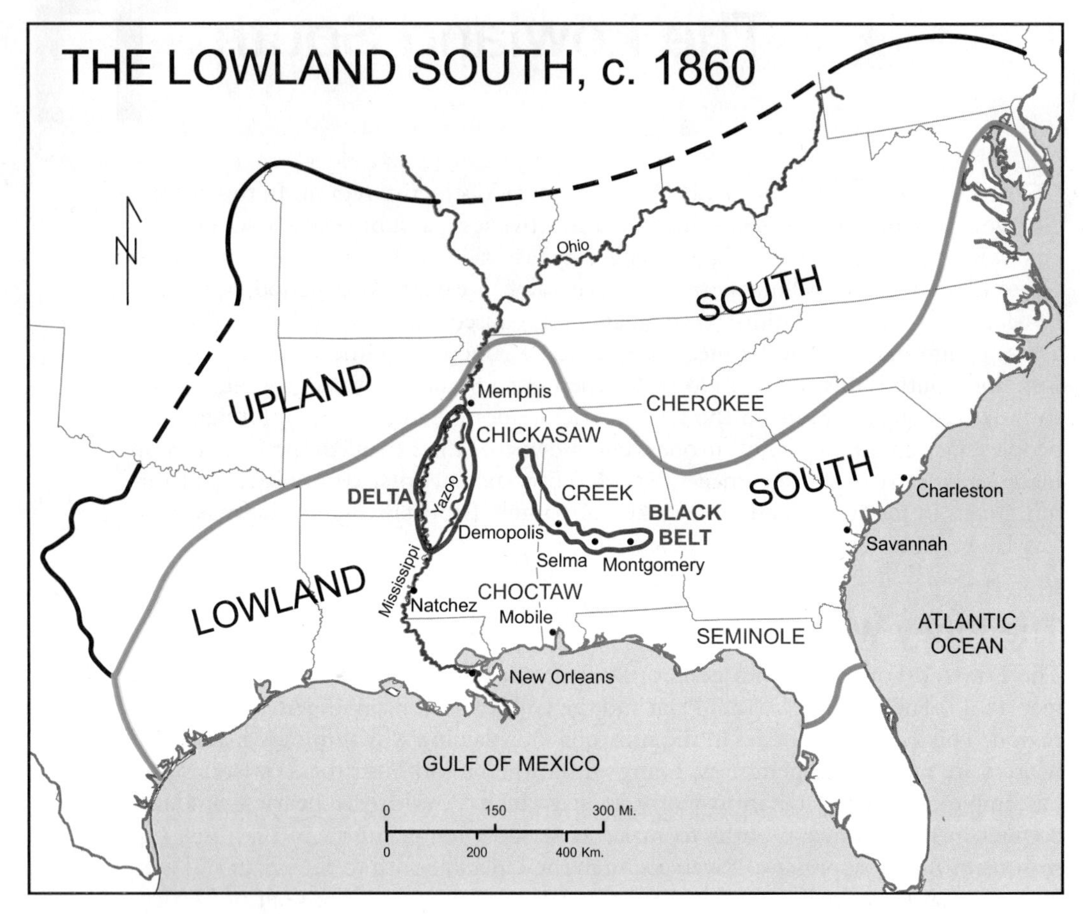

Map 11.1 The Lowland South, c. 1860. The region stretches from the Atlantic coast to East Texas, and omits much of Florida which was lightly settled and was to develop quite differently.

Sources: Rostlund, "The Myth of a Natural Prairie Belt in Alabama," 393, for the Black Belt; Adkins, "The Geographic Base of Urban Retardation in Mississippi, 1800–1840," 37, for the Delta.

the descriptions to say quite uniformly that, while the Black Belt had many patches of open prairie, it was basically a pine-oak woodland—which made it little different from the larger Lowland South. Rostlund concluded that it was a "myth" to call the Black Belt in historic times a natural prairie.[1]

The second large area with superior soils is the so-called Delta. The delta in question is not the birdfoot delta at the mouth of the Mississippi River. Rather, it is the delta formed by the Yazoo River where it converges with the Mississippi River in the state of Mississippi (map 11.1). It is the one that would also become the setting for the novels of William Faulkner (1897–1962). Rich alluvial soils cover this Delta. Slaves did the arduous work of clearing the Delta's woodland cover with axes and burnings, much as they had cleared the Black Belt. Except for people taking up choice riverine sites, however, the costs of clearing and draining land delayed settlement of the Delta until the 1870s and 1880s, but once settled, cotton became its major crop.[2]

"King Cotton"

In South Carolina and Georgia during colonial times, rice and indigo were the chief commercial crops, as seen in Chapter 6. Cotton was grown, but it was relatively unimportant. Two varieties of cotton had been introduced: Sea Island from the West Indies and Upland from Mexico. Sea Island, with its two-inch-long silky fibers, or "staples," had black slippery seeds that could rather easily be removed or ginned from the staples with a roller gin. But this superior cotton variety was restricted to the longer growing season of the barrier beach islands and the coastal lowland. Upland cotton, on the other hand, could spread inland on higher ground where the growing season was shorter, but its problem was that its one-inch-long staples were tightly intermeshed with green seeds in the cotton boll. Removal by hand of the green seeds from one pound of cotton amounted to approximately one day of slave labor.

Soon after the colonial period ended, two developments enabled cotton to become "king." The first was a heightened demand for raw cotton in the burgeoning cotton textile industry in the Manchester district of central England, served by the port of Liverpool. And the second was the call for a more efficient way to remove the green seeds from Upland cotton. Eli Whitney (1765–1825) is correctly associated with the invention of a more efficient cotton gin. In 1793 this young Yale graduate visited a Georgia plantation, learned of the ginning problem, and in 1794 patented a cylindrical device within which revolving wire teeth separated seed from fiber. With such a gin, it became possible to gin some fifty pounds of cotton in one day. Other gins were under development at the same time, and one, a saw gin (1796), was favored over Whitney's spiked-cylinder gin. Thus new markets and new technology, which besides gins included improved power looms and spinning machines, turned the Lowland South into a cotton kingdom, increasing slavery at the same time.[3]

About 1800, explains D. W. Meinig, the Piedmont in the South Carolina–Georgia area became something of an Upland cotton culture hearth (see map 11.2). From it, cotton diffused northeast into the Piedmont zone of North Carolina and west into Georgia's Piedmont. Rivers connected this upper country to port facilities in Charleston and Savannah, from which cotton was exported. Before 1820, cotton planters reached Alabama in a "pincer movement" south from Tennessee and north from Mobile. In the 1820s more cotton planters spread west into Alabama from Georgia, and in the 1830s the Black Belt emerged as a cotton district with prosperous planter communities at Montgomery, Selma, and Demopolis (map 11.1). Mobile, on the Gulf coast, provided cotton export facilities for the Black Belt. All this expansion had quite a tragic impact on the Cherokee, Creek, Seminole, Choctaw, and Chickasaw: Removal from and loss of title to their homelands was the plight of these Five Civilized Nations.[4]

Meanwhile, Meinig explains, cotton diffused from a second cotton culture hearth in the Natchez District (map 11.2). In that fertile area adjacent to the Mississippi River, planters in the 1810s developed a hybrid cotton, apparently from "creole black seed," introduced by the French from Siam, and from a cotton variety from Central Mexico. The result was a high-yielding, easy-to-pick cotton that was immune to common diseases. From the Natchez District, this new cotton diffused north into the lowlands of the Delta adjacent to the Mississippi River, where slaves of wealthy "river planters" cleared and drained the land. Cotton then spread across the Mississippi to the bottomlands of Arkansas and south down the Mississippi Valley to Louisiana. Diffusion continued to the west up Louisiana's Red River Valley and eventually well into East Texas. New Orleans and Galveston emerged as the ports from which cotton was exported. Meanwhile, the wealthy planters in Natchez built opulent Greek Revival houses atop bluffs above the Mississippi River, and in 1850 Natchez was reported to have more millionaires than did any other community in America.[5]

What seems to have happened, then, is that from two cotton hearths, the South Carolina–Georgia Piedmont and the Natchez District, cotton diffused within the Lowland South, rather rapidly after about 1815. In each direction there were environmental limits (approximate, not absolute) to this diffusion—the 210-day growing season line to the north (shown in map 11.2); the twenty-three-inch average annual precipitation line to the west (later to be exceeded with irrigation); and as

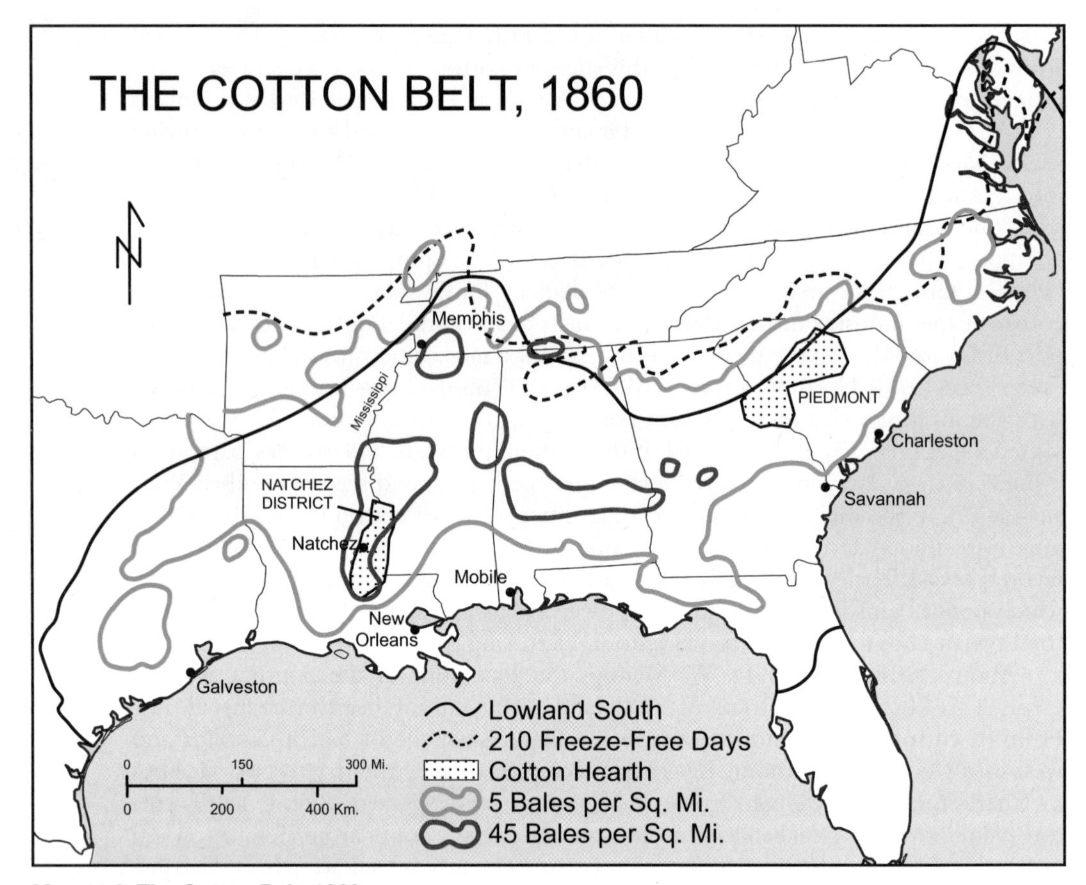

Map 11.2 The Cotton Belt, 1860.

Sources: Hilliard, *Atlas of Antebellum Southern Agriculture*, 71, for the Cotton Belt; 17, for the northern limit of the 210-freeze-free-days line; and Meinig, *The Shaping of America*, 2:288, for the two cotton hearths.

one approached the Gulf of Mexico to the south, excessive autumn rainfall which deteriorated fibers in the cotton boll. As one of his many contributions to our understanding of the antebellum Lowland South, geographer Sam Hilliard (see spotlight) mapped the extent of the Cotton Belt in 1860 (map 11.2). Most amazingly, from this Cotton Belt in 1860 was produced three-quarters of the world's cotton, and given the shift of cotton growing to the west in the Lowland South, three-quarters of the cotton grown came from the Gulf states. The importance of the Lowland South can be measured by the fact that in 1860, cotton represented half the total value of America's exports.[6]

In the antebellum period, then, cotton became king, the clear leader as a regional income generator and the probable leader in acreage devoted to one crop. But Hilliard makes the point that in this time before the Civil War, the acreage for *corn* in the Lowland South was about as great as it was for cotton. Much less attention was paid to the proper cultivation of rows of corn than to those of cotton, and this neglect explains why per-acre yields of corn were perhaps only half that of farms in the same period in the Old Northwest. Corn acreage, nevertheless, was high because corn was needed as feed for animals (mules were preferred to horses for fieldwork in the Lowland South) and for the food consumed by slaves and whites. From his analysis of the amounts of corn produced in the antebellum Lowland South, Hilliard concluded that the region, if not totally self-sufficient in the production of its own foodstuffs, was largely so. Besides corn, these foodstuffs included pork, the heavily

Spotlight

Sam Bowers Hilliard (1930–2011)

Sam Bowers Hilliard, leading the Mississippi Valley Plantations field trip, Natchez, Mississippi, July 19, 1986. Photograph by the author.

Sam Hilliard knew the historical geography of the American South—especially its changing agriculture—like few others. Born in Bowersville, Georgia, he earned an AB (1960) and an MA (1962) at the University of Georgia. At the University of Wisconsin–Madison, he worked under renowned historical geographer Andrew H. Clark, earning a second MA (1963) and PhD (1966). After short teaching positions at the University of Wisconsin–Milwaukee (1965–1967) and Southern Illinois University–Carbondale (1967–1971), Hilliard taught for twenty-two years at Louisiana State University, retiring as alumni professor in 1993. Sam and I worked closely together, Sam as local-arrangements chair and I as program chair, to sponsor the Sixth International Conference of Historical Geographers held July 19–26, 1986, in Baton Rouge and New Orleans. Sam made it a "rolling" conference with four days of paper sessions sandwiched between three field trips to look at Mississippi Valley plantations, the "Acadiana" swamp, and New Orleans streetcar landscapes. Following the conference, we coedited *The American South*, a volume containing thirteen papers presented at the conference. Following his retirement, Sam returned to his family farm in Bowersville in northeastern Georgia, where he died at the age of eighty.

preferred meat in Lowland South foodways. During much of the year, hogs foraged in the woods for acorns and wild fruits, but after corn harvest, they were turned loose in the corn (and sweet potato) fields to be fattened.[7]

So it was that as the people of the Lowland South moved west from the Atlantic-facing states to East Texas, cotton and corn dominated crop acreage. Cotton and corn are cultivated as row crops with much exposed soil between the rows. Their cultivation in a region of sandy soils with summers characterized by heavy downpours led to serious soil erosion. Within several decades, gullies three- to four-feet deep appeared in the Piedmont country. Cotton and corn also quickly exhausted soils of their nutrients. The response of some farmers and planters to soil erosion and soil exhaustion was to pick up and move west to fresh ground. Examples of land abandonment were numerous enough to prompt some historians and others to characterize the westward movement within the Lowland South as a "hollow" frontier.

The Plantation

We associate large plantations in the United States with the Lowland South. The literature on the region, however, makes it abundantly clear that, from the antebellum period down to the present, farms in the Lowland South have been more numerous than plantations. Farms were relatively small: 40 to 250 acres as an approximate range in sizes. They were typically family-owned and operated with few if any slaves in antebellum times. Farmers grew cotton, but to be self-sufficient they also grew corn and other foodstuffs. Nonetheless, plantations were what made the Lowland South different. Plantations were large landholdings of more than 250 acres; many measured one thousand to several thousand acres. They were run by the owner or the owner's overseer or both. In antebellum times the emphasis was, of course, on growing cotton for export. Gangs of unskilled slaves, one hundred or more on the larger operations, supplied the heavy inputs of needed labor, and the fact that slave labor was involved was pivotal in making cotton-growing profitable.[8]

Owning slaves was clearly a moral issue, but for white planters in the South, it was also an issue of economic survival. As geographer Carville Earle noted, "the economics of staple crops and labor costs," not "moral fiber," "explain[ed] the geography of slavery."[9] Cotton was a subtropical crop. It did best south of the 210-day freeze-free line—where the growing season lasted seven months or longer. During the long growing season, much labor was required to plant, hoe, and pick the cotton, and beyond the growing season, tasks included ginning and bailing. Exploiting slaves to do this often back-breaking labor under often oppressive climatic conditions was morally wrong. But for most Southerners the moral issues were subordinated to economic goals and realities. In 1860 blacks represented at least 40 percent of every state's population between South Carolina and Louisiana, and in the prime agricultural lands containing the large plantations, the percentage was much higher.

Geographer Merle Prunty Jr. explained in rich detail how the geography of blacks on a Lowland South plantation evolved in three periods: antebellum, postbellum, and twentieth century. The major elements in this changing geography drawn from Prunty are illustrated in figure 11.1, which shows a hypothetical plantation set in Alabama (or farther west) where the township and range system left landholdings carved into rectangles. In the antebellum period, the planter's house—built on a carefully chosen site and likely built as an I-style or dogtrot and only rarely in the more pretentious Greek Revival style—was the focal point. Near it were the slaves' quarters, the mule barn, the equipment sheds, and the ginhouse (see figure 11.1A). Beyond the plantation nucleus was the cropland, which was subdivided into several large fields, none of which was usually fenced. The area set aside for pasture, however, was fenced, typically in antebellum times, said Prunty, with stake-and-rider, "wooden worm," or willow wattle fencing. Woodland, which often represented rougher and less desirable ground, was also a major land-use category.[10]

The Civil War disrupted plantation production and marketing, but following the war the large plantations emerged basically intact. The end of slavery, however, significantly changed the internal geography of these plantations. Some freed blacks moved to towns, but those who remained on plantations became sharecroppers. Planters subdivided their cropland into twenty- to forty-acre plots and on them built houses for black families, who worked the plots (see figure 11.1B). Verbal contracts negotiated annually between planters and sharecroppers had the blacks supply the labor; pay rent for the land, houses, and mules; pay part of the cost of seed and fertilizer; and at the end of the growing season, share the crop in proportions already agreed upon. In the decades after the Civil War, cotton decreased in acreage while corn, grains, and hay crops increased. Raising cattle also gained importance, and for fencing pasture, barbed wire came into use. Connecting the dispersed houses of sharecroppers with the plantation nucleus added to the linear miles of plantation lanes and roads. And as an alternative to ginning cotton on the plantation, public steam-powered gins sprang up in towns.[11]

Sharecropping continued with strength until the 1940s and World War II. A key development in its decline was mechanization, notably the appearance of tractors, which Prunty notes came to be mass-produced in 1927. Mechanization necessitated fewer sharecroppers to keep the plantation running; the degree to which

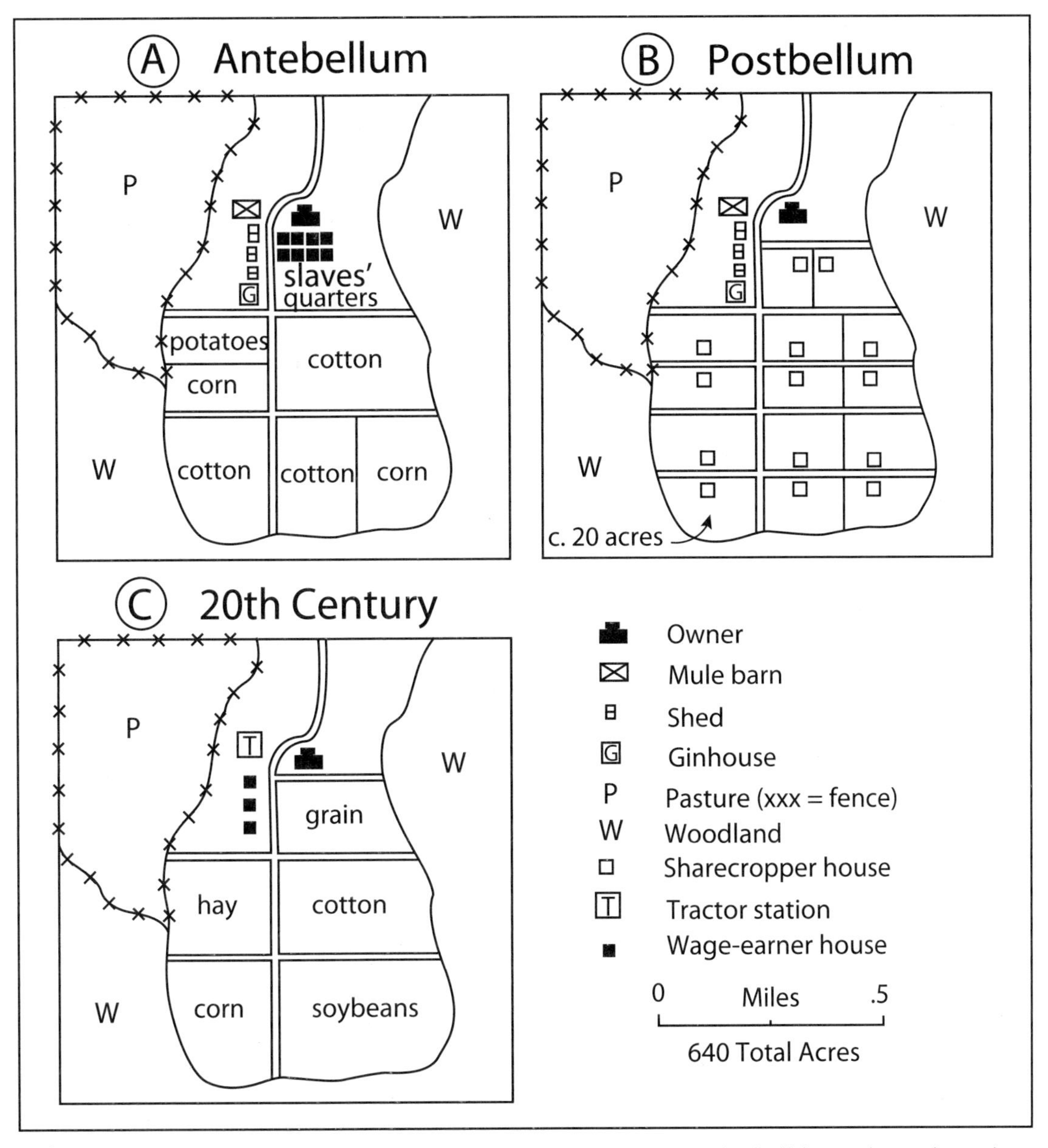

Figure 11.1 Lowland South Plantation Evolution. This hypothetical plantation in Alabama shows change in the antebellum, postbellum, and twentieth-century periods.

Source: Prunty, "The Renaissance of the Southern Plantation," 464, 467, 474, 482.

mechanization pushed blacks from rural plantations to urban places in the South, North, and West, however, seems to need greater scholarly attention. For the "neoplantation," Prunty's word for the new, mid-twentieth-century landholding, planters cobbled together the plots of sharecroppers into large fields reminiscent of antebellum times (see figure 11.1C). The sharecroppers' houses were removed or razed, and their access lanes and roads were plowed under for cropland. Near the planter's house, a tractor station was built, and near it were the houses of several black families whose wage-earning heads possessed the skills to operate tractors, harvesters, and other equipment. The crops grown—cotton, corn, soybeans, grains, hay—continued to be diversified and rotated. And pasture and woodland remained as land-use elements in a neoplantation controlled, as in antebellum times, from one nucleated plantation headquarters.[12]

The Confederate States of America

In the decades leading up to the Civil War, tensions between South and North intensified. Underlying these tensions was the issue of slavery. Both South and North sought a balance of power in the Senate between the number of slave and nonslave states. No official policy was put in place to require this balance, but informally sectional balance was maintained by admitting slave and nonslave states in pairs. When Tennessee became the sixteenth state in 1796, the slave and nonslave balance was eight and eight (see map 11.3). Between 1803 and 1859, seventeen new states were admitted, fourteen of them as direct pairs or at least as "more or less" pairs. The issue of the extension of slavery to the West had become explosive in 1821 over the admission of Missouri as a slave state. The cost to the South for Missouri's statehood was acceptance, in the Missouri Compromise of 1821, of the southern boundary of Missouri—latitude 36°30' N—as the future northern limit of slavery. In 1845, Texas was brought into the Union with no sectional confrontation; in 1848, Wisconsin became its pair. But between 1850 and 1859, the admission of California, Minnesota, and Oregon tipped the balance of power to the North. In 1860, on the eve of the Civil War, of the thirty-three total states, fifteen were slave and eighteen were nonslave (map 11.3), an imbalance that Southerners did not take lightly.[13]

From late 1860 through the first four months of 1861, eleven of the thirty-three states seceded. Abraham Lincoln's election on November 6, 1860, triggered the secession: Lincoln (1809–1865) and the North stood for national unification

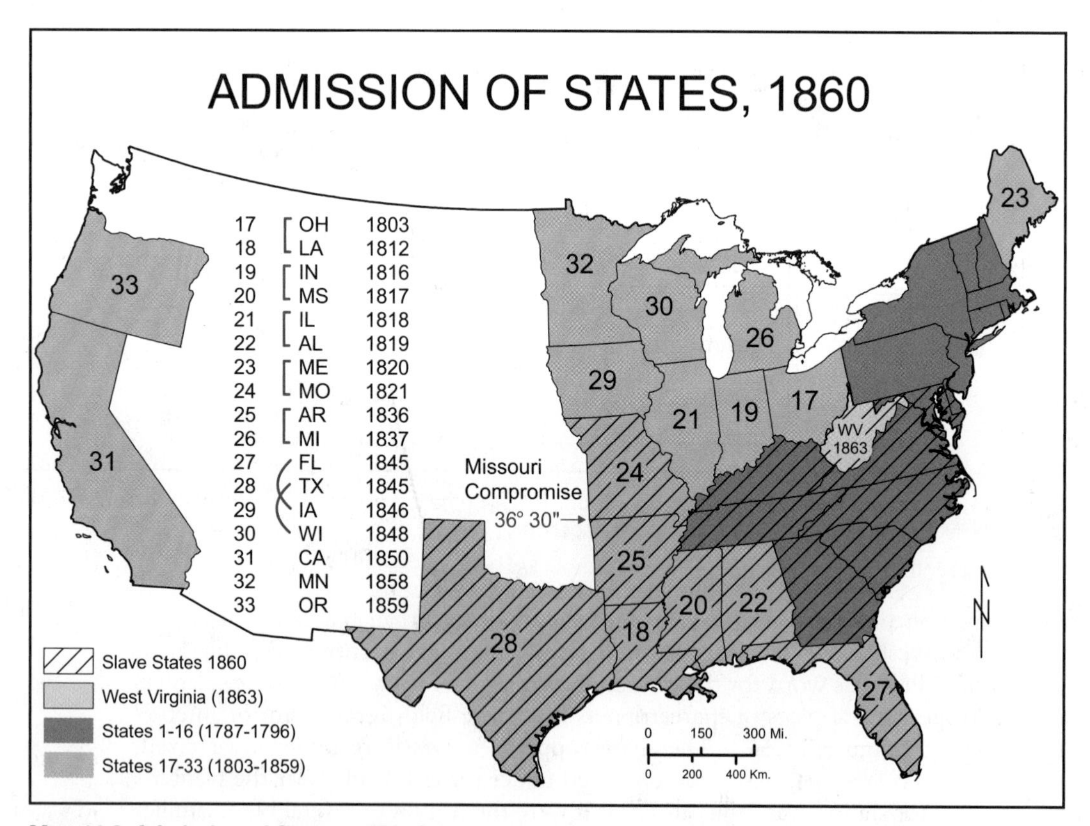

Map 11.3 Admission of States, 1860. On the eve of the Civil War, thirty-three states had joined the Union. The first sixteen had been admitted between 1787 and 1796. (West Virginia would break away from Virginia in 1863.) The next seventeen were admitted between 1803 and 1859. Fourteen of the seventeen were paired as slave-nonslave, as indicated. The Missouri Compromise line of 1821 (36°30' N) represented the northern limit of future slavery.

and preservation of the Union; the South, however, wanted independence in order to preserve its way of life, which included slavery. The seven so-called cotton states were the first to secede. South Carolina led on December 20, 1860, followed over the next forty-two days by Mississippi, Florida, Alabama, Georgia, Louisiana, and Texas (see map 11.4). On February 8, 1861, representatives of these seven states met in Montgomery, Alabama, where they formed the Confederate States of America (CSA). Two days later, they elected Jefferson Davis (1808–1889) president. After Southerners fired on Union-held Fort Sumter near Charleston on April 12, 1861—the first shots fired in the war—four additional states joined the Confederacy: Virginia, Arkansas, Tennessee, and North Carolina (map 11.4). And on May 20, 1861, the capital of the CSA was moved from Montgomery to Richmond, Virginia.[14]

Meinig explains that textbook maps (like map 11.4), when displaying the eleven Confederate states, fail to show the complexities behind secession. As he notes, the "tear" that ran through especially the Upland South to separate South from North was painfully "jagged." For example, Kentucky declared its neutrality, but people living around Bowling Green in western Kentucky had strong sympathies for the South (map 11.4). In eastern Tennessee around Greeneville, many strongly supported the North. Confederate sympathizers were strong in southwestern Missouri around Neosho. Indeed, in October 1861 the flag designed to represent the CSA had thirteen stars, not eleven, to include both Missouri and Kentucky. And, as we have seen, in June 1863, Upland Southerners in western Virginia broke away from Virginia to form West Virginia. To quote Meinig, "the breadth of the tear . . . was a profound

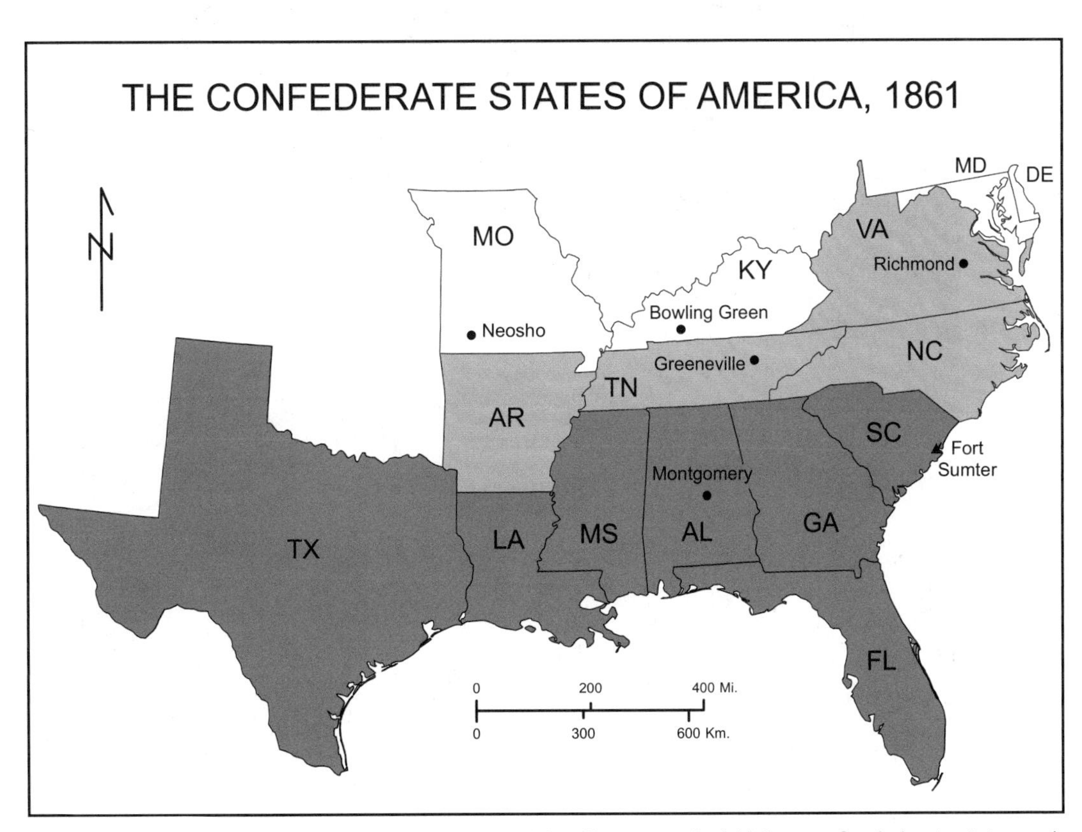

Map 11.4 The Confederate States of America, 1861. Shown are the initial seven Confederate states and the four to follow.

Source: Meinig, *The Shaping of America*, 2:485 and 488.

sorting that affected not only districts, towns, and countrysides but clans and families, and it took place from the Atlantic seaboard to the Plains."[15]

The Civil War lasted four years, from 1861 to 1865. Significantly, Southerners called it the "War between the States" while Northerners called it the "War of the Rebellion." The North had the advantage of far greater industrial development and a superior transportation infrastructure based on railroads. The North also had a larger population: over 20 million people compared to fewer than 13 million Southerners, some 3.5 million of whom were slaves. But Southerners were fighting out of desperation to preserve their way of life. During the conflict, 620,000 Americans died, which amounted to 2 percent of the population, and an additional 500,000 were wounded. The camera, invented about 1830, captured, for the first time for any war of consequence, examples of the horrors of battle, leaving vivid memories of this most tragic of American events (see figure 11.2). An important outcome was the official end to slavery: Lincoln had emancipated slaves in the Confederate states on January 1, 1863, and in December 1865 the Thirteenth Amendment formally abolished slavery in all the United States. But the collapse of slavery brought much chaos, and after 1865 the South sank to a level of poverty from which it would not recover for generations.[16]

The secession of eleven Confederate states is, of course, a defining regional characteristic of the Lowland South, much as being torn apart defines the Upland South. Comparing the Lowland South with the Upland South brings out additional regional

Figure 11.2 Dead Soldiers, Gettysburg, Pennsylvania, 1863. The bloated bodies of Confederate soldiers gathered for burial at the southwestern edge of Rose Woods in Gettysburg, Pennsylvania, are visceral reminders of the tragedies of war.

Source: Photograph by Timothy H. O'Sullivan, July 5, 1863. Downloaded from Library of Congress Prints and Photographs Division, Washington, DC, call number LC-B811-245 [P&P] LOT 4168.

attributes. The people of the Lowland South were heavily of British (English, Scotch, Welsh) descent, with large numbers of blacks, while Upland Southerners were Scotch-Irish, English, and German, with few blacks. In religion, Lowland Southerners of both races were mainly Baptist and, in time, also strongly Evangelical (in agreement with the New Testament gospels and salvation through atonement), while Upland Southerners were heavily Presbyterian and, in time, also Fundamentalist (believers in a literal interpretation of the Bible). Because of plantations, landholdings in the Lowland South were large compared to those in the Upland South. A single commercial crop of cotton was the focus before the Civil War in Lowland South agriculture, while livestock and small grains like wheat and corn had favor in the Upland South. Mules, the sterile offspring of a mare and a donkey, were favored for farmwork in the Lowland South, while horses were preferred in the Upland South. However, both regions were similar in that their inhabitants were basically conservative politically; they lacked urban centers, in a relative sense; and were basically rural.

Notes

[1] Rostlund, "The Myth of a Natural Prairie Belt in Alabama," 392–411.

[2] Regarding the Delta's late settlement, see Adkins, "The Geographic Base of Urban Retardation in Mississippi, 1800–1840," 36. The term "delta" also accurately characterizes the bottomlands on the Arkansas side of the Mississippi River and in northern Louisiana.

[3] See Aiken, "An Examination of the Role of the Eli Whitney Cotton Gin in the Origin of the United States Cotton Regions," 5–9; and Aiken, "The Evolution of Cotton Ginning in the Southeastern United States," 196–224.

[4] Meinig, *The Shaping of America*, 2:287, develops the two culture hearths of the South Carolina–Georgia Piedmont and the Natchez District. See Hilliard, "A Robust New Nation, 1783–1820," 167, for "pincer movement," and 168, for its cartographic illustration.

[5] Meinig, *The Shaping of America*, 2:286, introduces new material on the complicated topic of cotton experimentation and new varieties in the Natchez District.

[6] About the "chimerical" environmental limits of cotton, which farmers have grown in Tennessee, Kentucky, and Missouri, see Hart, "The Demise of King Cotton," 308. For the Cotton Belt in 1860, see Hilliard, *Atlas of Antebellum Southern Agriculture*, 71.

[7] See Hilliard, *Hog Meat and Hoecake*, 151, for corn acreage equal to cotton; 152, for neglect of corn compared to cotton; 153, for yields half those in Old Northwest; 157, for demands for corn; 9 and 158, for corn and foodstuffs self-sufficiency; 111, for pork preferred; 100, for cornfields opened to hogs.

[8] Hilliard contrasts farms and plantations in "Plantations and the Moulding of the Southern Landscape," 105–7.

[9] Earle, "A Staple Interpretation of Slavery and Free Labor," 65.

[10] Prunty, "The Renaissance of the Southern Plantation," 464, for an antebellum plantation example; 466, for fencing types.

[11] Prunty distinguishes between the more common "cropper" and the higher-status "tenant" plantations, shown diagrammatically in ibid., 467 and 474.

[12] Ibid., 482, for Prunty's "neoplantation" term and diagram; 483, for tractors mass-produced since 1927.

[13] Meinig has much to say on the admission of states. See his *The Shaping of America*, 2:458, for the existence of no fixed policy on pairing; 2:448–55, for admission as pairs; 2:504, for "more or less" pairing; 2:297, 449, 453, for the Missouri Compromise.

[14] Ibid., 2:299, for the indication that the North wanted to contain, not abolish, slavery; 2:473, for "cotton states"; 2:474–75 and 480–81, for the CSA.

[15] Ibid., 2:487. See also 2:481, for the examples of how the United States "would be ripped apart, jaggedly"; 2:482, for West Virginia and Kentucky; 2:483, for Kentucky, Tennessee, Missouri; 2:485, for the flag.

[16] Ibid., 2:502, for the names given for the war; 2:307 and 555, for the number of slaves; 2:528, for numbers killed and wounded; 2:513, for events that ended slavery.

Texas 12

Only in Texas can the state flag be flown at the same height as the American flag. This is because only Texas was a republic before its admission as a state. Concessions made to lure Texas into the union resulted in other unique conditions. For example, Texas was given control of its public lands. This meant that Texas could choose whether to implement the township and range system (which it did in unsettled West Texas). And it also meant that Texas had the option to create Indian reserves (it has only three small Indian reservations today). The state's one-time independence, its size (the largest until Alaska's admission in 1959), and its cosmopolitan population have prompted some to refer to Texas as an "empire." Reasons abound for treating Texas as a region. We look first at how "The Balcones and Caprock Escarpments" divide Texas into three landform areas. Next, we see how "Mexican Texas (1821–1836)" became the destination for thousands of "Anglos" and European-born. The Mexican strategy to populate Texas with non-Mexicans backfired, triggering independence and "The Republic (1836–1845)," as noted in the third section. And the lasting impress of Germans in "Germans and the Hill Country" concludes the chapter (with the recognition that the role Texas played in "The Cattle Kingdom" is discussed in Chapter 19).[1]

The Balcones and Caprock Escarpments

An escarpment is a break in slope which can mean a gentle decline or a high bluff. In Texas two escarpments divide the state into three landform areas (see map 12.1). Between the Balcones Escarpment and the Gulf Coast, the Coastal Plains are flat and low in elevation. Between the Balcones and Caprock escarpments are hills, plateaus, and country that is generally rugged—giving this part of Texas its acknowledged aesthetic appeal. And west and north of the Caprock Escarpment is the Llano Estacado, a Spanish term for "Staked Plain": an area that is high in elevation and sometimes extremely flat. Indeed, in places the Llano Estacado is so featureless that, say some historians, Spaniards used yucca stems or buffalo bones to stake out their route back to camp. A more plausible origin for the term "Staked Plain," however, is that *ciboleros*, or buffalo hunters, cured their buffalo meat by hanging it from lines strung between rows of wooden stakes taken onto the high plains for that purpose.

The Balcones Escarpment represents a fault zone that is between five and fifteen miles (eight to twenty-four kilometers) wide and extends in the form of a crescent for some three hundred miles (480 kilometers) between Del Rio on the Rio Grande and Waco (map 12.1). With justification, this escarpment has been likened to the fall line. Both represent breaks in slope (some three hundred feet [92 meters], for the Balcones) that separate hills underlain with older harder rock from plains underlain with younger softer rock. People settled along both—the fall line because of its head of navigation function and its waterfalls for power (Chapter 5), and the Balcones Escarpment because of the presence of artesian, or free-flowing, springs. Inland from the Balcones Escarpment, groundwater is stored in thick beds of porous limestone known as the Edwards Aquifer, and this water emerges along the escarpment in springs that, in a land of highly variable precipitation and frequent droughts, attracted people, as at Waco, Austin, San Marcos, New Braunfels, San Antonio, and Del Rio (see figure 12.1). The San Antonio River and its small tributary, San Pedro Creek, as noted in Chapter 2, are examples of spring-fed water courses.[2]

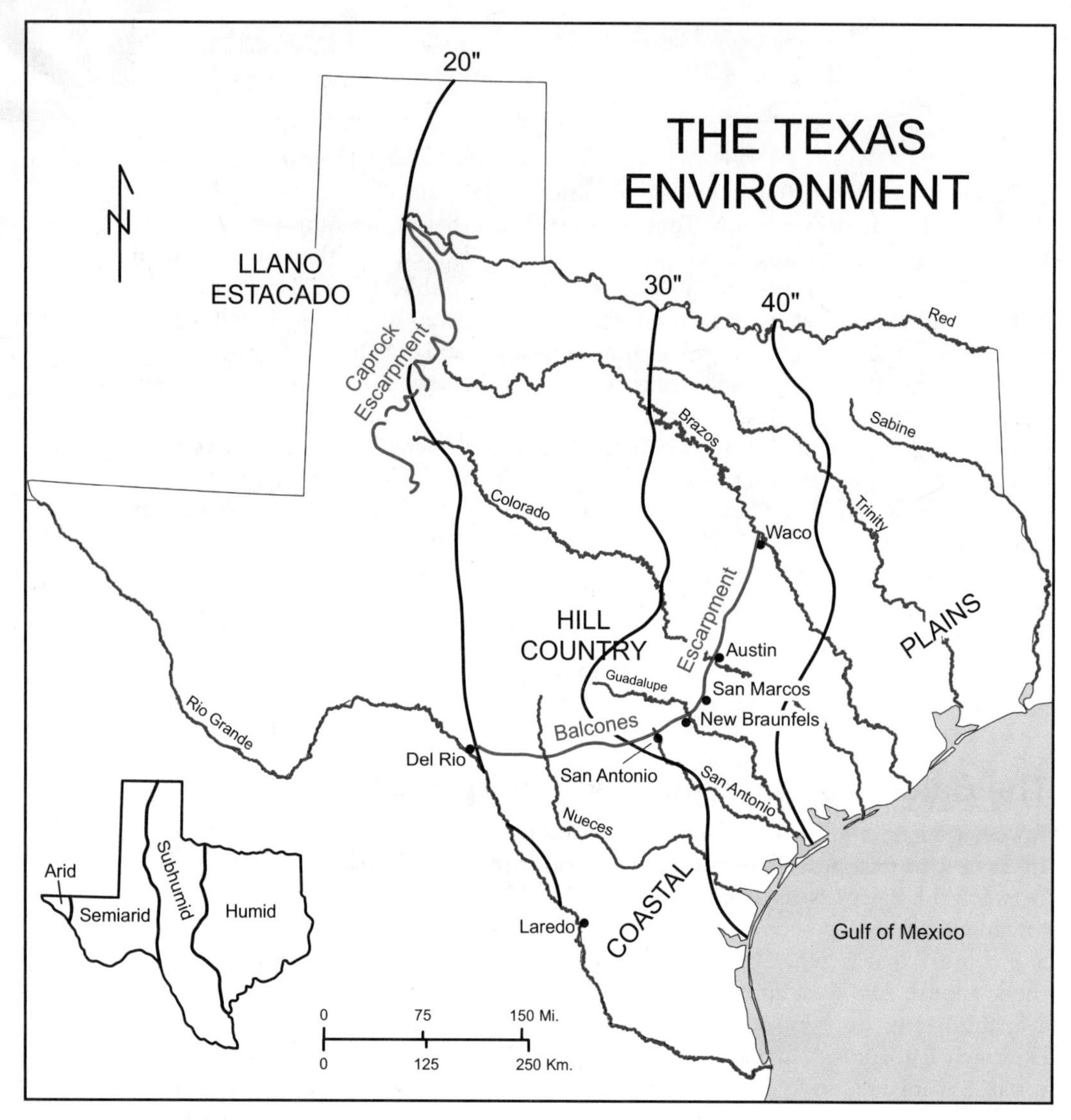

Map 12.1 The Texas Environment.

Source: Estaville and Earl, *Texas Water Atlas*, 5, for average annual precipitation lines (20, 30, and 40 inches).

Flowing from high ground to low ground across these landform zones are eight important rivers. Two of the eight form state or international borders—the Sabine between Texas and Louisiana, and the Rio Grande (in Mexico, the Río del Norte) between Texas and Mexico. Intermittent in their flow at their headwaters, these rivers become perennial downstream, rising and falling seasonally but in general too shallow annually to be navigable except near the gulf. Because precipitation amounts decrease across the state from east to west, the rivers become increasingly smaller and shallower in that direction. From east to west, these rivers are the Trinity, Brazos, Colorado, Guadalupe, San Antonio, and Nueces (map 12.1). As we shall see, the Nueces was the southern boundary of Texas during the Mexican period; Laredo on the Rio Grande was founded in the Mexican state of Nuevo Santander, not Texas. The eight rivers become like lines on a stage that set the framework for much of the human drama that unfolds.

Figure 12.1 Bird's-Eye View of San Marcos, Texas, 1881. Drawn by Augustus Koch, this view shows clearly the Balcones Escarpment, from which flows the San Marcos River. Locate the Hays County Courthouse in its Shelbyville-type square. Follow Austin Street (now LBJ Drive) north three blocks to the orchard at the break in slope. At that point today, one enters Texas State University, whose campus now covers the sloping ground.

Courtesy of Brock J. Brown, who used this image in teaching geography courses at Texas State University.

Geographer Terry G. Jordan points out that the stage for that drama in east and central Texas was set with an unusually complex vegetation cover (see spotlight). Eastern and central Texas are located at the southern end of a longitudinal zone where average annual precipitation amounts decrease from more than forty inches (one thousand millimeters) in the east to less than twenty inches (five hundred millimeters) in the west—a transition from humid to subhumid to semiarid (see map 12.1 inset). Reflecting this drop in precipitation amounts is a transition in vegetation from pine forests and oak-hickory woodlands to grasslands. Within the forests to the east, observes Jordan, were thousands of small patches of open prairie, while within the grasslands to the west stood thousands of groves of trees. Into this mixed forest-grassland mosaic in the period between about 1815 and 1840 came southerners, first to the southern banks of the Red River, then to the Austin Colony, and then across East Texas more generally. Drawing on firsthand accounts of settlers and travelers, Jordan asked how these southerners evaluated the mixed forest-grassland cover in selecting land on which to settle.[3]

Jordan found that, with few exceptions, southerners *chose* the prairies, provided these grasslands were located on the margins of the forests or woodland groves. As southerners evaluated where to settle, the advantage of choosing prairie was that

Spotlight

Terry Gilbert Jordan-Bychkov (1938–2003)

Terry G. Jordan-Bychkov. This photograph appeared in a memorial for Jordan-Bychkov in *Annals of the Association of American Geographers* 95, no. 2 (June 2005): 462. Courtesy of *Annals of the Association of American Geographers.*

Terry Jordan-Bychkov knew more about his home state of Texas than few will ever know about any state. To fulfill his humanistic goal to understand Texas, Jordan-Bychkov undertook archival research, and more importantly, in the tradition of his highly esteemed geography friend, Fred B. Kniffen, he crisscrossed all 254 counties in Texas to observe and to "read" the cultural landscape. From this came nearly a dozen books and more than fifty book chapters and articles about Texas—publications that after 1997 displayed his newly hyphenated surname to honor his marriage that year to Bella Bychkova. Born in Dallas to a German American, Hill Country father (and a professor of German literature at Southern Methodist University) and a mother of Texas southern stock, Jordan-Bychkov earned degrees at SMU (BA, 1960), the University of Texas–Austin (MA, 1961), and the University of Wisconsin–Madison (PhD, 1965), all in geography (with a double major in German at SMU). He taught at Arizona State University (1965–1969) and North Texas State University (1969–1982) before holding the prestigious Walter Prescott Webb Chair of History and Ideas at the University of Texas–Austin (1982–2003). His publications earned him the Association of American Geographers Honors Award in 1982, and in 1987 his colleagues elected him AAG president. For some, Jordan-Bychkov was overly outspoken, and on occasion he rushed into print a bit prematurely. But his loss to pancreatic cancer at the young age of sixty-five meant for geography the loss of one of its most committed supporters and truly gifted scholars.

---

crops could be planted and livestock pastured without clearing trees. The disadvantage was the difficulty of plowing the grasslands, a task that required a team of no fewer than four oxen and, before the availability of the steel-tipped plow, a heavy wooden plow. Proximity to trees meant access to lumber and fuel, and for shelter many chose to build their cabins on the edges of the trees. Prized more than prairies on the grassland-forest margins of Texas were the "Brazos weed prairies," apparently one-time Indian Old Fields strung along the Brazos in the Austin colony, where only burning, not plowing, was required before planting. Canebreaks (dense-growing reeds) along rivers also required only burning. From an earlier study of land evaluation by southerners, northerners, and foreign-born in the Old Northwest and Kentucky, Jordan came to the same conclusion—namely, that with few exceptions, colonists selected the prairie-forest fringe areas first, when such lands were available and affordable.[4]

Mexican Texas (1821–1836)

In 1821, the year Mexico won her independence from Spain, present-day Texas held perhaps five thousand people, excluding Indians. On this remote Mexican frontier, the largest population cluster, perhaps two thousand, lived in the capital, San Antonio, and in a string of settlements and ranchos that extended down the San Antonio River (see map 12.2). On the lower San Antonio, Goliad, the mission-presidio community that had been relocated from La Bahia on Matagorda Bay, was a second settlement cluster. Nacogdoches, located on the Old Spanish Road in the Piney Woods of East Texas, had a third cluster. By 1821, a few "Anglos," as people in the Southwest who are not of Spanish-Mexican or Indian descent are often known, had settled on the south bank of the Red River, initially at Jonesborough in 1815. And along the north bank of the Rio Grande in what would become Texas, Mexicans lived at Laredo and in other settlements and ranchos.[5]

Newly independent Mexico knew full well that the number of its citizens in Texas—Spain's population legacy—was too small to secure that vast frontier from westward-moving Anglos. Indeed, Spain itself had recognized this, and at the end of its regime, it had ended its closed-frontier policy by initiating a carefully defined *empresario* program to populate Texas. An empresario was an individual (not a grant of land) who, in return for land, contracted with the government to introduce a minimum number of colonists within a specified period. The intent was for the incoming colonists to accept the Spanish-speaking Catholic tejano society, to become loyal Mexican citizens, and to act as something of a barrier against Anglo intruders. In 1820 Spain had awarded Moses Austin, a Connecticut-born Anglo who had been a citizen of Spanish Louisiana and now lived in Missouri, an empresario contract in Texas. Moses Austin died, but at his deathbed, his son Stephen F. Austin promised to carry through with his father's award. Stephen F. Austin successfully reapplied with the new Mexican government to take his father's place.[6]

Austin's award straddled the basins of the lower Brazos and Colorado rivers (map 12.2). Initially the award stretched from the Old Spanish Road to a strip of land along the Gulf coast that Mexico wished to keep free of non-Mexicans. In 1823 Austin founded San Felipe as his headquarters on the banks of the Brazos. The Anglos who came streaming in to select the bottomlands and prairie-forest margins were, prior to 1836, mainly Upland Southerners from Tennessee and Missouri. In the Mexican period, Lowland Southerners also arrived with their slaves, but they were concentrated near the mouths of the Brazos and Colorado in the coastal strip that Austin had subsequently received. Mexico was officially a nonslave country; as such, the importation of slaves to Mexican Texas, as Meinig notes, was "technically forbidden" but nonetheless "tacitly allowed." By 1836 the Austin colony had become the largest and the most successful of the empresario awards, and it had become the focus of Anglos in Texas.[7]

Between 1824 and 1832, at least twenty-four empresarios had signed contracts, and by 1836 their lands covered nearly all of Texas from the Sabine to the Nueces, and from the Red River to the Gulf coast. Green DeWitt, an Anglo, introduced colonists, many from Missouri, to land bordering Austin's colony on the west. In 1825 DeWitt set up headquarters at Gonzales on the Guadalupe River, a town he laid out with a central plaza (map 12.2). Martín de León, a Mexican rancher from the state of Tamaulipas, received an empresario award, and in 1824 he founded Victoria as his headquarters on the lower Guadalupe. Near the mouth of the Nueces, Irish settlers established San Patricio in the late 1820s, and Irish also resettled an abandoned mission at Refugio forty miles (eighty kilometers) to the northeast. In 1831 Mexican officials awarded Friedrich Ernst, a German, land northwest of San Felipe within the huge Austin colony. Many Germans arrived, and in 1838 Ernst himself

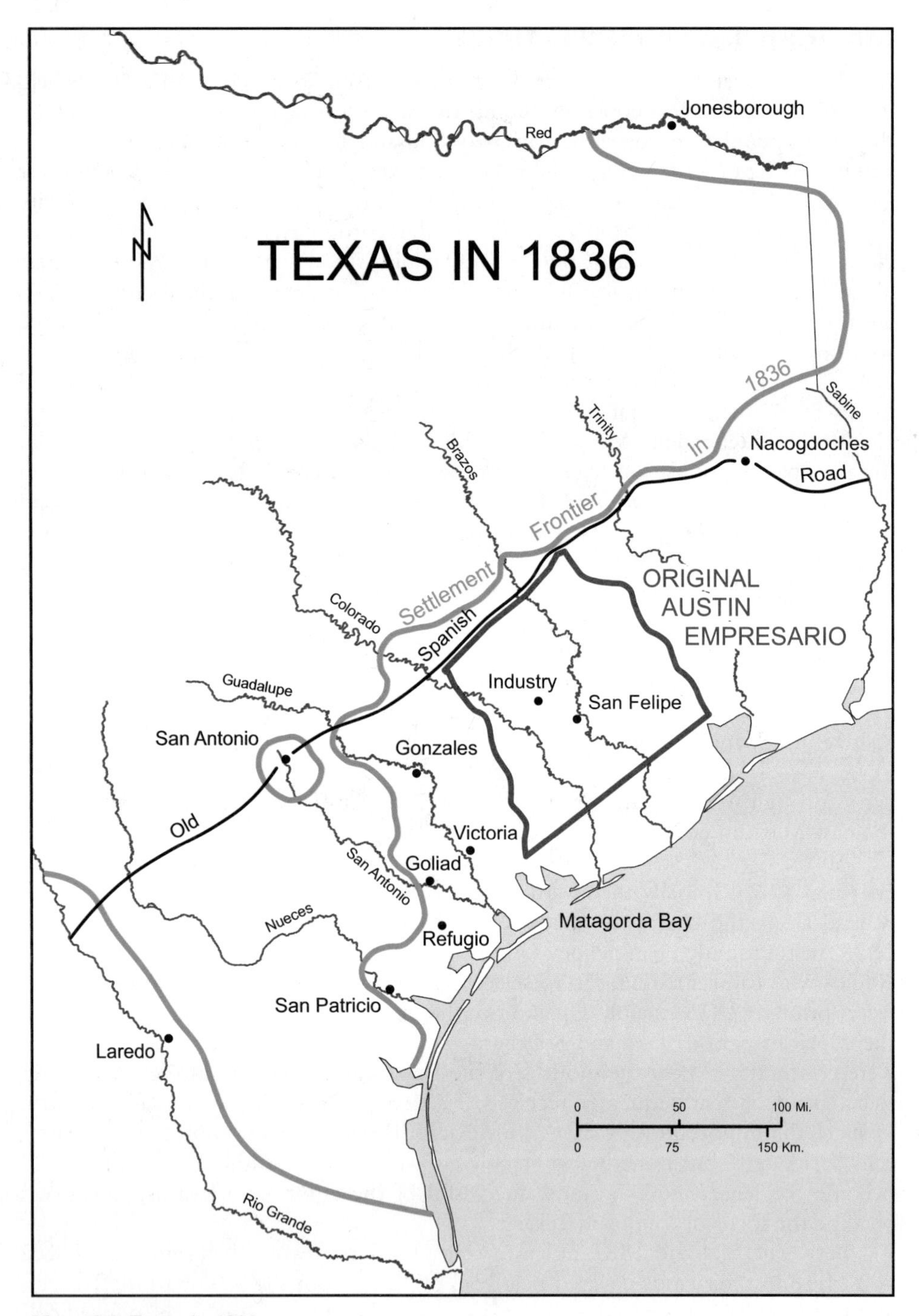

Map 12.2 Texas in 1836.

Sources: Jordan, "Population Origins in Texas, 1850," 102, for the settlement frontier line in 1836; Meinig, *Imperial Texas*, 30, for the original Austin empresario.

laid out Industry, the key German settlement. The German communities around Industry would come to be known as "German East" to differentiate that area from Germans who, beginning in 1844, settled "German West," located mainly in the Hill Country.[8]

By 1836 Texas held some thirty thousand non-Indians. Anglos, especially Upland Southerners, were the largest faction, Germans led among the foreign-born, and the Mexicans, who probably numbered about five thousand, had by now been outnumbered five or six to one. The thirty thousand were sorted in a rough way with Anglos anchored in the Austin colony in the east, a mixture of Anglos and foreign-born in something of a middle zone, and Mexicans in the west, notably at San Antonio and Goliad. As Mexicans came to be outnumbered, the tensions between them and Anglos increased. Mexicans complained that their arrogant Anglo neighbors were disloyal and had disrespected Mexican law by bringing slaves into a nonslave country. Anglos expressed as their grievances a judicial system that lacked trial by jury and a tariff system that levied duties on them. The stage was set for revolt.[9]

The Republic (1836–1845)

In late 1835 Anglo and Mexican Texans clashed at San Antonio, Goliad, and Gonzales. The conflict escalated, and on March 2, 1836, Anglos who met at Washington-on-the-Brazos in the Austin colony declared Texas to be an independent republic (see map 12.3). The line that separated Anglos from Mexicans was not entirely sharp, for in the short war that followed, says historian David Weber, perhaps 160 Mexican tejanos would side with the Anglos. In the war's three major battles, Mexicans under the command of General Antonio López de Santa Anna (who was also Mexico's president) defeated the Anglos at the Alamo in San Antonio and at Goliad, but at the third battle on the San Jacinto River, which marked the eastern boundary of the Austin colony, rebel leader General Sam Houston decisively defeated Santa Anna. Following that battle Anglos forced Santa Anna to sign the so-called Treaties of Velasco in which Santa Anna recognized an independent Texas—treaties that Mexico would never ratify. A year later (1837), Texans voted overwhelmingly to be annexed to the United States. Because Texas would come in as a slave state at a time when the slave-nonslave issue was explosive, the United States rejected the Texas petition, shocking both Texans and Mexico. Thus, for nine years Texas would remain a republic, during which time it would claim land west to the Rio Grande and north to the forty-second parallel (see map 12.3 inset).[10]

The new Republic of Texas made Austin its capital in 1839 (map 12.3). For Austin, officials chose a site on the banks of the Colorado River where the small community of Waterloo already existed; the new community name honored Stephen F. Austin who had died in December 1836. Officials chose Austin's location on the western edge of major Anglo settlement to encourage Anglo expansion in that direction. Meinig points out that some Texans also hoped that Austin—located between the Texas gulf ports and a distant Santa Fe—might also help divert the lucrative Santa Fe trade to Texas. However, Austin's selection had its detractors. One, Sam Houston, now president of the republic, was able to convene the government at his namesake town of Houston in 1842, and between 1842 and 1845 Washington-on-the-Brazos served as the de facto capital. In 1850 Texans voted to affirm Austin as the permanent capital.[11]

To bring much-needed colonists to Texas, the government of the republic revived Mexico's empresario program in 1841. Germans continued to be drawn individually and as families to German East in the area northwest of Houston (map 12.3). After 1844, Germans also arrived in organized group-colonization programs to German West.

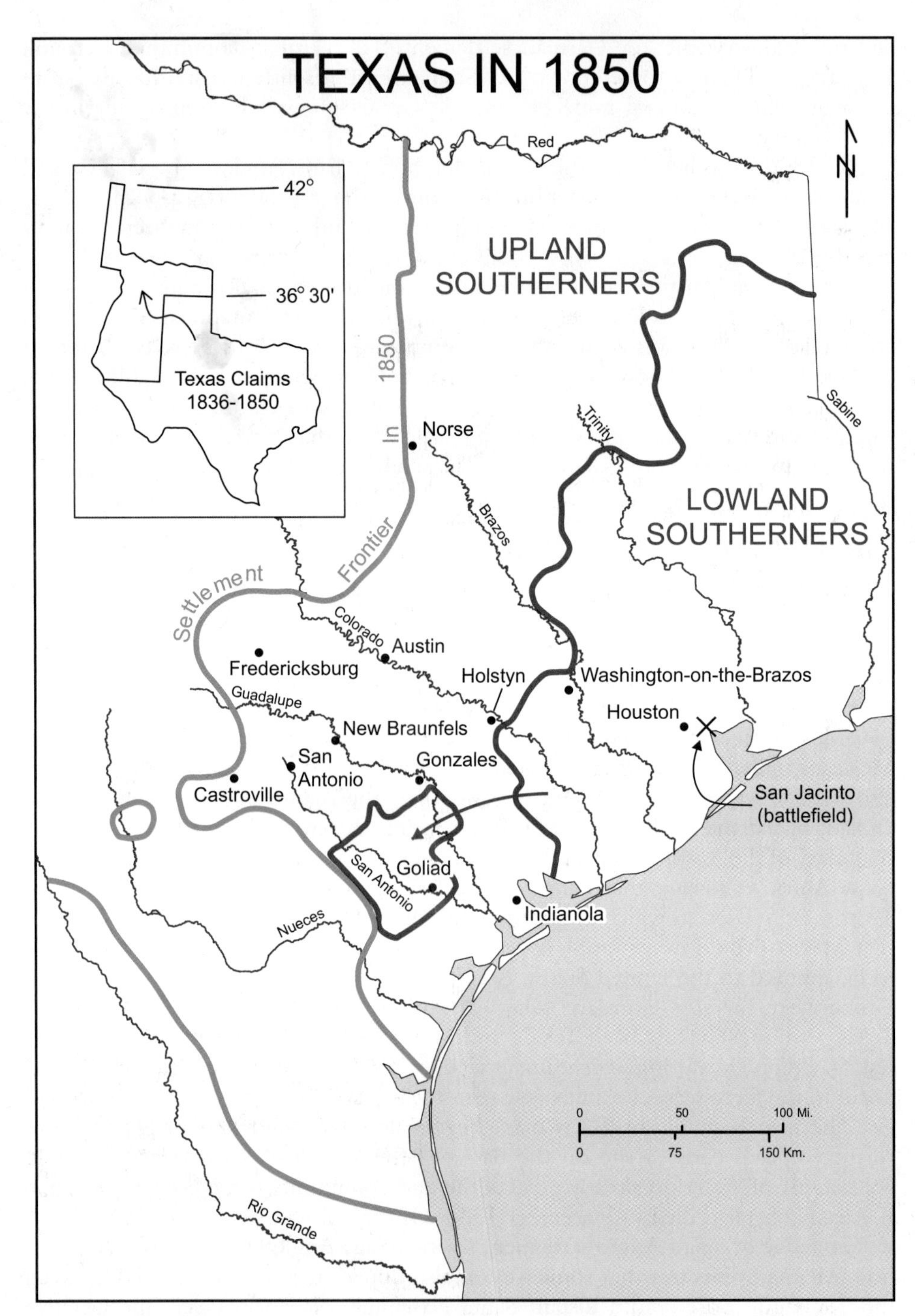

Map 12.3 Texas in 1850. Inset: In 1836, the Texas republic claimed land west to the Rio Grande and north to latitude 42°. With annexation in 1845, the United States accepted those boundaries. In 1850, however, it paid Texas $10 million to accept its present borders: the panhandle could not exceed 36°30', the northern limit for slavery, set in 1821 as Missouri's southern border.

Source: Jordan, "Population Origins in Texas, 1850," 102 and 87, for line of settlement frontier in 1850, and the location of Upland and Lowland Southerners.

In 1844 Henri Castro received a grant west of San Antonio, and at Castroville he introduced Alsatians from the French-German borderlands. A second group, the Verein Society, received land deep in the Hill Country, which it never settled, but it did establish Indianola as its port of entry on Matagorda Bay. And leading in the direction of its grant, the society founded two quite successful way stations at New Braunfels in 1844 and at Fredericksburg in 1846. In the 1840s German immigrants were now joined by Czechs, who founded their "mother colony" at Hostyn, and Norwegians, whose major settlement was Norse. In the 1850s Poles and Wends would arrive, adding to the cosmopolitan Texas population.[12]

By far the largest group to stream into Texas during the time of the republic, however, was southerners. By 1850, as shown in map 12.3, the line marking the Texas settlement frontier had moved significantly west from where it had been in 1836. Southerners were the major reason. Primarily Upland Southerners from states like Tennessee and Missouri settled the prairie lands in the west, and mainly Lowland Southerners and their slaves from states like Alabama and Louisiana came to dominate the Piney Woods and coastal prairies in the east (map 12.3). Slavery had been made legal under the republic, prompting a heavy influx of Lowland Southerners. Jordan's analysis of the Texas population in 1850 is revealing: Of 212,000 total Texans in 1850, 81.2 percent were either Southern Anglos (114,040; 53.7 percent) or slaves (58,558; 27.5 percent). By 1850 even the Germans (11,534; 5.4 percent) outnumbered the Mexicans (11,212; 5.3 percent), the latter identified by Jordan by their Spanish surnames, in the unpublished census schedules.[13]

The Republic of Texas ended when the United States annexed Texas as a state in 1845. In 1844 President James K. Polk had campaigned for office on a platform that called for the "re-annexation" of Texas. On March 1, 1845, before Polk's inauguration, outgoing president John Tyler signed a joint resolution of Congress that provisionally extended statehood to Texas. The Texas Congress unanimously approved this annexation resolution on July 4, 1845, and on December 29, 1845, Polk signed the act that admitted Texas. A sometimes overlooked reason this annexation was significant is that it marked the first time that any Mexicans became Mexican Americans. For Polk, a strong expansionist, acquiring Texas was not enough. Following his failure to buy California and New Mexico from Mexico, Polk provoked the Mexican War (1846–1848). A war of land aggression that was unpopular both in the United States and in Mexico, the outcome to this day has left feelings of bitterness among many in Mexico who are keenly aware of the unjust loss of territory to the United States. In the Treaty of Guadalupe Hidalgo, signed February 2, 1848, to end the war, Mexico was forced to cede, as David Weber wrote, "one-third of its territory, or a half counting Texas,"[14] to its more powerful neighbor. The 1848 treaty, then, marked the second event whereby Mexicans would become Mexican Americans, and the Gadsden Purchase in 1853 would be the third and final event.

Germans and the Hill Country

The Texas Hill Country lies north and west of the Balcones Escarpment. The hills are underlain with horizontal beds of limestone and sandstone once formed under an inland sea, and outcrops find the rock readily accessible. Live oak, post oak, cypress, and juniper trees (locally the juniper are called cedar) cover the hills. When Germans first settled this part of German West about 1845, they built distinctive one-and-one-half-story houses that have lasted to give the Hill Country a unique character (see map 12.4). Employing home-country building methods such as half-timbering and plastering, these Germans drew upon the local materials—oak for beams, cypress for shingles, and stone for walls—in their house construction. Hill Country German

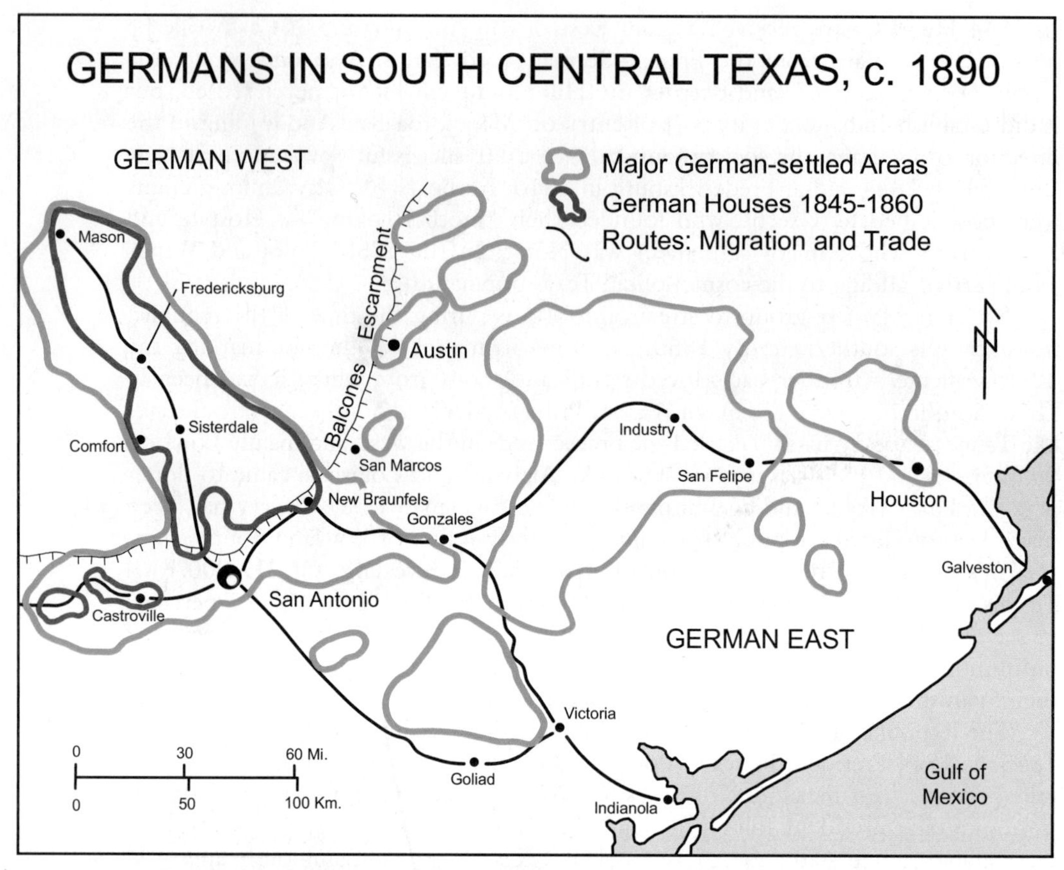

Map 12.4 Germans in South Central Texas, c. 1890.

Sources: Jordan, "German Houses in Texas," 25; Jordan, *German Seed in Texas Soil*, 42 and 46.

houses illustrate nicely all three themes in this book: the use of native materials as cultural ecology; the introduction of construction methods such as half-timbering as cultural diffusion; and the persistence of early relic houses as cultural landscape.

Terry Jordan tells us that the earliest houses built by Hill Country Germans were simple log cabins where oak logs were laid horizontally over stone foundations. Germans soon turned to half-timbering (in German, *Fachwerk*), wherein spaces between supporting oak beams used in constructing walls were filled with stone (see figure 12.2). (In Germany, brick would commonly have filled the interspaces.) And as half-timbering continued, walls came to be built entirely of stone. The building of one-and-one-half-story houses persisted to about 1860, and because Germans often plastered over and then whitewashed the outer (and inner) walls of houses, whether walls in a given house are made by half-timbering or stone is often hidden. To help hold the plaster to a half-timbered wall, wire would be strung between nails that stood out from the oak beams. An additional early German building method was the use of casement windows that swung open and were pulled back to close, unlike the more common sash windows that move up and down to open and close.[15]

While building these early houses, notes Jordan, Germans borrowed a number of architectural features from their southern Anglo neighbors. For example, to cope with the intense heat in summer, Germans added open covered porches to the fronts of houses (see figure 12.3). To save space inside their houses, Germans built outside stairs to reach the upper half-story—a practice that would have been unthinkable in Germany, given its winter ice and snow. Some Germans also adopted the southern

Figure 12.2 Goldbeck-Faltin House, Comfort, Texas, 1978. In 1854 brothers Fritz and Theodor Goldbeck built a log house in Comfort. In 1856, August Faltin bought the house and added a second front room built in *Fachwerk*. The side view of this house, which faces Seventh Street near the corner of Main, shows a rear "lean-to" also built in *Fachwerk*.

Photographs by the author, April 15, 1978.

penchant for building separate outbuildings for kitchens and smokehouses, to avoid heat and potential fire. However, kitchens and other rooms built onto the back sides of houses in "lean-to" or "lean-on" fashion, sometimes as part of the original house but more generally as later additions, may not have had southern origins. Jordan notes that "lean-tos" existed in Germany.[16]

Figure 12.3 "Sunday" House, Fredericksburg, Texas, 1978. This residence, located on East San Antonio Street, shows two Anglo architectural features: the open covered porch and the outdoor stairs that lead to a loft. Some German families who lived on farms owned two-room Sunday houses in town and used them for shopping on Saturday and church on Sunday.

Photograph by the author, April 15, 1978.

In the Hill Country, then, one-and-one-half-story houses built between about 1845 and 1860, with walls made from half-timbering or from stone, both of which were often plastered over, survive as a landscape signature (see figure 12.4). Jordan emphasizes that these houses did not represent German house types; houses in Germany were a full two stories. Rather, these houses represented a combination of German building methods and borrowed Anglo architectural features. Their distribution is limited to the Hill Country with its limestone and sandstone quarry sources (map 12.4). The exceptions are those German-populated communities including New Braunfels, San Antonio, and Castroville that lie just below the Balcones Escarpment. There, German stone masons, because of fellow German demands, built stone houses within a dozen or more miles of quarry sources, despite the difficulty and expense of transporting the stone. Significantly, an absence of stone on the coastal plains of German East found Germans there building wood-framed houses that typically have not survived.[17]

After about 1860, Germans in the Hill Country continued to build with stone, but their houses were built a full two stories in height. And after about 1885, according to Jordan, German attributes in house construction disappeared altogether. The tradition of building with stone, nevertheless, has given rise to an attractive Hill Country architectural style. Houses and commercial buildings today are often constructed of limestone that appears almost white. Standing two full stories in height, the street-facing broad sides of these structures literally radiate brightness. Their typically low-angle roofs constructed with metal sheeting often extend shed-like to cover a second-story front balcony that itself gives shade to the ground floor. These attributes are seen in figure 12.5; so are the symbolic statements of Texan pride that find the Texas flag and the American flag flying at equal heights and the ubiquitous lone star within a dark circular field found high in the gable end.

Figure 12.4 Lindheimer House, New Braunfels, Texas, 1972. Built about 1852 on Comal Avenue, this house exemplifies the one-and-one-half-story houses built by Germans between about 1845 and 1860. The plaque records this house to be a Texas Historical Landmark (1936); it is also listed on the National Register of Historic Places (1970).

Photograph by the author, April 16, 1972.

Figure 12.5 Hominick Homes Design Center, Fredericksburg, Texas, 2008. Texas Hill Country houses and commercial structures today are built of stone, have low-angle metal roofs, and often stand two full stories. This commercial building on East Main Street exemplifies the style.

Photograph by the author, October 27, 2008.

Notes

[1] D. W. Meinig's *Imperial Texas* is geography's notable example of referring to Texas as an empire.

[2] The facts about the Balcones Escarpment, also the comparison with the fall line, are from Petersen, "Along the Edge of the Hill Country," 20–30.

[3] Jordan, "Vegetational Perception and Choice of Settlement Site in Frontier Texas," 244–57.

[4] Ibid., 252–53, for the prized "Brazos weed prairies" and canebreaks. Jordan's earlier study that focused on the Old Northwest and Kentucky between 1800 and 1850 is "Between the Forest and the Prairie," 205–16.

[5] The estimate of five thousand in 1821 may be excessive. David J. Weber, in *The Spanish Frontier in North America*, 299, notes that political upheaval saw the Texas population decline from four thousand in 1803 to two thousand in 1821 (as stated in Chapter 2 of this volume). In *The Mexican Frontier, 1821–1846*, 160, Weber reports 2,500 in Texas in 1821.

[6] Meinig, *The Shaping of America*, 2:36–41, for empresario Austin.

[7] Jordan, "The Imprint of the Upper and Lower South on Mid-Nineteenth-Century Texas," 669, for Upland Southerners as the clear majority in the Austin colony before 1836; 670 and 672, for Lowland Southerners and slaves were found largely near the Brazos-Colorado river mouths. Quotations are from Meinig, *The Shaping of America*, 2:38.

[8] For at least twenty-four empresarios covering much of Texas, see Weber, *The Mexican Frontier, 1821–1846*, 163 and 165 (map). The year 1838 for Industry is from Jordan, *German Seed in Texas Soil*, 41.

[9] Weber, *The Mexican Frontier, 1821–1846*, 177, gives 35,000 in 1836, of which 3,500 were tejanos.

[10] Weber says "perhaps" 160 tejanos sided with Anglos, in *The Mexican Frontier, 1821–1846*, 252.

[11] Stephens and Holmes, *Historical Atlas of Texas*, 29, for Waterloo and Austin's impermanence as capital in the 1840s; Meinig, *Imperial Texas*, 42, for the imperial concept that Texas might tap the Santa Fe trade.

[12] For reviving the empresario program in 1841, Jordan, *German Seed in Texas Soil*, 26. Jordan identifies "mother colonies" such as Jonesborough, San Felipe, Gonzales, Industry, and Hostyn in "Annals Map Supplement Number Thirteen."

[13] For the 1850 manuscript census schedule breakdowns, see Jordan, "Population Origins in Texas, 1850," 87.

[14] Weber, *The Mexican Frontier, 1821–1836*, 274.

[15] Jordan makes these points in "German Houses in Texas," 24–26.

[16] Jordan, in personal correspondence dated April 27, 1978, noted that "lean-tos" were generally added later, and that they existed in Germany.

[17] Jordan, in *German Seed in Texas Soil*, Chapter 6, contrasts German East with German West.

The New Northwest 13

Our use of the term *New Northwest* requires some explaining. Technically the New Northwest lay west of the Mississippi River, while the Old Northwest lay east of that river. In reality, because the remoteness of Wisconsin and that part of Minnesota lying east of the Mississippi caused them to lag in settlement, they came to be thought of as New Northwest and not Old Northwest. In this chapter our focus is on all of Wisconsin and Minnesota as well as the upper peninsula of Michigan (see map 13.1); we leave the Dakotas part of the New Northwest for inclusion in the Great Plains. Few regions in the United States demonstrate more starkly the correlation between resource availability and the attraction of people. We look first at how "Lead, Copper, and Iron Ore" attracted miners. We turn next to how the always important resource of soils attracted farmers to "A Wheat Frontier." Under "Frontier Footsteps," we follow the moves made by one frontier family, the Ingallses, as they were chronicled in the well-known Little House books, and we also follow the footsteps of census takers for insights to doing research. Finally, we see how "White Pine" was the resource that drew lumbermen west to a timber industry in the upper lakes country.

Lead, Copper, and Iron Ore

In areas north of the Ohio and Missouri rivers, continental glaciers reworked surface landforms some ten thousand years ago, as noted in Chapter 3. In the New Northwest, one result was that ice scoured the Superior Upland around the western and southern sides of Lake Superior (map 13.1). In their scouring, glaciers removed surface soils and rock and deposited this overburden farther south in the form of "drift"—thick layers of clay, sand, and gravel sorted by glacial meltwater. Rather strangely, however, an area the size of southern New England that was centered on southwestern Wisconsin altogether escaped, according to accepted theory, the several glacial advances. This so-called Driftless Area seems to have been completely surrounded, yet unaffected, by ice. The Driftless Area, thus, has sharper relief features in the form of rugged hills; shallow if fertile soils; an absence of lakes, swamps, and "erratics" or boulders; and, of course, an absence of drift.

Found in the Driftless Area, lead was the first mineral to attract European American miners and settlers to the New Northwest. Given the area's absence of drift, lead ore could be found right on or within fifty feet (seventy meters) of the surface. These lead deposits occurred where the three states of Iowa, Illinois, and Wisconsin come together (map 13.1). Long before 1800, French settlers on the Iowa side of the Mississippi River mined lead; Julien Dubuque, a major figure, left his name on the Iowa community. Because lead melts at low temperatures, the French merely made bonfires, heaped on the lead ore, and the next day collected the molten metal. Indians followed the French example: They took the molten lead from under the fires and formed it into round balls called "pigs," which they exchanged for goods at French trading posts. By 1811, Americans were mining lead on islands in the Mississippi, from which they spread to the Illinois side. In 1822 they founded Galena, whose name means "lead ore" in Latin, near the mouth of the Fever River. That same year at the mouth of that narrow river, steamboats arrived to carry lead ore down the Mississippi to markets, and Galena became the industry's gateway and shipping point, its riverbank mansions built with industry profits. Up the Fever River

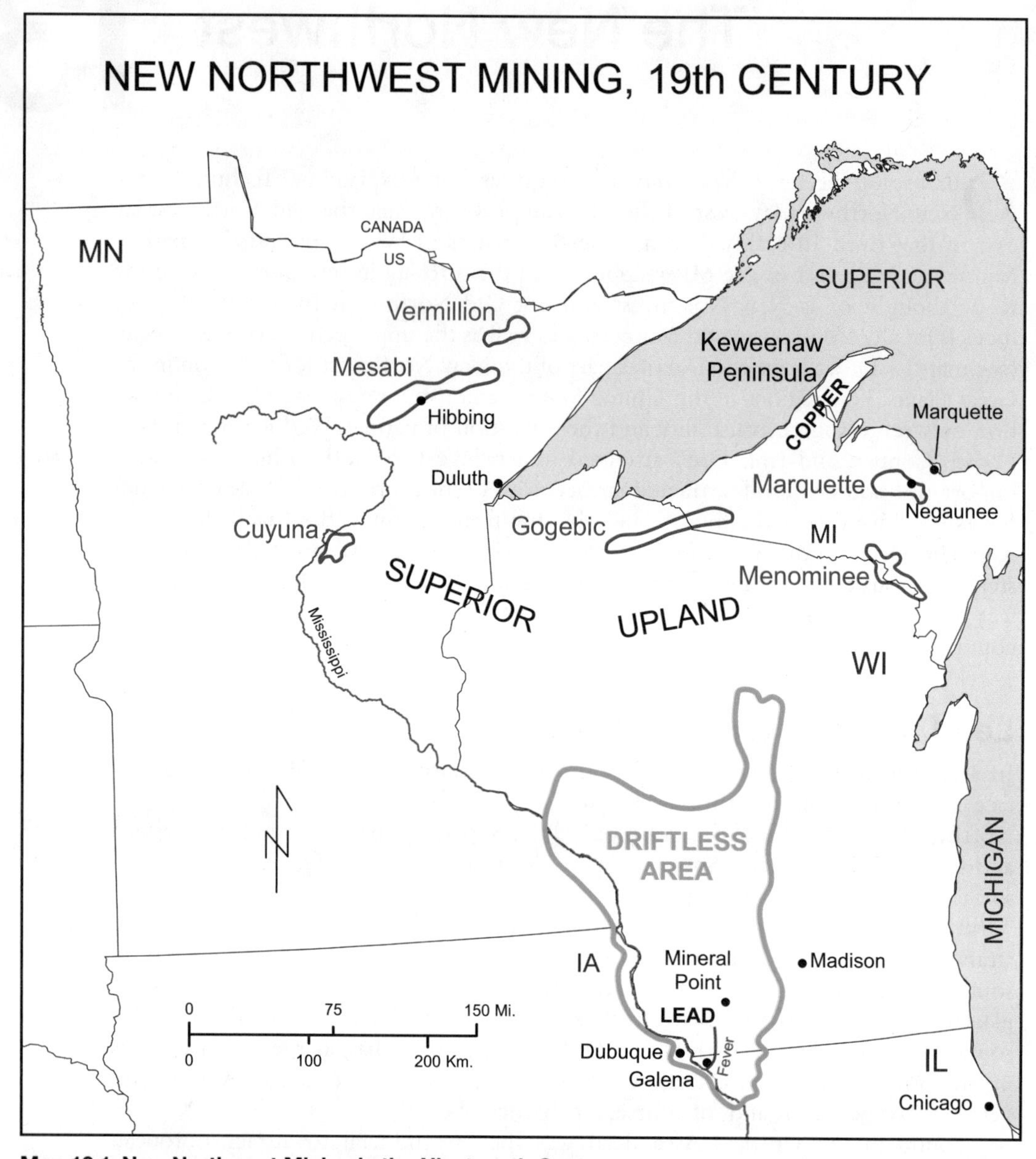

Map 13.1 New Northwest Mining in the Nineteenth Century.

Source: Raisz, *Landforms of the United States*, for the Driftless Area.

in Wisconsin, lead mining began in 1823, with Mineral Point the eventual focus. Joining the rush of at least ten thousand people to the lead district were Cornish miners from Cornwall in southwestern England; in 1845, they made up half of Mineral Point's 1,500 people. Lead mining in the tristate district peaked in 1845 as surface lead deposits gave out and the expense of digging shafts and draining groundwater for deeper deposits increased. Many miners turned to farming while others left for places like California's gold rush in 1849; mainly Cornish miners kept the mining productive until about 1860.[1]

Copper was the second mineral to attract European Americans to the New Northwest. It is rare for nearly pure copper to exist in nature anywhere in the world. Thus when Spaniards in Florida in the early 1500s found Native Americans wearing ornaments of nearly pure copper, they tried—without success—to learn where this metal came from. The only place in North America where nearly pure copper could have come from was Michigan's Keweenaw Peninsula (map 13.1). The Keweenaw Peninsula juts northeast for fifty miles (eighty kilometers) from Michigan's Upper Peninsula into southern Lake Superior. Continental glaciers had scoured the area, exposing surface veins of sometimes nearly pure copper. Reports of sporadic European American mining of Keweenaw copper date from the 1760s, but the first systematic mining did not begin until 1844. Thereafter, Michigan led the nation in copper production until the inevitable happened—namely, that the rich surface deposits gave out, and the expense of shafting to reach deeper veins curtailed production. After 1887, Montana, with its mines at Butte, became the new copper mining leader, although Keweenaw mines produced copper to about 1970. Copper did attract people, yet unlike the lead district to the south, poor soils and a short growing season made opportunities quite limited for copper miners to become farmers.[2]

While searching the upper lakes country for additional minerals to mine, Indians in 1845 led a party of white people to deposits of iron ore atop the Marquette Range (map 13.1). Iron ore became the third mineral to bring in settlers. Again, continental ice had removed the overburden from the top of the low Marquette Range that was located a dozen miles south of Lake Superior, exposing rich deposits of iron ore on or near the surface. Inexpensive open-pit or strip mining operations commenced, and first roads and then railroads were constructed to haul the ore the dozen miles to the new lake port of Marquette. Blast furnaces at Negaunee used charcoal made from trees cut locally for limited iron and steel production. In 1855 completion of the first Soo Canal around the rapids of the St. Mary's River at Sault Ste. Marie, between lakes Superior and Huron, facilitated transporting the iron ore to the lower Great Lakes. By 1870, after a quarter century, mining declined in a relative sense, again because of the depletion of rich surface ore and the expense of exploiting deeper deposits. Local cutover land provided only limited opportunities for agriculture.[3]

Meanwhile, additional low mountain ranges containing iron ore were discovered in the scoured Superior Upland. They included the Menominee and Gogebic ranges on the Michigan-Wisconsin border and the Vermillion, Cuyuna, and Mesabi ranges in Minnesota (map 13.1). The largest and most famous of these, the Mesabi, focused on the community of Hibbing and is to this day a major source of iron ore in the United States. As in the case of the Marquette Range, iron ore mined in these newer areas was shipped on ore boats from ports like Duluth on Lake Superior through the series of improved canals at Sault Ste. Marie to iron and steel plants at, for example, Chicago, Detroit, and Cleveland on the lower Great Lakes. Meanwhile, coal mined in the Appalachian field made its way by railroads to the lower lake ports. The fortuitous linking by cheap water transportation of iron ore around Lake Superior with coal in Appalachia gave rise in the second half of the nineteenth century to an iron-and-steel industry that catapulted the United States into a leading position in world industry.

A Wheat Frontier

Almost as clear as the correlation between glaciation and mining is the correlation between the environment and agriculture. Soils—notably the thick and fertile deposits of drift—were the basic resource that attracted farmers; where soils were good, farmers sought land along the prairie-broadleaf tree interface, which channeled their agricultural frontier diagonally to the northwest (see map 13.2). Transitions in precipitation and temperatures explain why the original vegetation changed. First, average annual precipitation amounts decreased from roughly forty inches along Lake Michigan in the east to about twenty inches along Minnesota's western border. Reflecting this was the transition from a mixed forest (broadleaf and needleleaf) in the east to tallgrass prairie in the west. Second, temperatures decreased from south to north, a decrease accentuated by the higher elevations of the Superior Upland. Within the mixed forest, the transition found broadleaf deciduous trees (e.g., maple and oak) in the south being overtaken by needleleaf evergreen trees (e.g., white pine and spruce) in the north. That needleleaf trees are better adapted to colder temperatures seems to be the reason. Special note should be made of an area that settlers knew as the Big Woods, which stretched from the "elbow" of the Minnesota River east across the Mississippi River into Wisconsin (map 13.2). In this park-like environment, groves of tall maple and oak were interspersed within tallgrasses, providing homesteaders with a source of timber and fuel, land ready to plow, and good hunting.

Fertile soils and their cover of mixed broadleaf trees and prairie, then, are what became the setting for a wheat frontier that found farmers, between the 1830s and

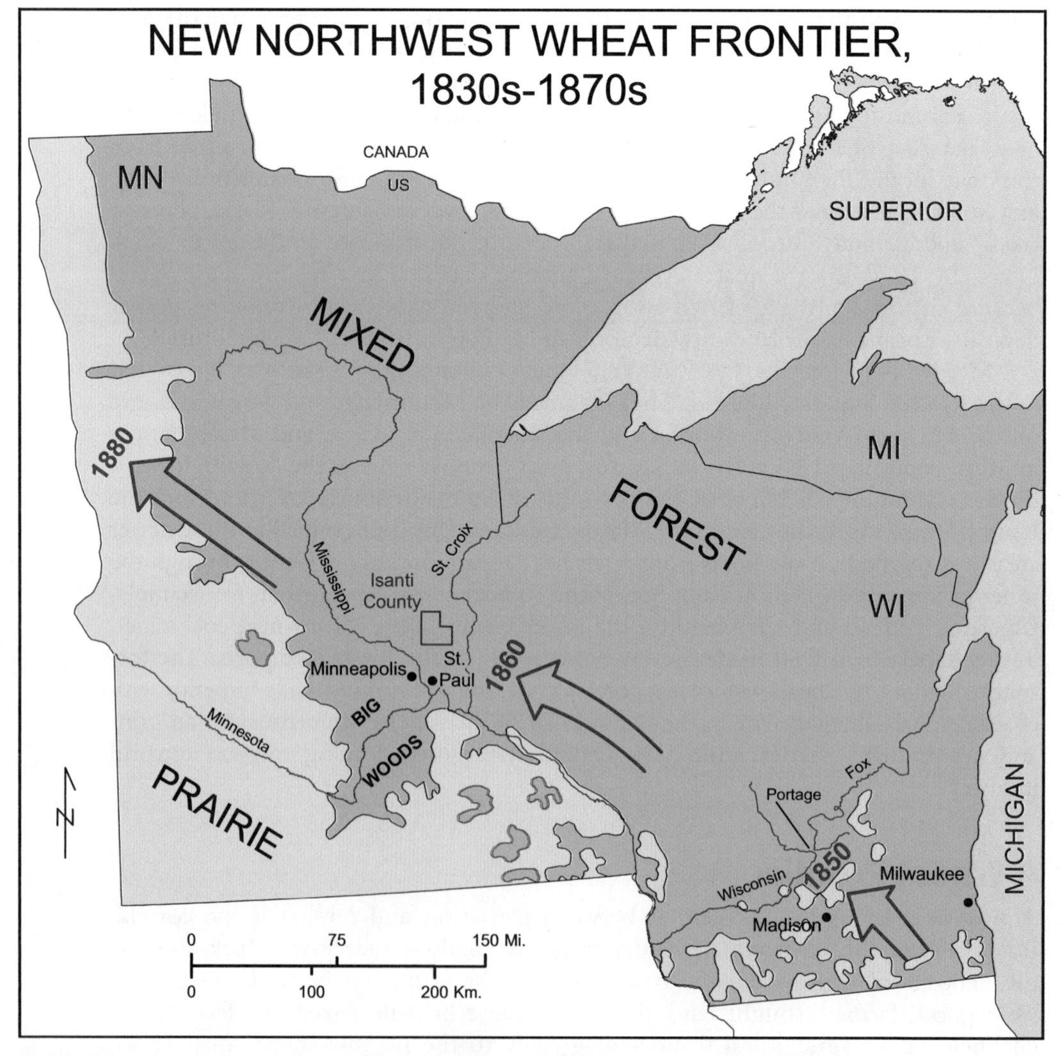

Map 13.2 New Northwest Wheat Frontier, 1830s to 1870s.

Source: Brown, *Historical Geography of the United States*, 317 and 329, for the original mixed forest–prairie line.

1870s, advancing from southeast to northwest, first through Wisconsin and then through Minnesota (map 13.2). Land in southeastern Wisconsin was opened when Indians of several tribes relinquished their strongholds following the Black Hawk War of 1832. The farmers who came were a mixture of native-born European Americans, notably Yankees (see Chapter 9), and immigrants from Europe, notably Germans, English, Irish, and Norwegians; agents representing the state of Wisconsin actively recruited foreign-born arrivals in New York City in the early 1850s and after the Civil War. In purchasing land already surveyed into a grid of townships and ranges, colonists sought especially those parcels that combined prairie and woodland. On this land they planted wheat, their preferred grain, and corn, because it did so well. Initiated in the mid-1830s, this farming frontier advanced from southeastern Wisconsin northwest to the Fox-Wisconsin portage route by 1850, and by 1860 agricultural settlement approached Wisconsin's western border at the St. Croix River. Wisconsin's population totals reveal how rapidly people arrived: 12,000 in 1836; 31,000 in 1840; 305,000 in 1850; 776,000 in 1860.[4]

In 1850 Minnesota's small population of six thousand was concentrated in the junction area of the Minnesota and Mississippi rivers (map 13.2). That year, treaties signed with the Sioux Indians opened "Sooland" (as the area west of the Mississippi was known) to European Americans. Although many Sioux remained in Sooland, in 1851 a surge of white settlers commenced to spread northwest across Minnesota. As in Wisconsin, the colonists represented a combination of native-born old stock, importantly Yankee, and newly arrived foreign-born, notably Germans, Norwegians, and Swedes, who again were actively recruited in New York City by state officials representing Minnesota. And as in Wisconsin, farmers sought land that contained a prairie-woodland mixture on which they grew primarily wheat and corn. The suddenness of Minnesota's growth is seen in its total population increase: 172,000 in 1860; 440,000 in 1870; 781,000 in 1880—by which time the wheat frontier had reached Minnesota's western border.[5]

In two important ways, Minnesota did differ from Wisconsin. First, the Homestead Act of 1862 was passed too late to affect land alienation in Wisconsin, but just in time to have a big impact in Minnesota. This act allowed an individual of twenty-one years of age or older to claim a quarter section (160 acres) provided that the person pay a small filing fee (ten or fifteen dollars) and live on and make improvements to the land for five years. Imagine the pull factor: 160 acres virtually free for the prospective emigrant who was, say, the second or third son in a Swedish farm family where in Sweden the oldest brother, under primogeniture, would inherit the entire family farm. The historian Marcus Lee Hansen tells us that, because of its timing, the Homestead Act had the heaviest impact on the "great triangle" lying between the upper Mississippi and Missouri rivers and the Canadian border—including, of course, Minnesota. The second way in which Minnesota differed from Wisconsin concerned the role played by railroads in attracting settlers. Advancing west from Wisconsin and north from Iowa, railroads reached Minnesota beginning in 1862, and as they built through the state, bringing with them new colonists, they advertised the sale of their alternate-section federal land grants. The idea was to settle farmers along railroad rights-of-way and earn revenue from the freight these farmers would ship. The plan worked: Settlers came, they bought land, and they marketed their products by rail.[6]

Farmers, then, were quick to settle Wisconsin and Minnesota, and given the speed of their arrival, territorial status and statehood came rapidly: in 1836 and 1848 for Wisconsin, and in 1849 and 1858 for Minnesota. Madison, in part because of its early centrality, was chosen to be Wisconsin's capital, and as happened in only a half-dozen other states (Texas, for one), it was also awarded the state university. Meanwhile, Milwaukee, the state's largest urban center, served as the main gateway for immigrants, many of whom came via the Erie Canal and by sail or steam through the Great Lakes. Most immigrants who were headed to Minnesota before about 1860 arrived by steamboats going up the Mississippi River. The Falls of St. Anthony marked the head of navigation on the Mississippi. St. Paul, founded below the falls on the east bank about 1823, served an agricultural hinterland that faced east; St. Paul's head start was behind its selection as Minnesota's capital. Minneapolis grew out of the village of St. Anthony located right at the falls. Tapping the falls for power, Minneapolis by 1848 was a milling site, first for lumber and then wheat, which gave it the nickname, "Flour City." Awarded the consolation prize of the state university, St. Anthony/Minneapolis grew to be the larger of Minnesota's "twin city" urban node.

Frontier Footsteps

We are able to follow the frontier footsteps of one much appreciated and very real New Northwest family through the writings of Laura Ingalls Wilder. Many readers will know the Little House books and, loosely based on them, the television series that began in 1974. Laura Ingalls Wilder wrote eight Little House books that document how her frontier family lived, and by way of example, how frontier families could be movers. Her books feature four sisters, Mary, Laura, Carrie, and Grace, born in 1865, 1867, 1870, and 1877, respectively; and their parents, Pa and Ma Ingalls, between roughly 1860, when Pa and Ma were married in Concord, Wisconsin (thirty-five miles west of Milwaukee), and 1885, when Laura married Almanzo Wilder in De Smet, South Dakota (see figure 13.1). Laura, who died in 1957 at the age of ninety, did not begin to write her Little House books until she was nearly sixty-five, and as they were published in the 1930s and early 1940s, her main source was her own remarkable memory. Encouraging Laura to undertake the project was her only daughter, Rose Wilder Lane, herself an author, and the degree to which Rose helped her mother continues to be debated.[7]

Pa and Ma were of Yankee stock with roots in that Yankee exodus stream discussed in Chapter 9. Pa, Charles Philip Ingalls (1836–1902), was born in Allegheny County, New York, to parents from New Hampshire and Vermont. Pa's family stopped over in Illinois en route to Wisconsin. Ma, Caroline Lake Quiner (1839–1924), was born in Milwaukee County, Wisconsin, to parents who had been married in Connecticut before moving west to Ohio, then Indiana, and finally Wisconsin. In 1862 Pa and Ma moved from eastern Wisconsin to a homestead in the Big Woods near Pepin, Wisconsin, in the bluffs above the Mississippi River (see map 13.3). Mary and Laura were born there. In 1868 the Ingallses and the family of Uncle Henry (Ma's brother) moved by covered wagon to near Rothville, Missouri. In 1869 Uncle Henry returned to Pepin County, and the Ingalls family continued southwest to land near Wayside, Kansas, in Indian Territory. Carrie was born there. The family returned to Pepin County and stayed there from 1871 to 1874 before heading west to a farm on Plum Creek some two miles outside of Walnut Grove, Minnesota. A son, Freddie, was born in Walnut Grove, but he died nine months later. In the summer of 1876, while en route to Burr Oak, Iowa, the Ingalls family made a stopover at a farm near Zumbro Falls, Minnesota, to visit Pa's older brother, Uncle Peter. The family lived in Burr Oak from 1876 to 1877, and Grace was born there. Pa and the family then returned to Walnut Grove for two years before making a final major move in 1879 to Dakota Territory and a railroad camp at Silver Lake that a year later evolved into De Smet.[8]

Altogether, the Ingallses made seven major moves as a family between 1868 and 1879, two of them to return to places they had already been (Pepin County and Walnut Grove), and along the way there were countless lesser moves. The reasons for

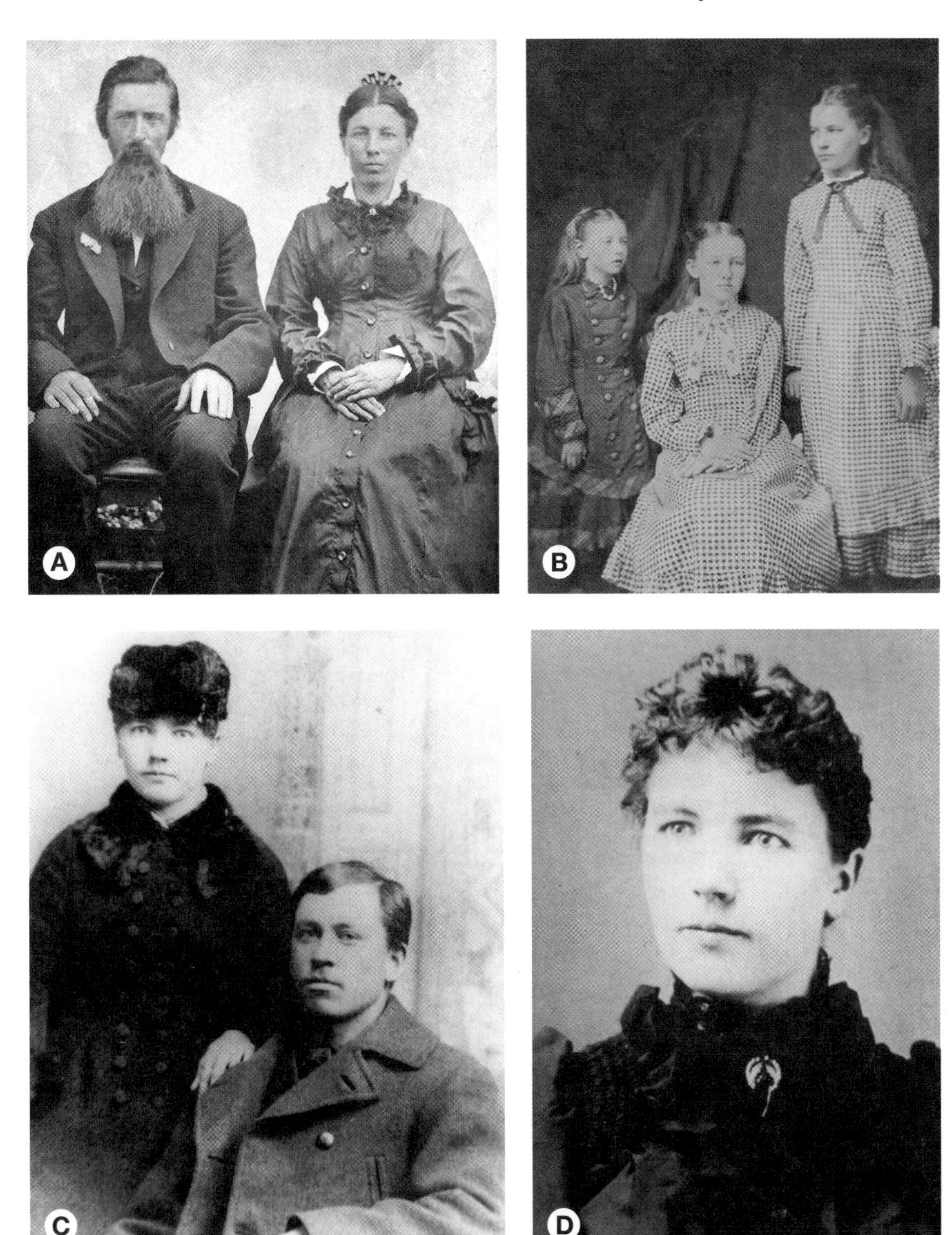

Figure 13.1 The Ingalls family, 1880, 1881, 1885, and c. 1891. (A) Pa and Ma Ingalls, 1880. (B) (left to right) Carrie, Mary, Laura, 1881. (C) Laura and Almanzo after their marriage, 1885. (D) Laura, c. 1891.

Photographs A, B, D, courtesy of the Laura Ingalls Wilder Memorial Society, De Smet, South Dakota. Photograph C used with permission of the Laura Ingalls Wilder Home and Museum in Mansfield, Missouri.

the major moves seem not to have been written about by Laura, but they appear to be reasonably clear. Pa wanted to farm, and in 1868 he and Uncle Henry were moving to what they hoped would be better land that they had purchased sight-unseen in Missouri, a venture that for unknown reasons did not work out. Pa may have found the Kansas prairie to be desirable land, but the southern strip of Kansas where he had squatted was part of Indian Territory, land titles were uncertain, and the Osage braves who visited his log cabin were definitely not welcoming. Near Walnut Grove,

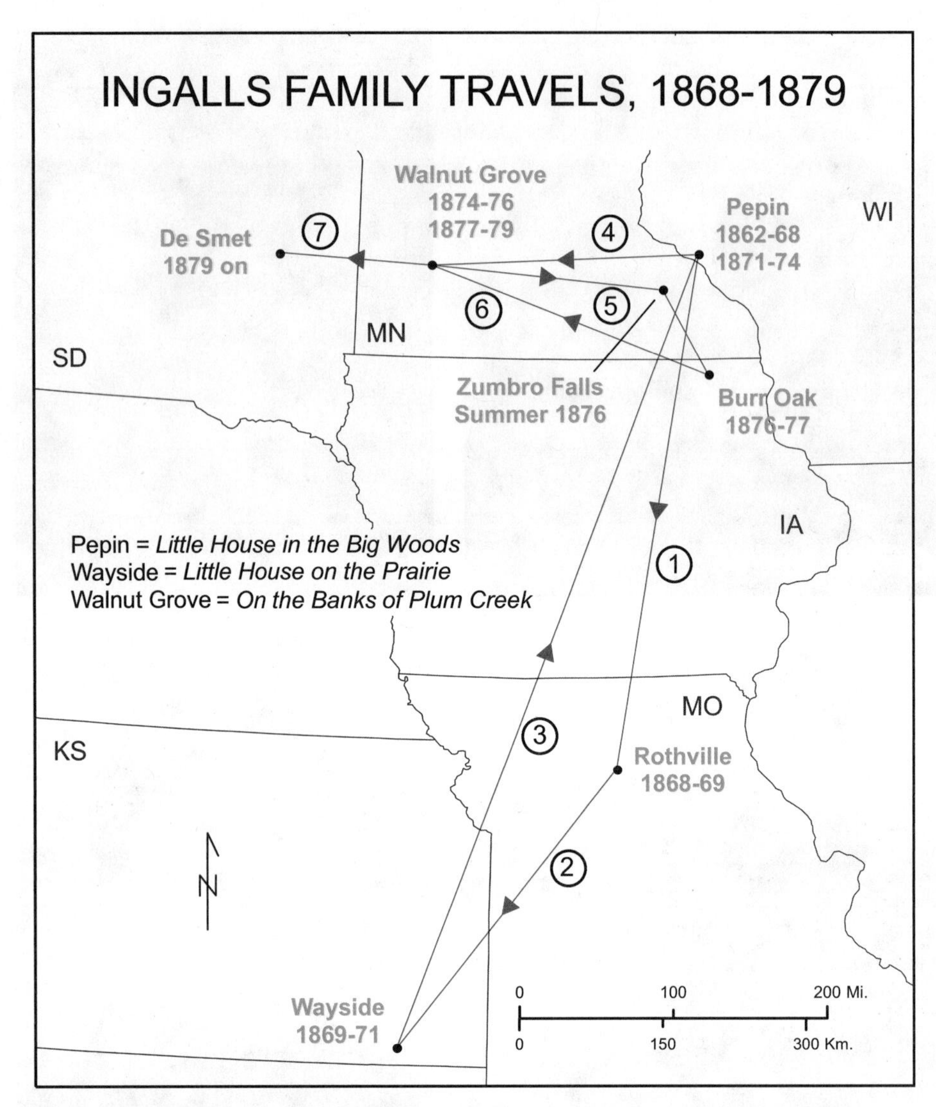

Map 13.3 Ingalls Family Travels, 1868–1879. The circled numbers indicate the seven moves made by the Ingalls family.

Source: Zochert, *Laura*.

Pa did get title to a homestead, and next to Plum Creek he built a cabin out of milled timber hauled from town. But in the consecutive summers of 1875 and 1876, he lost his crops of wheat to grasshoppers, which forced him to sell his homestead and to make an unhappy move back east to Burr Oak. In Burr Oak, Pa worked at a hotel for an acquaintance from Walnut Grove, at a gristmill, and on a farm as a laborer. Even so, he was "haunted" by debt, and he could barely pay the money he owed as rent for a small house. Back in Walnut Grove from 1877 to 1879, Pa worked as a store clerk and a carpenter, and for a while he ran a butcher shop, but a steady job did not materialize. When Pa's sister, Aunt Docia, said that her husband, Hiram, knew of a railroad bookkeeper-storekeeper job out in Dakota Territory, the family made its final major move.[9]

Pioneer life was full of hardships, and the Ingalls family experienced many of them. All members of the Ingalls family periodically had illnesses, and a bout with measles compounded by a stroke left Mary blind at the age of fourteen. Grasshoppers took Pa's wheat, and unexpected blizzards took the lives of fellow Walnut Grove villagers. The Ingallses were so desperately poor that Pa's tax bill of $2.10 owed in Walnut Grove could not be paid until after his return from Burr Oak. Compounding his financial troubles were the family's more-than-typical number of moves, which forced Pa to promise Ma that if she would follow him to Dakota Territory, that would be their "last move." Then again, the Ingalls family enjoyed some good times. In Walnut Grove, Burr Oak, and De Smet, the family belonged to the Congregational Church, which meant singing, mutual support, hand-me-down clothes, and books to borrow. There were the memorable, if very infrequent, visits with relatives. Laura loved going to school when circumstances permitted, and at school she always had many friends. Laura also saw great beauty in the prairie grasses and flowers as the family trekked from place to place. And there was Pa's fiddle, which he played many evenings, and which on occasion prompted dancing.[10]

Over the years, Pa had faithfully taken responsibility to file for title to his land claims—everywhere but in Kansas. In Kansas, the narrow strip of Indian Territory where Pa and other white families had squatted about 1869 still belonged to the Osage Indians; no one living there filed for land until the summer of 1871, by which time the Ingalls family had already returned to Wisconsin. Without a land claim document, finding where that Little House had been located required some detective work. Margaret Clement, a bookseller and one-time teacher from Independence, Kansas, the Montgomery county seat, became that detective. In the 1870 census, Clement learned that Asa Hairgrove, the census taker, had recorded the Ingallses, including newborn Carrie, as living in Rutland Township in Montgomery County. She knew that census takers almost always recorded families and houses one after another (in seriatim). To be able to follow the census taker's path, she constructed a map from courthouse records that showed family quarter-section claims, once they had been filed for, in Rutland Township. By comparing her map and Hairgrove's probable path, she found, among the names of Ingalls family acquaintances whom Laura had written about, two unfiled-for quarter sections in the southeast corner of the township. In the field, Clement learned that only one of these two quarter sections had once had a hand-dug well, and a county atlas made about 1881 also showed that the same parcel once had a cabin. Given the absence of a well and apparently a cabin on the other parcel, Clemens concluded that she had found the Ingalls place, and on this quarter section today is located the *Little House on the Prairie* interpretive center (see figure 13.2).[11]

Clement was probably unaware that she was following good historical geography procedure when she located the Ingalls family parcel in Kansas. In a short article, geographer Michael Conzen (see spotlight) laid out how a researcher could follow the footsteps of a census taker in order to combine data from the original census schedules (population and agriculture) with data on plat (landownership) maps to reconstruct an area's geography. Conzen had applied this technique in a master's thesis written in history at the University of Wisconsin–Madison, where, for 1860, he reconstructed the geography of the thirty-six-square-mile township of Blooming Grove, which has part of Madison in its northwestern corner. From the original population and agricultural census schedules, Conzen obtained data on family households, places of birth of individuals in these households, the value of a farm's real estate, and the number of acres held. And from a contemporary plat map (Ligowski's map of 1861), he determined the sizes and shapes of land parcels, the names of land parcel owners, and the road networks that connected these farms. Assuming that census takers logically visited households in seriatim, Conzen could follow the routes taken and combine census data with map data. The fit, noted Conzen, was not always perfect, but when names and numbers of acres in the census coincided with landowner names and parcels on the map, one had complete accuracy (Conzen

Figure 13.2 Little House on the Prairie near Wayside, Kansas, 2008. The sign announces that this log cabin may resemble the onetime Ingalls log cabin. Little House sites attract thousands of visitors; the two girls shown in period dresses were tourists from Minnesota.

Photographs by the author, July 23, 2008.

recognized that the names of property owners on maps sometimes differed from those of farm operators in the census).[12]

Conzen suggested how merging census data with plat maps to produce farm-level reconstructions could be of value. Three examples will suffice: First, by merging value of real estate data in the 1860 census with properties on the Ligowski plat map, Conzen showed that farm values per acre did predictably increase with proximity to the urban center of Madison. Second, Conzen noted that if one were to map the places of birth of individuals, one could identify possible ethnic clustering of, say, Norwegians or Yankees, valuable information for understanding a number of

Spotlight

Michael P. Conzen (1944–)

Michael P. Conzen, while taking a close look at a car bumper sticker at Crossroads Cathedral, Shields Boulevard, Oklahoma City, Oklahoma. Photograph by the author, April 12, 1981.

Graduate students who took my seminars in historical geography had total respect and admiration for Michael Conzen. First, Conzen was resourceful, using, for example, bank correspondent linkages (clearing house data on financial activity) to map how America's urban system matured between 1840 and 1910, or passenger transport frequency information for railroads to analyze Midwestern regional growth between 1850 and 1910. Second, Conzen's maps (with "MPC" always appearing in a bottom corner), which showed such things as New York City leading the urban hierarchy or how Chicago was linked regionally, are truly masterful. Conzen is an urban historical geographer who specializes in nineteenth- and early-twentieth-century America, and he backs it up with a significant secondary interest in geographical cartography. The number of his publications in these and related areas is staggering—as of 2011, 26 books and edited works, 121 articles and book chapters, 11 edited monograph series, and a great deal more. Born in Grappenhall, England, Conzen earned BA (1966) and MA (1970) degrees at Cambridge University. At the University of Wisconsin–Madison, he earned three additional degrees: an MSc in geography (1968), an MA in history (1970), and a PhD in geography (1972). From 1971 to 1976, Conzen taught at Boston University, and since 1976 he has been a faculty member at the University of Chicago. Because of his scholarship, Conzen is a veritable giant in American historical geography.

patterns, including what crops or animals raised were favored by which group. Third, if data from the census of agriculture (1859–1860) on the number of bushels of wheat or "Indian" corn produced were merged with farm units, one could map crops as a land-use item, and the level of per-acre yields could also be differentiated across space. The procedure Conzen outlined is useful in conducting archival research and can result in good, original large-scale (or small-area) scholarship.[13]

White Pine

White pine was truly the lumberman's delight: These trees stood 250 feet (75 meters) tall, measured five to seven feet (1.7 to 2.3 meters) in diameter, and yielded two hundred board feet per log of even-grained, easily planed, durable softwood. This most sought-after tree was part of that mixed broadleaf-deciduous-hardwood and needleleaf-evergreen-softwood forest that covers much of eastern Canada and extends south into the northern parts of Maine, New York, Michigan, Wisconsin, and Minnesota (see map 13.4). Between the 1820s and the 1890s, a commercial lumbering frontier swept from east to west across the area, beginning in Maine and ending in Minnesota. Michael Williams, geography's expert on *Americans and Their Forests* (the title of his major synthesis on the subject), tells us how the systematic exploitation of white pine evolved geographically.[14]

From Maine to Minnesota, the initial assault on white pine found lumbermen penetrating rivers and exploiting river basin watersheds. Operations took place in winter as

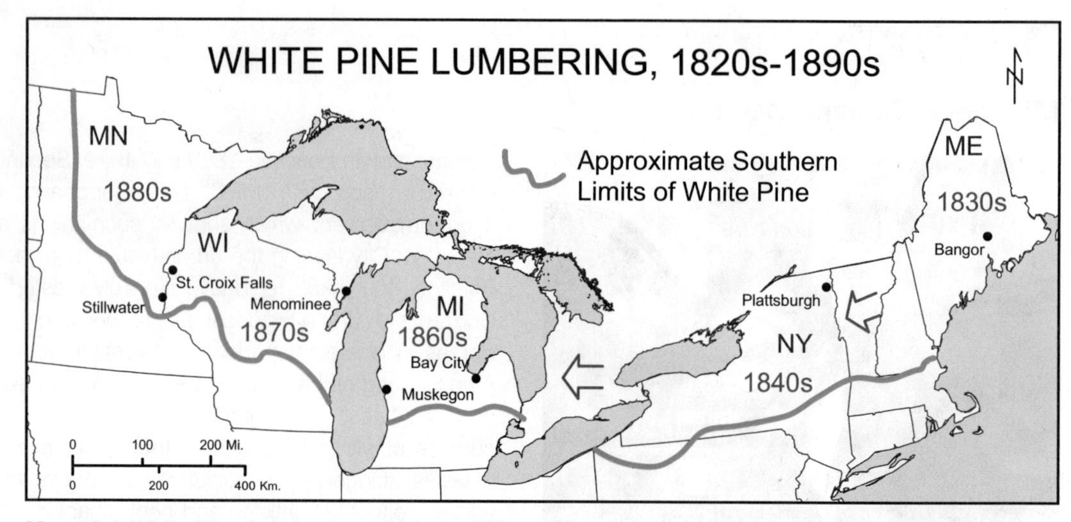

Map 13.4 White Pine Lumbering, 1820s to 1880s. Shown are approximate decades of peak activity.

lumberjacks felled trees with axes and used oxen to haul the logs over snow to rivers. Come spring thaws, they would float the logs downstream to water-powered sawmills found along rivers and at river mouths. Bangor, Plattsburgh, Bay City, Muskegon, Menominee, St. Croix Falls, and Stillwater were all important river-oriented sawmill towns (map 13.4). Two significant innovations introduced in major ways in the 1870s, railroads and steam-driven sawmills, then changed the geography of forest exploitation. Railroad spur tracks widened the cutting areas that could be tapped by lumberjacks, who had begun to fell trees with crosscut saws. And new steam-driven sawmills could be positioned closer to cutting sites along railroad main lines and at some distance from the rivers that had transported logs and provided waterpower sites. Railroads and steam-powered sawmills also increased the pace of logging operations; logging became a year-round activity, and whereas white pine trees were once cut selectively in the forest, large areas came to be "clearcut" of all trees, leaving behind only stumps.[15]

Lumbering was in its heyday in northern Wisconsin in the 1870s and in northern Minnesota in the 1880s and1890s. By the end of the century, white pine trees were essentially exhausted, and this lumbering frontier had run its course. In places, the uncontrolled depletion of white pine left behind only a wasteland of stumps. Short growing seasons and poor soils precluded these cutover lands from becoming agricultural, and cold temperatures would make it years before a forest cover of any kind could regenerate. The mismanagement of white pine, more than any other resource example, led quite directly to the conservation movement. In 1905, under President Theodore Roosevelt, the US Forest Service was created, and Gifford Pinchot, dean of the School of Forestry at Yale University, was appointed its director. Pinchot was a pragmatist who would strive to manage, not exhaust, forests as a renewable natural resource. And as lumbermen turned to the new commercial lumbering frontiers of yellow pine in the American South and Douglas fir in the Pacific Northwest, they would be forced to use better management practices.[16]

Notes

[1] For lead mining, see Nesbit and Thompson, *Wisconsin*, 106–17; also see Brown, *Historical Geography of the United States*, 311–16.

[2] Krause, *The Making of a Mining District*, 51, for the Keweenaw's uniqueness as a world copper deposit; 20, for the likelihood that it was the copper source in a widespread Indian trade. For the significance of years 1844 and 1887, see Brown, *Historical Geography of the United States*, 303–5.

[3] Brown, in *Historical Geography of the United States*, 305–8, draws on the research of geographer J. Russell Whitaker in discussing the Marquette Range.

[4] Ibid., 319, 321, 328, 330, 337, where Brown underscores how, in Wisconsin and Minnesota, colonists sought a combination of prairie and woodland; 321 and 323, for a careful study by Joseph Schafer, which Brown cites at length, of land selection in four counties in Wisconsin's southeastern corner. Recall that southerners also selected primarily land with a prairie-woodland cover, in Texas (see Chapter 12).

[5] Geographer Robert Ostergren's *A Community Transplanted* is a landmark case study of the transatlantic experience of Swedes migrating between 1835 and 1915 from the villages of Rättvik parish in Dalarna province to farms in Isanti County, located between the St. Croix and Mississippi rivers in Minnesota (map 13.2).

[6] For the "great triangle," see Hansen, *The Immigrant in American History*, 73. Geographer John Rice notes that railroad companies received more federal land in Minnesota (26 percent of the state) than in any other state. See his "The Effect of Land Alienation on Settlement," 61 and 62.

[7] For this chapter section, I draw on the carefully researched *Laura*, by Donald Zochert.

[8] Zochert, *Laura*, 2, for Pa's roots; 7, for Ma's roots; 23, for the move to Missouri; 26, for the move to Kansas; 69, for movement to Plum Creek and Walnut Grove; 86 and 97, for Freddie; 95, for Uncle Peter; 99, for Burr Oak; 113, for Walnut Grove again; 131, for South Dakota.

[9] Ibid., 23, for the Missouri purchase; 35 and 41, for Kansas Osage Indians; 84 and 90, for grasshoppers; 99, for the Burr Oak house; 120, for "haunted"; 119, 124, 129, for Walnut Grove employment; 130, for Aunt Docia.

[10] Ibid., 129–30, for Mary blinded; 83, for grasshoppers; 128, for blizzards; 117 and 150, for the tax bill; 131, for "last move"; 71–79, 117, 146, for Congregational Church; 140, for Pa's fiddle.

[11] Ibid., 31–33, 43, 50, for how the Ingalls's Kansas parcel was found.

[12] See Conzen, "Spatial Data from Nineteenth Century Manuscript Censuses," 337–43.

[13] Conzen maps "Farm Value per Acre, 1860," in "Spatial Data from Nineteenth Century Manuscript Censuses," 342.

[14] Williams, *Americans and Their Forests*, especially Chapter 7 on the "Lake States."

[15] Ibid., 201–14.

[16] Brown gives the heyday decades for Wisconsin and Minnesota in *Historical Geography of the United States*, 308.

The Humid East 14

At this point we pause to elaborate and compare. Under "Indian Territory and Oklahoma," we elaborate on the continuing and ever-tragic story of how Native Americans were dispossessed of their lands. We then compare "Square Farms and Long Lots" to reinforce the point that how people organize their space—in this case, their farms—really *does* matter. We then review the ideas of geographer Carville Earle, who compared basic differences between the "Corn, Cotton, and Wheat Belts." Finally, as a transition between the humid East and the dry West, we celebrate the completion of the transcontinental railroad in 1869—and the need it created to keep track of time more systematically—under "The Pacific Railroad and Time Zones."

Indian Territory and Oklahoma

The Louisiana Purchase in 1803 prompted Thomas Jefferson and others to hope that somewhere west of the Mississippi River, a permanent western sanctuary could be created for Native Americans. In the four decades after 1803, mainstream political thinking held that Indians should be removed from east of the Mississippi to the Louisiana Purchase. But when Congress created the Arkansas Territory in 1819 and the state of Missouri in 1821, the choices of where to relocate Indians narrowed (see map 14.1). In the 1830s authorities finally formalized an Indian Territory that lay west of the states of Missouri and Arkansas (Arkansas became a state in 1836). The area designated as Indian Territory stretched from the Platte River south to the Red River, which marked the border with Mexico. Tribes indigenous to this area, and "immigrant" tribes brought in from elsewhere, were relegated to reservations, some of which were laid out as latitudinal strips that extended west into buffalo country. Between the upper Neosho and Platte rivers, however, Indian titles to recently awarded reservations were soon compromised because of an assault on these lands by white squatters and railroad and townsite speculators. In 1854, when Congress created the Kansas and Nebraska territories north of latitude 37° north, what was left for an Indian Territory was only the southern half of the lands in question—modern Oklahoma without its Panhandle.[1]

Into the area south of latitude 37° north in the 1830s went the so-called Five Civilized Nations: the Cherokees, Creeks, Choctaws, Chickasaws, and Seminoles. Passed during Andrew Jackson's presidency, a law called the Indian Removal Act (1830) forced these Indians to relinquish their lands in several southeastern states (map 14.1). All five nations were removed to Indian Territory, but the drama of what happened to the Cherokees received the greatest publicity. The Cherokees had acculturated to the point that they lived in log cabins, wore the clothing of white people, grew cotton, and owned many slaves. Pointing to treaties that gave them firm title to their lands, they fought displacement, but in the end they lost all appeals clear up to the Supreme Court. A ruthless and tragic dispossession followed. In a forced migration carried out by US Army troops in 1838 and 1839, the main body of the Cherokees followed a "Trail of Tears" to Fort Gibson and their new Ozark reservation, losing en route by some reports one-quarter of their people to starvation and other hardships. (A Trail of Tears pageant reenacted annually near Tahlequah, the Cherokee capital in Oklahoma, keeps memories of the event alive.) Similar tragic stories of removal could be told by the Creeks, Choctaws, Chickasaws, and Seminoles,

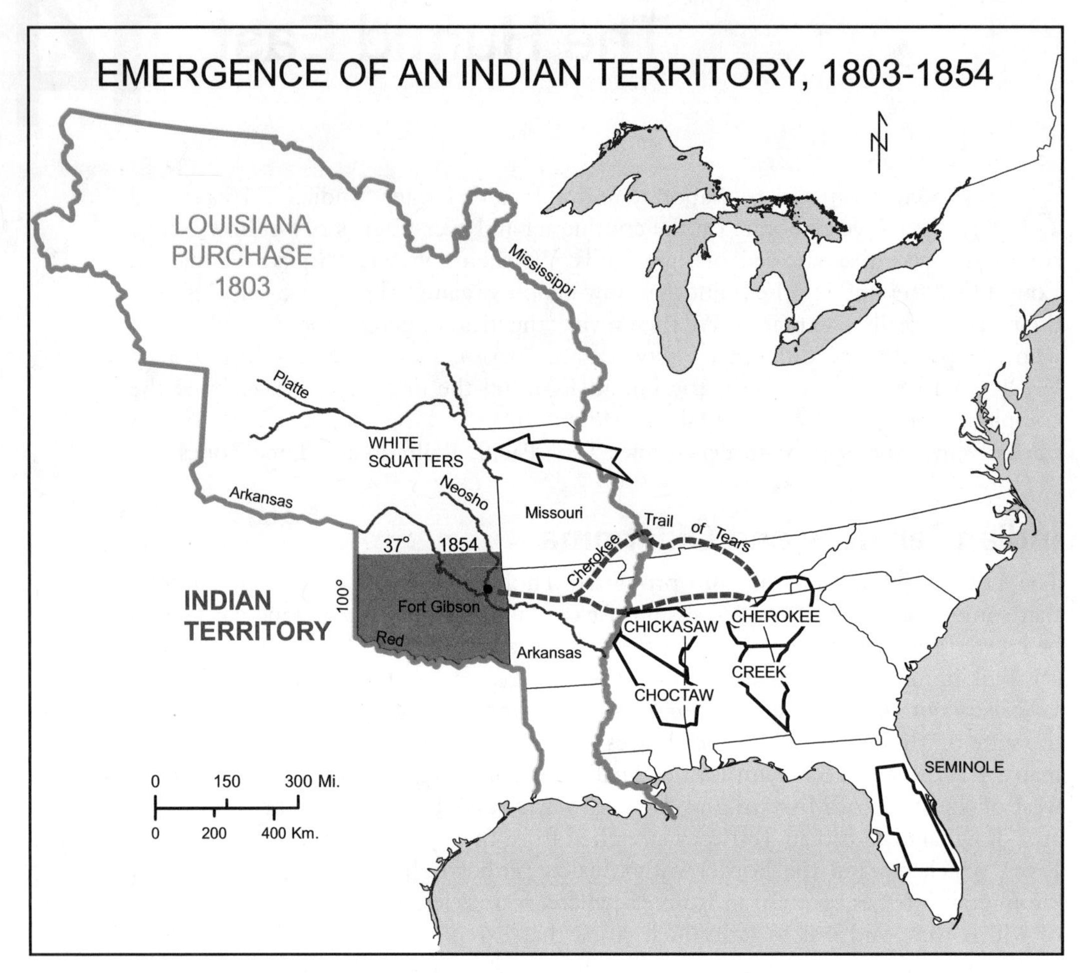

Map 14.1 Emergence of an Indian Territory, 1803–1854.

Source: Morris, Goins, and McReynolds, *Historical Atlas of Oklahoma*, 20, for the Five Civilized Nations and the Cherokee Trail of Tears.

each of whom had a Trail of Tears along which many lives were lost. By 1845 seventy thousand Indians from all five nations were estimated to live in Indian Territory.[2]

The Five Civilized Nations came to occupy only the eastern half of Indian Territory (see map 14.2). During the Civil War the majority of these five peoples sided with the Confederate states, in part because of their ownership of slaves. When the war ended, the victorious Union states, as punishment for Confederate allegiances, forced the five nations to give up the lands they had not occupied in western Indian Territory. Authorities then awarded some thirty "uncivilized tribes," including the Cheyenne, Arapaho, Comanche, Kiowa, and Apache, reservations in this western half of Indian Territory. Between both groups in the middle of the territory were some two million acres that remained "unassigned." Like the rest of Indian Territory, these unassigned lands had been (or soon would be) surveyed with the township and range grid, and railroads would be built across them. In the unassigned lands, the Santa Fe Railroad laid its track in 1887, and it maintained stops with water needed for steam engines at Guthrie, Oklahoma Station (Oklahoma City), and Norman. What happened next in this unassigned enclave within central Indian Territory would begin the process of the unraveling of all of Indian Territory—a process that would shatter any hopes held by many that a largely Indian commonwealth or state would come to exist.[3]

In the 1880s many white people were bent on opening Indian lands to non-Indian settlers. Known as Boomers, these individuals successfully lobbied in

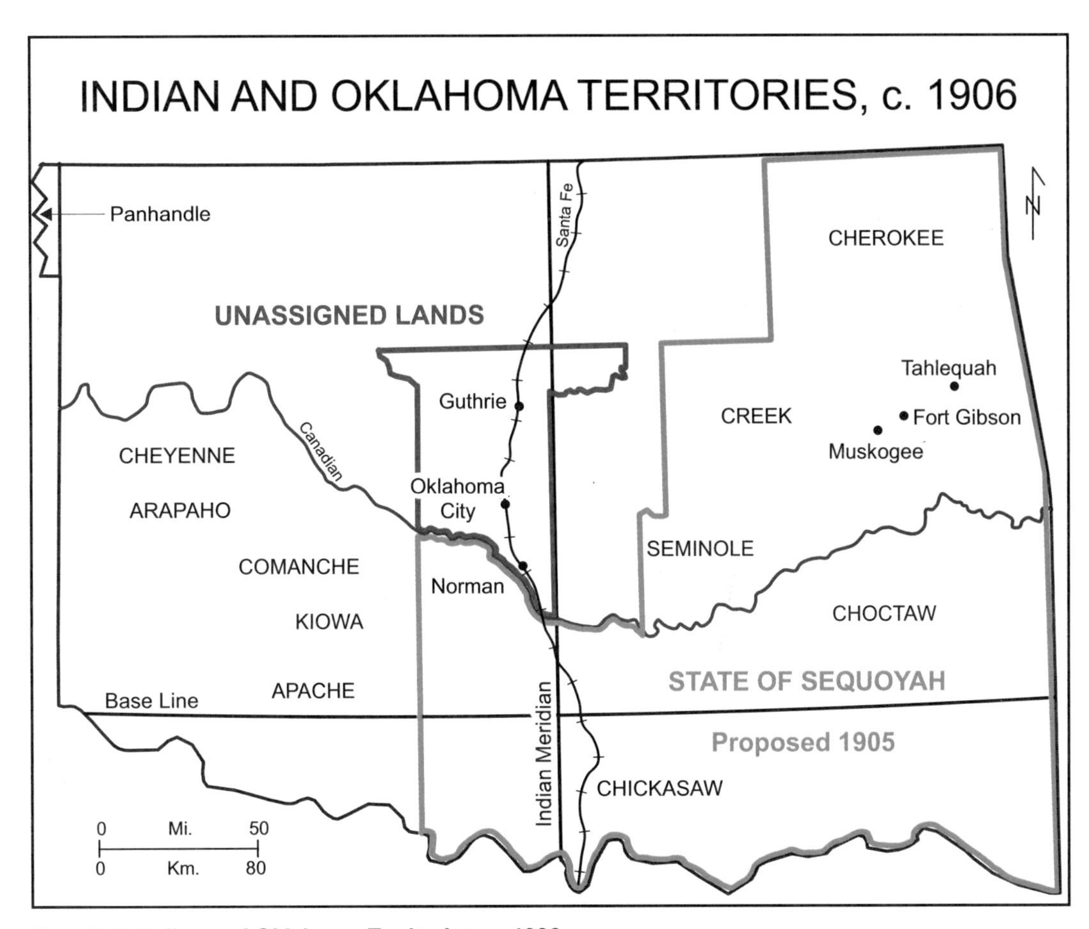

Map 14.2 Indian and Oklahoma Territories, c. 1906.

Source: Morris, Goins, and McReynolds, *Historical Atlas of Oklahoma*, 33 and 55, for the unassigned lands and the proposed state of Sequoyah.

Congress for lands to be opened, and with an eye to increased passenger and freight revenue, railroad companies weighed in with their support. The annual Indian appropriations bill that President Grover Cleveland signed on March 2, 1889 (his term of office would end on March 15), had a rider that gave permission to a future president to open the unassigned lands to non-Indians. Benjamin Harrison quickly became that future president. At noon on April 22, 1889—the date Harrison chose for a land run to open the unassigned lands—guns were fired, bugles were blown, and some ten thousand 160-acre homesteads were thrown open as essentially free land to claimants. By sundown, an estimated fifty thousand people from at least thirty-two states and territories had rushed on foot, on horseback, in horse-drawn wagons, and by train—in all, nine trains from the north and three from the south, and all were required to travel no faster than those making the run on horseback or in wagons—to identify and secure the center stake in the pre-surveyed 160-acre tracts (see figure 14.1). Those "bad guys" who jumped the guns were known as Sooners. When geographer Leslie Hewes was quite advanced in age, he fascinated his geography colleagues in a University of Oklahoma colloquium when he told the story of how his father had made the land run on "Eighty-Niner Day." Out of a feeling of generosity, the father had paid a Sooner one hundred dollars to relinquish his 160-acre claim in the Mount Hope locality near Guthrie, the homestead where Hewes was born in 1906 and where he spent his early childhood.[4]

Figure 14.1 Black Homesteader Family near Guthrie, Oklahoma, 1889. Nearly one thousand black people were among the fifty thousand who made the land run in 1889. They were especially numerous in the Langston area northeast of Guthrie. This remarkable photograph shows a black family and the family's dugout dwelling in the Unassigned Lands.

Photograph apparently by Harmon Swearingen (Swearingen Collection 107), courtesy of the Western History Collections, University of Oklahoma.

When the dust settled late on April 22, 1889, Guthrie contained perhaps ten thousand of the estimated fifty thousand who made the run. Guthrie, everyone knew, was projected to be the new capital of an envisioned Oklahoma Territory. In one day a tent city comprised of claimants to town lots in this future capital came to cover two square miles (see figure 14.2). A law then in effect mandated that no newly founded town in the public domain could exceed 320 acres, and to comply, four Guthries immediately came to exist, each with a mayor and its own laws (see map 14.3). These four Guthries lasted until August 1, 1890. Meanwhile, on May 2, 1890, Congress passed the Organic Act that created Oklahoma Territory out of the unassigned lands (to which the Panhandle, then called No Man's Land, was added), and Guthrie, as anticipated, was made the territorial capital. When Guthrie's streets were surveyed and straightened in December 1891, Oklahoma Avenue was laid out to lead east up a gradual incline to a ten-acre Capitol Park, land with a vista designated to contain the future state capitol building. The streets on either side of Oklahoma Avenue were named Cleveland and Harrison to honor the two presidents responsible for Oklahoma Territory. Because trees were in short supply in this prairie environment, many of Guthrie's early buildings were made using brick and sandstone; Belgian-born architect Joseph Pierre Foucart would rejoice if he could see how his building designs continue to make downtown Guthrie distinctive today (see figure 14.3).[5]

In the 1890s, from their rented offices in buildings around Guthrie, the several appointed governors of Oklahoma Territory oversaw the territory's areal growth parcel by parcel. The Dawes Act, passed in 1887, had set a new direction for federal Indian policy. The objective was to make Indians similar to white people by allotting in severalty (as private property) to each Indian head of household a homestead of 160 acres (80 acres for single Indians) from Indian reservation lands. The thousands of acres of surplus lands then were to be opened to non-Indian homesteaders, and in western Oklahoma, this was to be accomplished by land runs and

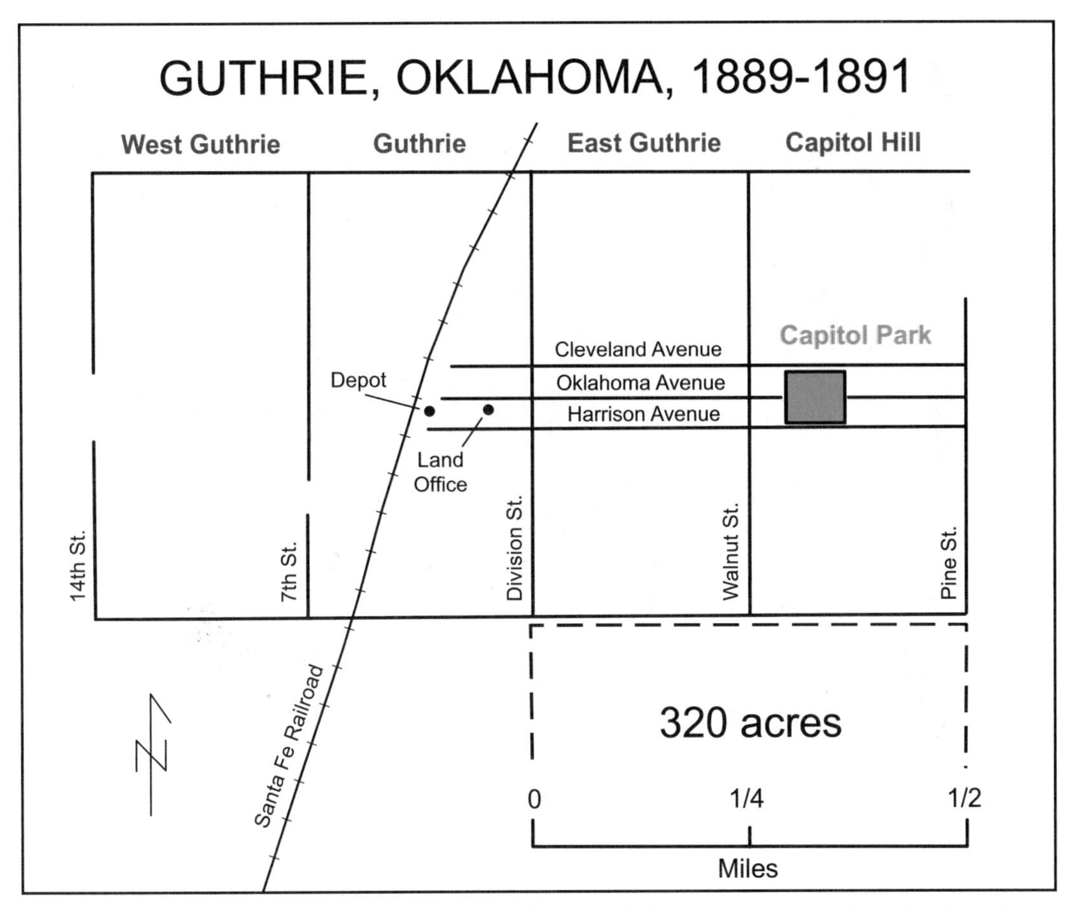

Map 14.3 Guthrie, Oklahoma, 1889–1891. Each of the four Guthries was seven-blocks wide. Homesteads of 160 acres were registered for at the Land Office, a wood-framed building that no longer exists. Modern Guthrie is not much larger than the two square miles shown.

Figure 14.2 Prospective Settlers Arriving in Guthrie, Oklahoma, April 22, 1889. Their arrival was apparently by train.

Harmon Swearingen is credited with taking this image on the day of the land run. Swearingen Collection 147, courtesy of the Western History Collections, University of Oklahoma.

Figure 14.3 The Victor Building in Guthrie, Oklahoma, 2013. Joseph Foucart's penchant for beveled corner entrances, Romanesque arches, and façade ornamentation are displayed in this brick structure, built in 1893, on the northwest corner of Harrison Avenue and First Street. Born of French ancestry in Birel, Belgium, in 1848, Foucart died in Muskogee, Oklahoma, in 1917.

Photograph by the author, March 13, 2013.

lotteries. The Indians themselves were opposed to this shift from tribal to individual land ownership. Nonetheless, in all thirty of western Oklahoma's reservations, lands were allotted, first to the Indian residents, and then through land runs and lotteries to non-Indians; when the transition to private ownership was completed, a given reservation would shift in political status from Indian Territory to Oklahoma Territory. Land runs much like the 1889 land opening in the unassigned lands took place between 1891 and 1895, and by 1901, allotments in western Oklahoma were finished. Each year, as the reservations were being broken up, officials in Guthrie urged Congress to allow those in Oklahoma Territory to start the process of becoming a state, but Congress, eager to make only one state of the "Twin Territories," Indian and Oklahoma, took no action.[6]

In the still intact Indian Territory in eastern Oklahoma, members of the Five Civilized Nations were at first exempt from the allotment policy of the Dawes Act of 1887. But in 1893 Congress required that they, too, accept allotments, which, because of Indian opposition, found the Dawes Commission itself making assignments. Authorities completed the allotment process in eastern Oklahoma in 1902. Unlike what happened in western Oklahoma, however, in eastern Oklahoma there were no surplus lands to be opened to non-Indian homesteaders: Oklahoma historian Arrell M. Gibson says that every acre of land in eastern Oklahoma was allotted to tribal citizens, which included freed blacks. And if forcing allotments on the Five Civilized Nations was not enough, the Curtis Act, passed in 1898, erased the governments of these five peoples. As these events were taking place, both Indians and whites who lived in Indian Territory expressed their strong desires to form a new state. In the summer of 1905, in a convention held in Muskogee, they drafted a constitution for the state of Sequoyah, and that fall they voted six to one in favor of its creation. Fort Gibson would be the capital (map 14.2). Knowing that a state of Sequoyah would vote strongly for Democrats, and wanting to make only one state out of the Twin Territories, the Republican Congress in Washington, DC, took no action.[7]

Out of these quite complicated circumstances, there emerged just one state of Oklahoma. On June 16, 1906, Congress passed the Oklahoma Enabling Act, which allowed the people of the Twin Territories to join in writing a constitution. From November 1906 to April 1907, and then during a week in July 1907, the elected delegates met in Guthrie, where they drafted a state constitution based heavily on the example of Nebraska. On September 17, 1907, people in the Twin Territories held an election wherein they approved the constitution and elected Charles N. Haskell, a Democrat from Muskogee, governor. On November 16, 1907, Congress in Washington, DC, approved both the constitution and the election of Haskell and other state officials. Theodore Roosevelt's signature made Oklahoma the forty-sixth state on that November day. The fact that more Native Americans today live in Oklahoma than in any other state should not be surprising.[8]

With statehood, Guthrie became the new capital. In 1909 a brown brick Convention Hall was finished in the ten-acre Capitol Park, and turning on the new electric lights for the first time thrilled the public and the legislators who met there. But state government in Guthrie would be short-lived. Guthrie's residents, including the vitriolic editor of the influential *Daily State Capital*, were heavily Republican. Moreover, Guthrie's population was perhaps one-third black, due importantly to the efforts of one Edward P. McCabe, a black territorial deputy auditor who successfully recruited black southerners to Guthrie and had a major role in founding the all-black town of Langston, located near Guthrie. Haskell and most other state officials, on the other hand, were Democrats and were prejudiced against blacks. Determined to move the capital out of Guthrie, Haskell and the Democrats submitted for a statewide vote a proposal to relocate the capital. On June 11, 1910, Oklahomans heavily favored Oklahoma City over Guthrie. Despite the outrage in Guthrie, the removal was upheld in the courts. Guthrie's Convention Hall then became the home of Oklahoma Methodist University until it too moved to Oklahoma City in 1918, where it became Oklahoma City University. Happily, in 1919 Guthrie sold Convention Hall and the 10-acre Capitol Park to the Scottish Rite Masons, and the temple that this Masonic order built—in grand Greek Revival style using limestone brought by rail from Bedford, Indiana—looks much like what Oklahomans would have approved of as a capitol building (see figure 14.4). Nonetheless, Guthrie stagnated—some would argue to its good fortune—because a relative lack of urban renewal that surely would have occurred had the community grown, plus the fact that Guthrie's many early buildings were built of brick and sandstone, have left the city of about thirteen thousand (in 2010) very well preserved. Billing itself as the "Williamsburg of the West," Guthrie is well positioned to sell its history to tourists.[9]

Square Farms versus Long Lots

In Oklahoma, those future farms that attracted many eager settlers were 160-acre *squares*. Squares were the legacy of the township and range system that came to dominate America's rural landscape. By contrast, the farms of French settlers in Canada and in parts of the United States were *rectangles*. Known as long lots, the ribbonlike farms of the French varied greatly in size: As seen in Chapter 3, long lots in Canada were typically one hundred acres or less, and in Louisiana and Illinois Country long lots could contain several hundred acres. Geographer Carleton P. Barnes, a specialist

Figure 14.4 The Scottish Rite Temple, Guthrie, Oklahoma, 2013. Limestone quarried in Bedford, Indiana, was used in constructing this Masonic temple between 1921 and 1929. The site was to have held Oklahoma's state capitol building.

Photograph by the author, March 13, 2013.

in agricultural land use at the US Department of Agriculture between 1929 and 1962, wrote a short but significant article in 1935 that compared these two cadastral, or landownership, patterns. And what Barnes found was that French long lots had major economic advantages over American square farms.[10]

The essence of what Barnes argued is captured in figure 14.5. Frame A shows two square miles of 160-acre square farms and their eight farmsteads—typical in the American system. Frame B shows two square miles of 106.6-acre long-lot farms and their twelve farmsteads found hypothetically in back tiers of a French-settled area where houses line both sides of a road (see figure 3.3). The obvious differences between frames A and B are the sizes (160 vs. 106.6 acres) and shapes (squares vs. rectangles) of the farms. Perhaps less obvious is the important fact that *two* linear miles of road are required to service the eight square farms in frame A but only *one* linear mile of road is needed to service the twelve long-lot farms in frame B; if the road servicing the American squares was drawn vertically down the middle, it would still stretch for two linear miles. Thus for the American square farms it takes twice as much road (and road maintenance after construction) than for the French rectangular farms. Also duplicated, as Barnes points out, are the costs of constructing and maintaining electric power lines and drop poles. Barnes also lists the costs for services such as mail delivery, school bus transportation, and snow plowing, all of which are greater where there are square farms because there is twice as much road to cover. A final point made by Barnes is that with mechanization, there is greater efficiency in plowing or harvesting a mile-long field, to which can be added that fewer turnarounds at the ends of fields mean less wasted land at fence rows.

For those who might assume that 160 acres is the ideal size for a farm (a questionable assumption at best), frame C in figure 14.5 shows 160-acre long lots at four to a square mile. Here, the advantage of having only one linear mile of road to service the new 160-acre farms in a two-square-mile area is no different. But once

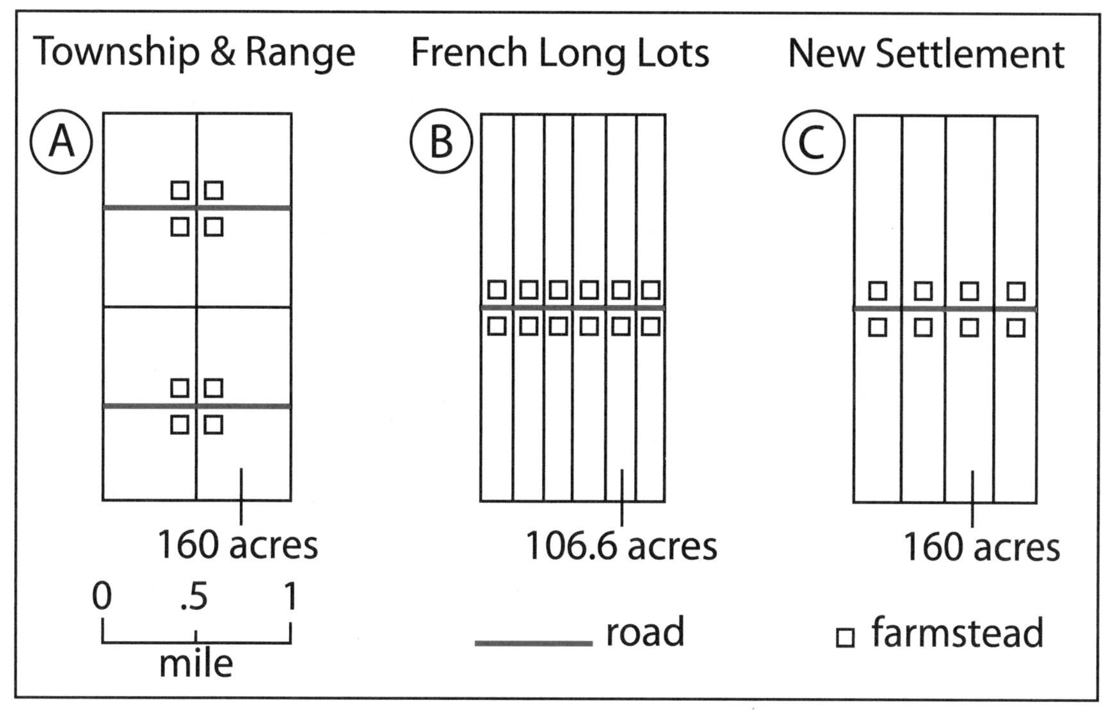

Figure 14.5 Square Farms versus Long Lots.

Source: Barnes, "Economies of the Long-Lot Farm," 299.

the township and range grid has already been implemented, knowing this advantage is too late to be of use: To resurvey and start over would be extremely difficult, if not impossible. America's square farms are here to stay. So the implication for what Barnes points out is that if new settlement were being undertaken somewhere, farms laid out as long-lot rectangles would have a clear cost advantage over farms laid out as squares. Barnes gives only two advantages that square farms have over rectangles: Square farms require less fencing, and on a square farm it takes a farmer less time to reach the opposite side of his or her property. And the big point to remember is that geography matters.[11]

Corn, Cotton, and Wheat Belts

In the humid eastern half of the United States, we observed where and when three agricultural "belts" developed. In Chapter 10, we saw how, in five areas north and south of the Ohio River in the early 1800s, farmers grew corn to be fed to hogs and cattle; marketing animals—and not the corn itself—was an innovation that spread west to Nebraska and north to Minnesota to create the Corn Belt. In Chapter 11, we saw how strains of cotton evolved soon after 1800 in two hearth areas and from them diffused to form a Cotton Belt. And in Chapter 13, we saw how, beginning in the 1830s, a wheat frontier moved northwest from southeastern Wisconsin to northwestern Minnesota. This was the incipient Spring Wheat Belt; spring wheat is sown in the spring, grows over the summer, and is harvested in the fall. A Winter Wheat Belt centered on western Kansas would also develop; winter wheat is planted in the fall, is in the ground over winter, and is harvested in the spring (see map 14.4). Earle (see spotlight) had a keen interest in these three *staple* (basic) crops of corn, cotton, and wheat, along with the staple crops of tobacco and wet rice. And Earle used

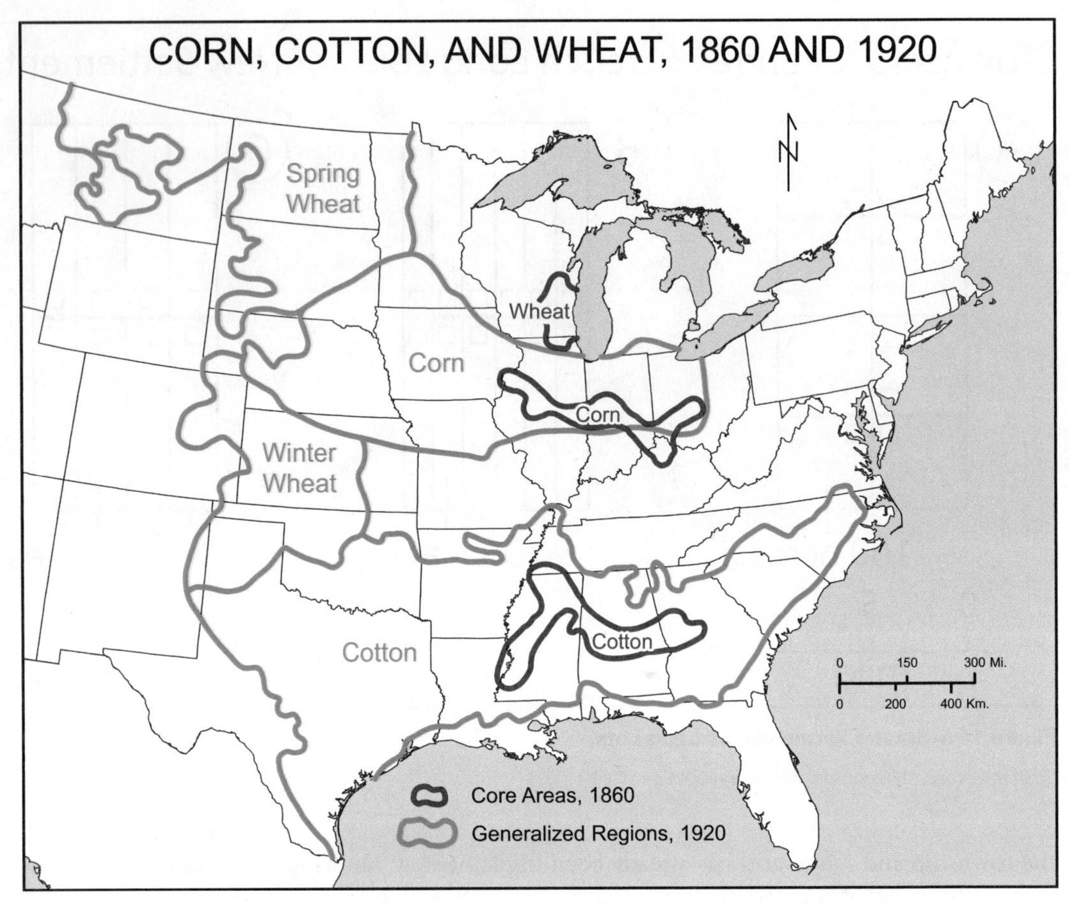

Map 14.4 Corn, Cotton, and Wheat, 1860 and 1920.

Sources: Earle, "Beyond the Appalachians, 1815–1860," 177, for core areas in 1860; Meyer, "The National Integration of Regional Economics, 1860–1920," 314, for generalized regions in 1920.

"staple theory" to analyze how it was that different pre-1860 societies developed in the three agricultural belts.[12]

Staple theory is an economic interpretation of how labor systems that develop around staple crops shape different societies. Of central importance is the fact that staple crops have quite different labor demands. Cotton requires a long (eight months) growing season, during which time labor demands are high for planting, for the many necessary cultivations after planting, and for harvesting—all by hand in antebellum times. Before 1860 relatively cheap slave labor provided the high number of labor days required by southern planters if they were to be successful economically. Corn required some four months of high labor input between April and early July; corn harvests could extend into the winter months beyond the first hard freeze, so labor for harvesting could be spread out. Corn, then, compared to cotton (and wheat), had a medium number of labor days required. Slave labor before the Civil War, noted Earle, would have been ideal in the Corn Belt, yet wage labor hired to supplement family labor in the four peak months worked well enough economically. Wheat required heavy labor inputs only for planting in spring and for some ten to fourteen days at harvest. For wheat, then, wage labor worked well so long as it was available in that "bottleneck" time at harvest.[13]

Earle uses these differences in staple-crop labor days to interpret society in the three agricultural belts, especially the differences in degree of urbanization in the

Spotlight

Carville V. Earle (1942–2003)

Carville V. Earle. Courtesy of Elizabeth Earle, Baton Rouge, Louisiana, May 2013.

Carville Earle questioned conventional historical interpretations, and he sought answers to big questions in American history. Self-admittedly more a geographical historian than a historical geographer, Earle was an "armchair" scholar, for unlike Terry Jordan-Bychkov and many others, it was not his practice to go to the field for information. Rather, Earle drew on an array of primary and secondary sources—economic, social, political, demographic—to answer some big questions about what shaped America's human affairs and its geography. His culminating academic achievement, *The American Way*, published in 2003, the year he died, was a plea for use of "macrohistorical cycles"—specifically, eight fifty-year periods or long waves of political economy between roughly 1600 and 2000—to explain the macrohistory of the United States. Earle earned his degrees, all in geography, at Towson State College (Towson, Maryland, BS, 1966), the University of Missouri (MA, 1967), and the University of Chicago (PhD, 1973). He taught at the University of Maryland–Baltimore County and Miami University in Oxford, Ohio, before finishing his career at Louisiana State University (LSU), where he held the title of Carl O. Sauer Professor of Geography. At LSU, Earle was editor between 1994 and 1996 of geography's flagship journal, the *Annals of the Association of American Geography*. He will long be known for the quite brilliant, sweeping theoretical challenges he proposed.

three areas. For cotton to be grown economically in the South, as noted, planters found it necessary to own slaves. Slavery created a society with marked disparities between rich and poor, and because the large class of the poor—the slaves—were not consumers, the need for urban places to provide goods and services was quite restricted. The South was notably rural with gradual urbanization coming only late in the colonial era. On the other hand, corn and wheat, both high-bulk and high-weight grains, generated a demand for transportation services and storage facilities that translated into a need for urban centers. Moreover, low labor demands, especially where wheat was grown, argued Earle, allowed wage earners in agriculture plenty of opportunity to also work in industry, and the availability and income of these workers fostered the growth of urban centers. Thus, urban centers of varying sizes did develop in the corn and wheat belts.[14]

The Pacific Railroad and Time Zones

In 1869 Americans celebrated when the rush to complete the transcontinental railroad ended. A race between two companies, the Union Pacific (UP), which was building west from Omaha, and the Central Pacific (CP), which was building east from Sacramento, received great public attention. Prompting the race to rapidly lay track were government incentives in the forms of land grants and per-mile subsidies. On May 10, 1869, workers hammered in the last spike (a golden one) at Promontory, Utah Territory, to connect the two lines (see map 14.5). Known as the Pacific Railroad, 1,800 miles of single track now connected Omaha and Sacramento; the UP and the CP made the Mormon community of Ogden, located near Promontory, their division point. D. W. Meinig underscores the point that the Pacific Railroad, in reality, represented only two of four operational segments of a full transcontinental linkage: East of Omaha, rail networks already in place linked Omaha with Chicago for a third segment, and an even denser network linked Chicago with Atlantic ports for a final segment (map 14.5). Other transcontinental railroads would follow in the 1870s and 1880s, but the Pacific Railroad, because it was first, symbolized this American triumph of mastery over space.[15]

A well-known photograph taken at the "last spike" ceremony at Promontory captured the engines of the UP and the CP facing each other, while in the foreground the two chief engineers of each rail line clasped hands (see figure 14.6). Until recently Charles R. Savage of Salt Lake City was typically cited as the author of this photograph. Geographer William D. Pattison, while analyzing a large collection of lantern slides in the archives of the American Geographical Society (AGS) in New York City in 1960, found that Andrew Joseph Russell (1831–1896) was the real photographer. From 1940 to 1941, the AGS had acquired the collection of glass plate lantern slides of Stephen J. Sedgwick (1820–1920), a resident of Newtown on Long Island. Sedgwick made his living by giving public lectures. In 1869 Sedgwick accompanied the UP photographic corps, of which Russell was in charge, to Promontory to acquire illustrative materials for future lectures about the Pacific Railroad. His collection of lantern slides, known at the AGS as the Combes Collection, attracted Pattison's critical attention. In that collection was the famous last-spike photograph. Evidence that Russell, not Savage, took this last-spike image is verified by comparing handwritten titles on the negatives and by the fact that this image is united in serial numbering with the other Russell photographs.[16]

Interestingly, travel by railroads prompted the need for standardized times within time zones. Before standardized times, communities in the United States set their times locally when the sun was directly overhead. When it was noon in Chicago,

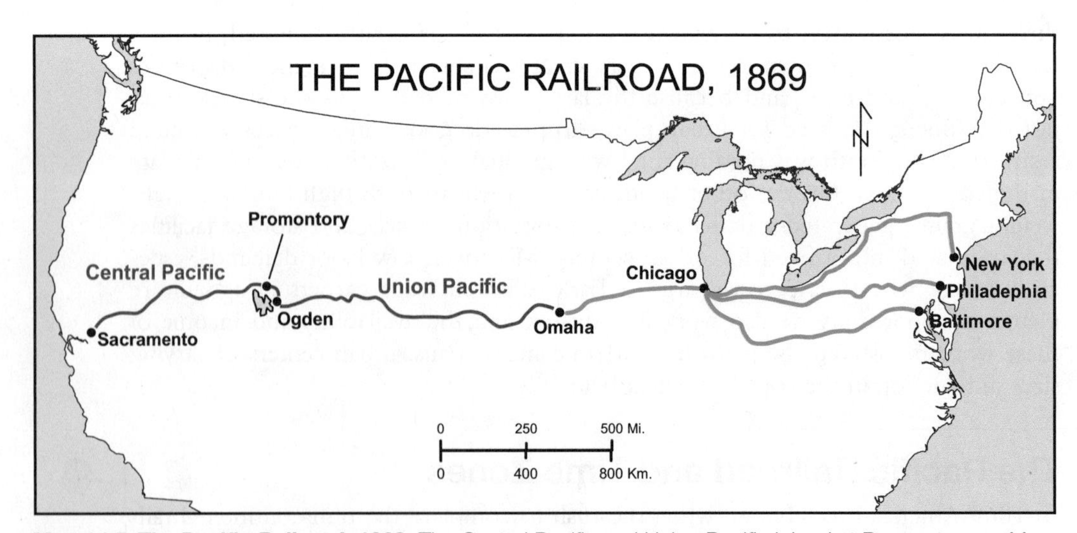

Map 14.5 The Pacific Railroad, 1869. The Central Pacific and Union Pacific joined at Promontory on May 10, 1869.

Source: Pattison, "The Pacific Railroad Rediscovered," 26.

Figure 14.6 Andrew J. Russell's "Last Spike" Photograph, Utah, May 10, 1869. In *Nothing Like It in the World*, between 192 and 193, Stephen Ambrose notes that the Central Pacific engine's smokestack (L) burned wood, was round, and had a screen to catch sparks, while the Union Pacific engine's smokestack (R) burned coal and was straight.

Source: Photograph obtained from the American Geographical Society Library, University of Wisconsin–Milwaukee Libraries.

it might be 11:50 in St. Louis or 12:18 in Detroit. Variations in local times created problems in printed railroad schedules, and travelers found themselves resetting their watches with frequency. Responding to the problem, on November 18, 1883, railroad companies in the United States and Canada standardized time into four zones: Eastern, Central, Mountain, and Pacific. In 1884 a worldwide system of twenty-four time zones was put in place, each fifteen degrees of longitude apart. The British Royal Observatory at Greenwich, England, located not too far up the Thames River from London, marked the beginning point at longitude 0°. In the United States, the boundaries of the four time zones, for reasons of commerce and convenience, have been redrawn many times, yet their creation by railroads in 1883 is testimony to the power that railroad companies then held.[17]

Notes

1 D. W. Meinig reviews with compassion the creation of Indian Territory and the tragic story of the removal of Native Americans to that territory, in "Shoving the Indians Out of the Way," in *The Shaping of America*, 2:78–103. See ibid., 92, for Indian Territory between the Platte and Red rivers; 102, for its "crumbling" north of the upper Neosho.

2 Ibid., 2:98, for seventy thousand in 1845. The dates for Cherokee removal (1838 and 1839) are in Hewes, "The Oklahoma Ozarks as the Land of the Cherokees," 270.

3 For original insights to mixed-blood and full-blood Indian ownership of slaves by members of all five nations in Indian Territory, see geographer Doran, "Negro Slaves of the Five Civilized Tribes," 335–50.

[4] See Hewes, "Making a Pioneer Landscape in the Oklahoma Territory," 588–603, for his family's role in the Mount Hope locality. Hewes spoke at the University of Oklahoma on October 24, 1994. Arrell M. Gibson, *The Oklahoma Story*, 155, notes that the unassigned lands contained about ten thousand potential 160-acre claims, and that some fifty thousand made the land run.

[5] See Morris, Goins, and McReynolds, *Historical Atlas of Oklahoma*, 48, for an estimated ten thousand people at Guthrie; 54, for the Organic Act; and 62, for No Man's Land. See Alley, "City Beginnings in Oklahoma Territory," 39, for townsite limits of 320 acres; 52, for August 1, 1890, as the date provisional governments ended; and 51, for the fact that thirty-two states and territories had representatives in Guthrie on April 23, 1889.

[6] See Gibson, *The Oklahoma Story*, 152, for the Dawes Act; 154, for the completion of allotments in 1901; 163, for the shifting from Indian to Oklahoma territories; and 182, for the urgings of people for statehood.

[7] Gibson, *The Oklahoma Story*, 182–83, including the fact that "every acre of land in the eastern half of Oklahoma was allotted to tribal citizens"; also see Meinig, *The Shaping of America*, 3:173–74, for a discussion of the Dawes Act and the state of Sequoyah.

[8] Gibson, *The Oklahoma Story*, 184–86, for events leading to statehood.

[9] Between 1973 and 1995, I took my University of Oklahoma students to Guthrie twenty-seven times on half-day field trips. Donald Keith Odom, a Guthrie public school history teacher with extensive knowledge about early Guthrie, led my students around his community on each of these trips.

[10] Facts about Barnes are in Marschner, "Carleton P. Barnes, 1903–1962," 233–34; comparisons of square farms and long lots are in Barnes, "Economies of the Long-Lot Farm," 298–301.

[11] Barnes, "Economies of the Long-Lot Farm," 300 and 301, notes twice that his findings have implications for those undertaking new settlement.

[12] For Earle on staple theory, see his *Geographical Inquiry and American Historical Problems*, 8. Earle compares in some detail the pre-1860 corn, cotton, and wheat belts in "Beyond the Appalachians, 1815–1860," esp. 180–87.

[13] Earle, in "A Staple Interpretation of Slavery and Free Labor," 60–62, compares the costs of possible slave labor versus wage labor in the Corn Belt; esp. 61, where he uses the term "bottleneck."

[14] Earle, *Geographical Inquiry and American Historical Problems*, 89, for his position that uneven consumer demand "stunted" urban development; 89–90, 141, for the nature of wheat and corn as crops generating urban development.

[15] Meinig, *The Shaping of America*, 3:24, for the four operational railroad segments.

[16] See Pattison, "The Pacific Railroad Rediscovered," 25–36; and Pattison, "Westward by Rail with Professor Sedgwick," 1–15.

[17] For time zones, see Meinig, *The Shaping of America*, 3:263.

THE DRY WEST III PART

Spanish Americans and New Mexico 15

We turn now to the West and New Mexico. Dryness is what distinguishes the West from the East, and under the heading "Images of the West," we review how perceptions held by Americans of this dry area evolved after 1803. We then look at "Mexican New Mexico," the first major western destination for Anglos after Mexico gained its independence in 1821. What drew people to New Mexico was "The Santa Fe Trade," as explained in the chapter's third section. Finally, we look at "Spanish Americans," the only Spanish colonial subculture to survive in the Borderlands and the people who have made this part of the West a region.

Images of the West

The term *the West* has been a relative one because it shifted west as Americans moved west. To clarify which West was meant, Americans used modifiers such as "Middle" West, "Trans-Mississippi" West, "Rocky Mountain" West, and "Far" West. But, stripped of any modifiers, what is meant when people today refer to the West? A strong case can be made that the West begins with the twenty-inch rainfall line or, for ease of location, the one-hundredth meridian (see map 15.1). That line, fuzzy as it is, given that it is based on averages of annual precipitation amounts, separates the subhumid (twenty to thirty inches) East from the semiarid (ten to twenty inches) West. Significantly, the twenty-inch isohyet correlates with two other important phenomena. It is a natural vegetation line because it marks the change from tallgrass prairie to shortgrass steppe. And it is a population line, for it separates "more than six people per square mile" from "fewer than six people per square mile." As historian Bernard DeVoto wrote, "The West begins where the average annual rainfall drops below twenty inches. When you reach the line that marks that drop—for convenience, the one-hundredth meridian—you have reached the West."[1]

When Thomas Jefferson purchased the Louisiana Territory in 1803, thus doubling the size of the United States, what image did he and others have of Louisiana and the West? According to geographer John L. Allen, in 1803 Jefferson and Americans more generally held two prevailing images about Louisiana and the West. First, they thought that the area west of the Mississippi was a "garden," lush and fertile and inviting for an agricultural people. Second, they believed that an all-water transportation route crossed this verdant land to link the Mississippi with the Pacific Ocean. Americans knew about the wide Missouri River and how it flowed into the Mississippi. Due to the discovery of the Columbia River by American-born Robert Gray in 1792, they also knew that a second major river emptied into the Pacific. The thought was that a single low mountain range along the Pacific coast separated the two rivers. To find what was seemingly an easy portage between the Missouri and the Columbia, and to be able to claim this western land for the United States, Jefferson sent an expedition led by Meriwether Lewis and William Clark.[2]

Jefferson picked Lewis, his private secretary, to lead the exploratory expedition. Lewis, in turn, asked Clark, a fellow Virginian with whom he had served in the Army, to be his coequal leader. The two were remarkably compatible, and they also complemented one another, Lewis being the better planner and natural historian yet possessed of an irritable and brooding personality, and Clark being the better boatman and navigator and of a more genial and even temperament. The two men formed their partnership at the Falls of the Ohio (Clark lived in Clarksville, Indiana), where

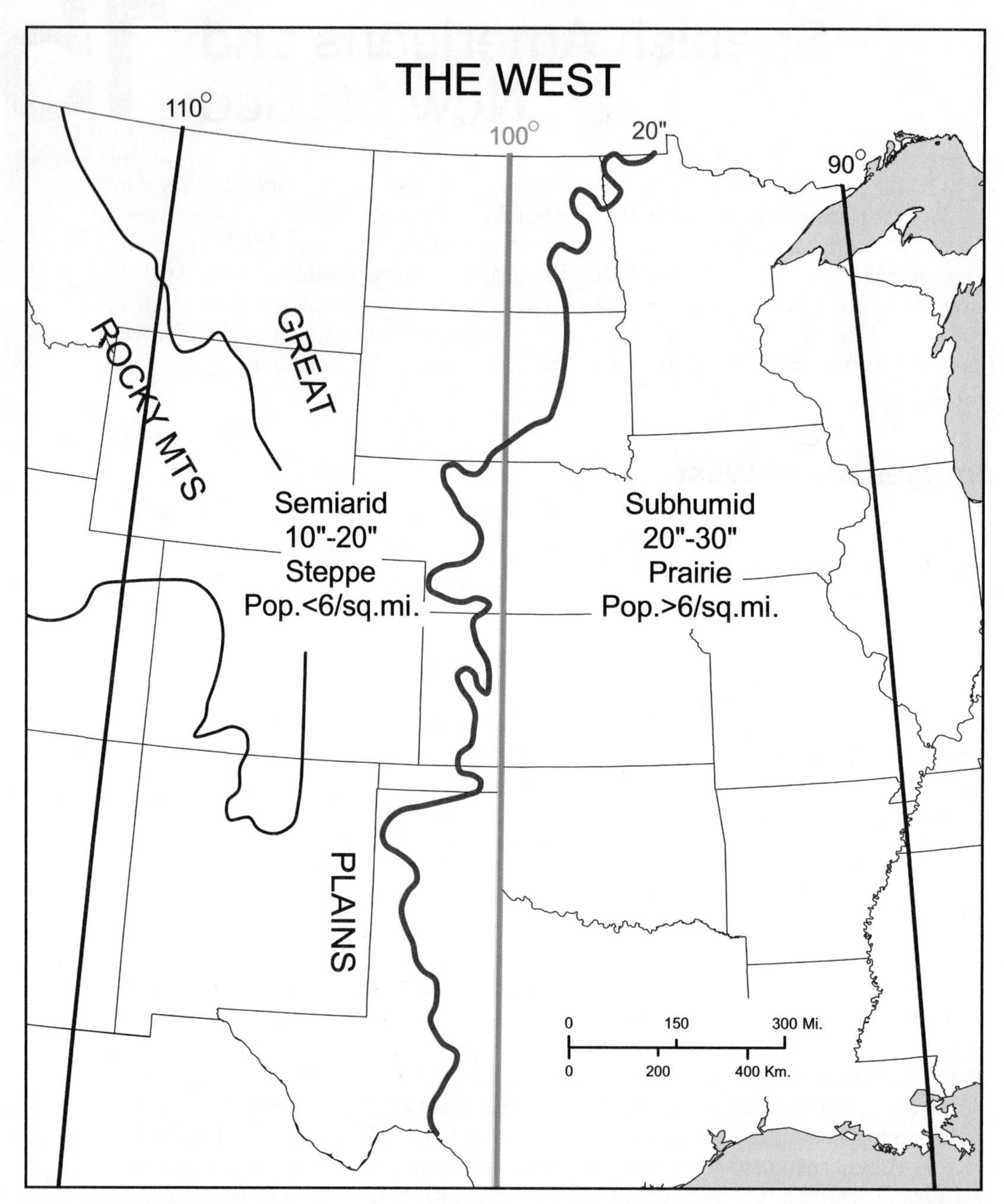

Map 15.1 The West. Many geographers and historians accept the argument that the West begins with the twenty-inch rainfall line.

they recruited many who made up the party of some forty persons (the number fluctuated). At Camp Wood, located at the mouth of Wood River across the Mississippi River from St. Louis, the party headquartered in the winter of 1803 to 1804 as Lewis and Clark made their plans. Their expedition would last for more than twenty-eight months between 1804 and 1806, and it would cover some eight thousand miles (12,900 kilometers).[3]

Departing on May 4, 1804, the expedition followed the Missouri to its second "great bend," where they wintered among the Mandan Indians at a post called Fort Mandan (see map 15.2). The next spring the party followed the Missouri and its

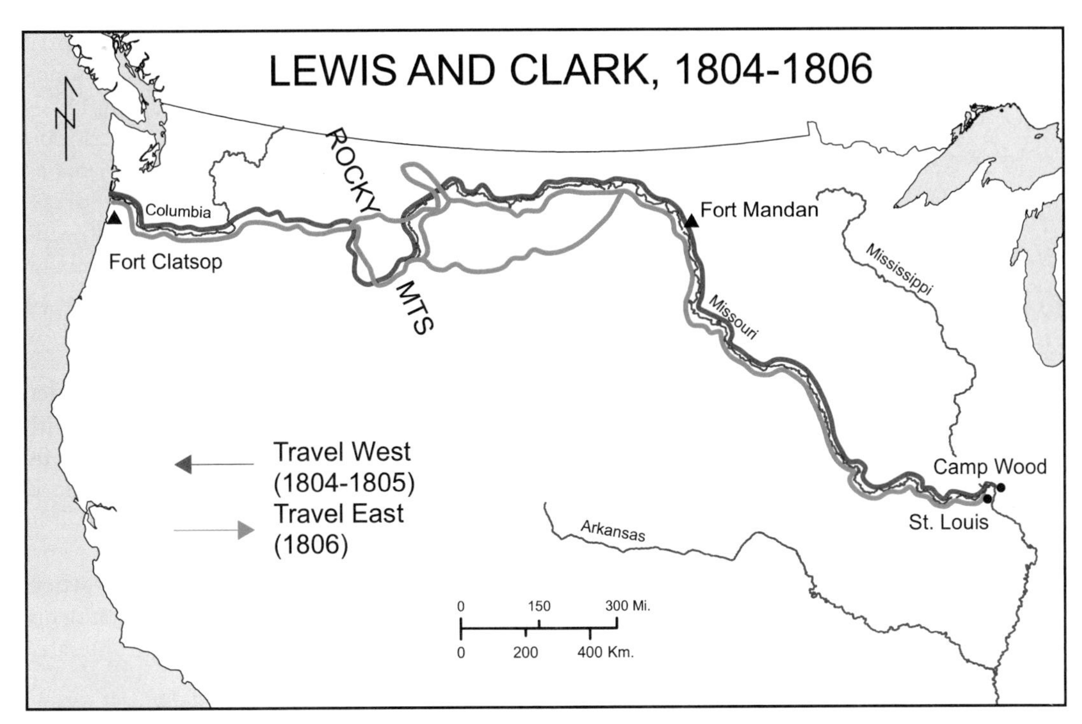

Map 15.2 Lewis and Clark, 1804–1806.

Routes west and east are based on *Time Map* by Ed Gabel, in Collins, "Domesticated Daring."

tributaries into the high and rugged Rocky Mountains, which they crossed with great difficulty in western Montana. West of the Rockies, the Columbia and its tributaries led them to the Pacific, where they wintered at Fort Clatsop, named for local Indians and built near the mouth of the Columbia in Oregon. In the spring and summer of 1806, the party retraced much of its route, returning to St. Louis on September 23, 1806. Amazingly, there was only one casualty—a man who died of appendicitis early in the expedition. Besides mapping the country and documenting many plants and animals (at Monticello, Jefferson proudly displayed the head of a buffalo that was given to him), the United States now had a basis for claiming the lands just explored. But overtures made to some Indians were not successful, and the perceived water route that Jefferson thought would be convenient for reaching the Pacific turned out to be a wide and brutal portage over the high Rocky Mountains.

After about 1810, the image of the West, particularly the Great Plains region lying east of the Rocky Mountains, gradually became one of a desert, at least among many better-read Americans. Zebulon M. Pike planted the seeds for this image. Pike, an Army officer commissioned in 1806 to explore the headwaters of the Arkansas River, had unofficial orders to learn more about the Spanish in New Mexico. In February 1807 he and his small party were easily captured by Spaniards at Pike's camp on the Conejos River (near La Jara, Colorado), briefly imprisoned in Santa Fe, and then taken to Chihuahua before being released in June 1807 at the international border between Los Adaes and Natchitoches (see map 2.3). In his journal, published in 1810, Pike described the land he had crossed in Kansas, Colorado, and New Mexico as treeless and dry, and he compared parts of it to the sandy deserts of Africa. A second military officer, Stephen H. Long, was far more influential than Pike in promoting a desert image. In 1819 and 1820, Long explored the upper Mississippi River and country that led west to the Rocky Mountains. With him was botanist and geographer Edwin James. In 1823 James reported that in the country they had crossed, wood and water were scarce, and that the area would be uninhabitable by a people depending on agriculture. Long's map, published in 1823, had "Great American Desert" written across the country drained by the Mississippi. Geography textbooks in the 1840s and 1850s came to refer to the Great Plains as "Great American Desert" and "American Sahara."[4]

Deserts are arid areas defined by their lack of precipitation; the rule-of-thumb definition is that deserts receive less than an average of ten inches of precipitation annually. The Great Plains area east of the Rockies, however, was not arid, but semiarid (ten to twenty inches). Gradually, the semiarid nature of the Great Plains—characterized by shortgrasses resulting from scant yet adequate precipitation amounts—replaced the desert image, held especially by educated New Englanders. Josiah Gregg, a trader on the Santa Fe Trail, in his *Commerce of the Prairies* (1844), described the land he crossed as available for pasturage if not for agriculture. Railroad surveyors in the 1850s compared the high plains to the steppe lands of the Ukraine, where soils were deep and fertile. The area, their reports suggested, was potentially arable, and was certainly good for year-round grazing. Gradually the area mapped as desert grew smaller and smaller, and by about 1870 it disappeared altogether. Geographer Martyn J. Bowden asked whether the myth of a desert slowed settlement in the Great Plains, and his conclusion was that it did not, given that the desert concept was held by the "literati" especially of New England and not by potential colonists.[5]

Mexican New Mexico

The Mexican era, which lasted from 1821 to 1848, was short but significant. In 1819, just two years before Mexican independence, the Adams-Onís Treaty established the boundary between the United States and Spain. This new border would separate the United States from Mexico until 1845 (the annexation of Texas) and 1848 (the Mexican Cession). Map 15.3 shows that, except in East Texas, the new boundary was drawn well north of Spain's four frontier outpost-clusters: Texas, New Mexico, Pimería Alta (Arizona), and California (also on map 2.3). Whereas Spain's mercantile policy had prohibited interaction between its frontier settlers and *americanos* (the term used for Anglos and all foreigners in the Mexican period), independent Mexico now opened her frontier. A flood of southerners quickly reached Texas, as we saw in Chapter 12. And in the American West, New Mexico, the oldest and by far the most populous outpost-cluster, became an immediate magnet for attracting Anglo traders. What, then, was Mexican New Mexico like?[6]

Located at the southern end of the Rocky Mountains, New Mexico was high and dry. Two exotic rivers, the Rio Grande (known to New Mexicans as the Rio del Norte) and the Pecos, formed at high-elevation snowpacks to flow south from the Rockies. *Exotic* means "foreign"; an exotic river is one that rises in a relatively humid environment before flowing through a dry environment—to which it is foreign. Such rivers, which also include the Colorado, understandably draw people to their banks. By 1821 more than thirty thousand New Mexicans had been drawn to watercourses in the upper basin of the Rio Grande. Since their arrival beginning in 1598, New Mexicans had come to distinguish between a higher and colder Rio Arriba and a lower and drier Rio Abajo; *arribeños* had more moisture while *rivajeños* had a longer growing season (map 15.3). In both New Mexican subregions, *nuevomexicanos* lived in scores of villages located along the Rio Grande, the Pecos, and their many tributaries.[7]

The villages of nuevomexicanos were compact "plazas" during times of warfare with nomadic Indians, but in more peaceful times they became linear affairs strung

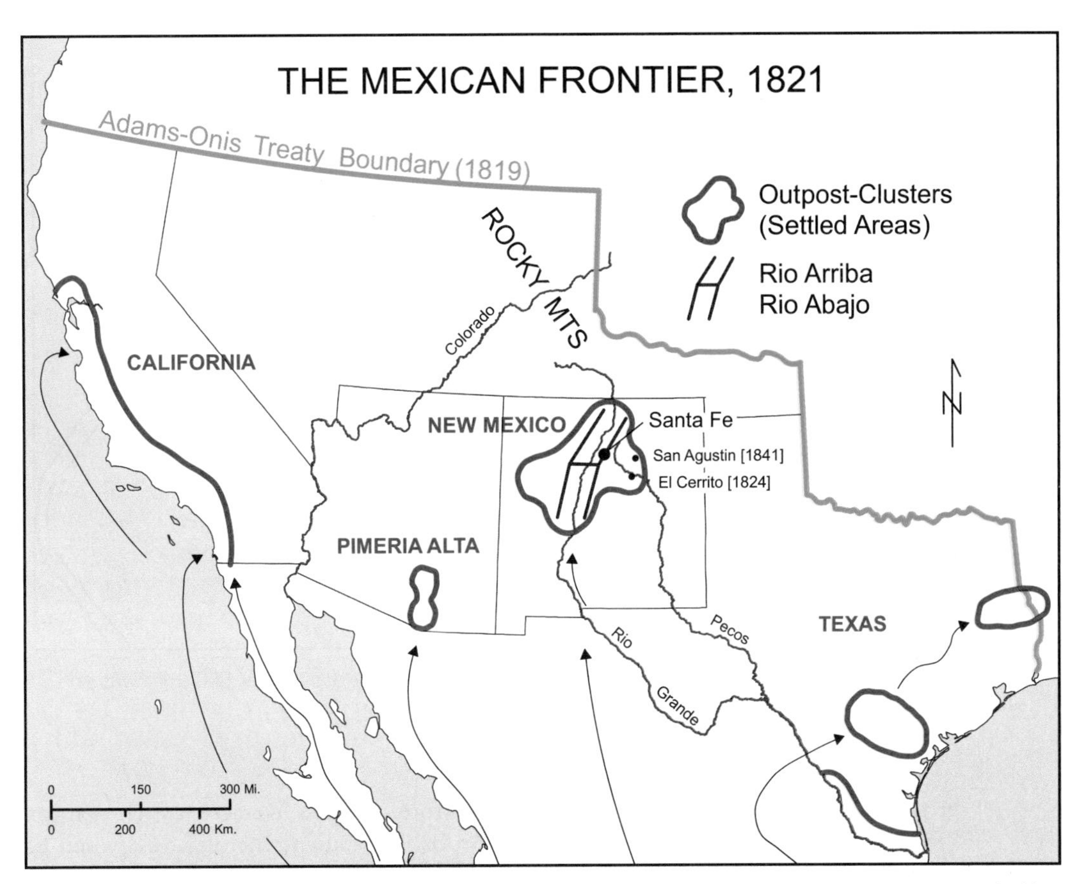

Map 15.3 The Mexican Frontier, 1821. See also map 2.3. The four outpost clusters were connected with Mexico to the south but not with each other. Arrows represent people moving north.

out along irrigation ditches and roads. Along rivers, villagers constructed simple dams made from piled rocks and felled trees, and these dams diverted water into irrigation ditches. As these ditches flowed downstream, they gradually swung away from their rivers, leaving flat land between the ditch and river. Villagers carved the flat land into long lots. As noted in Chapter 2, in a simple yet amazingly efficient system, irrigation water would flow by gravity from the irrigation ditch through furrows across a field to the river, and because long lots fronted on the irrigation ditches, each villager had access to the region's precious supply of water. In Mexican times, Americans were struck by an absence of fences around the long lots. The explanation, said Gregg, was a contrast in cultural values: In the United States, farmers built fences because *farmers* were obliged to keep livestock out of their fields, yet in Mexico, restricting livestock was the *livestock owner*'s responsibility, which explained the ubiquitous presence of the shepherd. Livestock, notably sheep, grazed on the mesa lands above the agricultural valleys; sheep and woolen products were of greater importance than agriculture in Mexican New Mexico's economy.[8]

One village example is El Cerrito (see map 15.4). Settled soon after 1824, El Cerrito's location on the Pecos only fifty miles (eighty kilometers) below that river's headwaters meant a perennial supply of water for a place that received only fourteen inches of precipitation annually. Moreover, El Cerrito's elevation of 5,700 feet (1,735 meters) gave it a five-month growing season (May 10 to October 10) and several isolated stands of Ponderosa pine that early villagers quickly harvested for vigas. The flatland found within tight *ancones* (bends) of the Pecos at El Cerrito

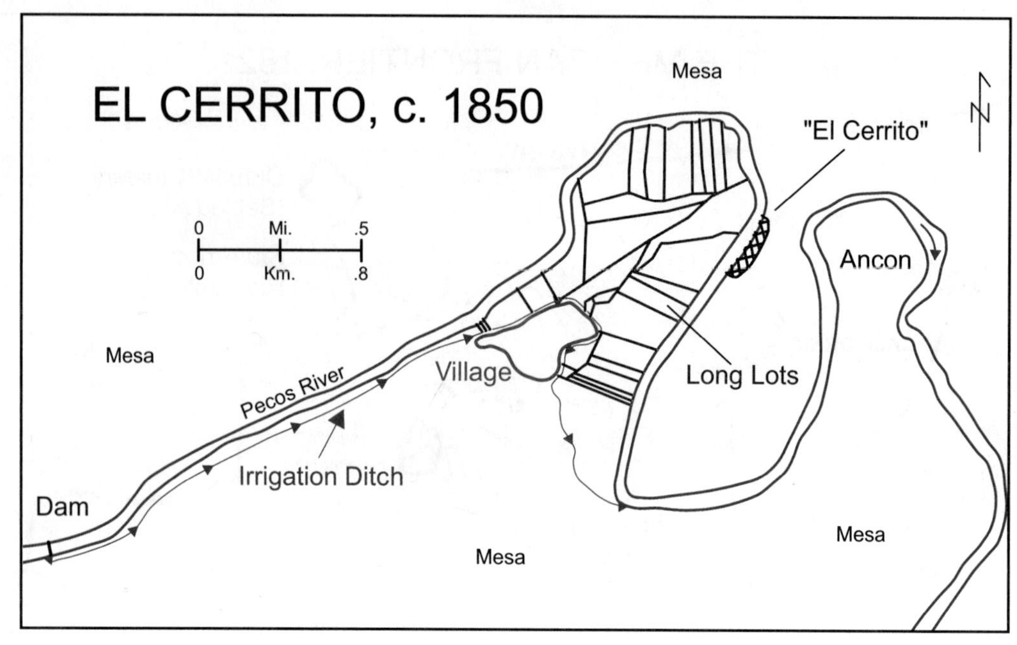

Map 15.4 El Cerrito, c. 1850. Named for an isolated "little hill" in the Pecos Valley, El Cerrito was an unusually compact village, owing to site limitations. Long lots are shown for 1922.

provided only limited irrigable land for agriculture, and the five-acre site chosen for the village was likewise constricted. So it is probably fortunate that in 1841, when El Cerrito's population peaked at 321, half the villagers left for new farmland at the new village of San Agustin (map 15.3). After this hiving off, some 150 *cerriteños* survived comfortably in a village of thirty one-story adobe houses, a Roman Catholic church (the first church built when the village was founded), and a grade school (built in 1882). Like villagers elsewhere in the Mexican period, cerriteños were poor. The sale of "Indian" corn and wheat grown on long lots in the roughly 110 acre (45 hectare) ancón below the village netted some income. But like most nuevomexicanos in the Mexican era, cerriteños were principally involved in raising livestock, and their main income came from sheep and cattle (also wool and hides) raised on the commonly held higher mesa lands of their San Miguel del Vado land grant.[9]

New Mexico's largest community in the Mexican era was its capital, Santa Fe, with a population of five thousand in 1850. When Zebulon Pike saw Santa Fe in March 1807, he likened its rows of one-story adobe houses facing dirt streets to a fleet of flat-bottomed boats descending the Ohio River. In the midst of these rows of houses was an open and unbeautified plaza with corners that pointed weakly in the cardinal directions (see map 15.5). Facing the plaza on the north was the Palace of the Governors (now a museum), behind which stretched the walls and buildings of a large presidio (of which little remains). Facing the plaza on the east was the Roman Catholic church, called La Parroquia (now the site of Saint Francis Cathedral), but by the Mexican era, the half of the plaza that faced La Parroquia had been encroached upon by houses. Since its founding, Santa Fe had been the terminus of the Camino Real that led 1,500 miles (2,400 kilometers) south to Mexico City. And during the Mexican era Santa Fe came to be the terminus of the 750-mile-long (1,200-kilometer-long) Santa Fe Trail that led to Missouri. Anglo traders in Santa Fe used the Exchange Hotel (the present location of La Fonda Hotel) as their headquarters while they traded their goods from their wagons or from rented space around the plaza.[10]

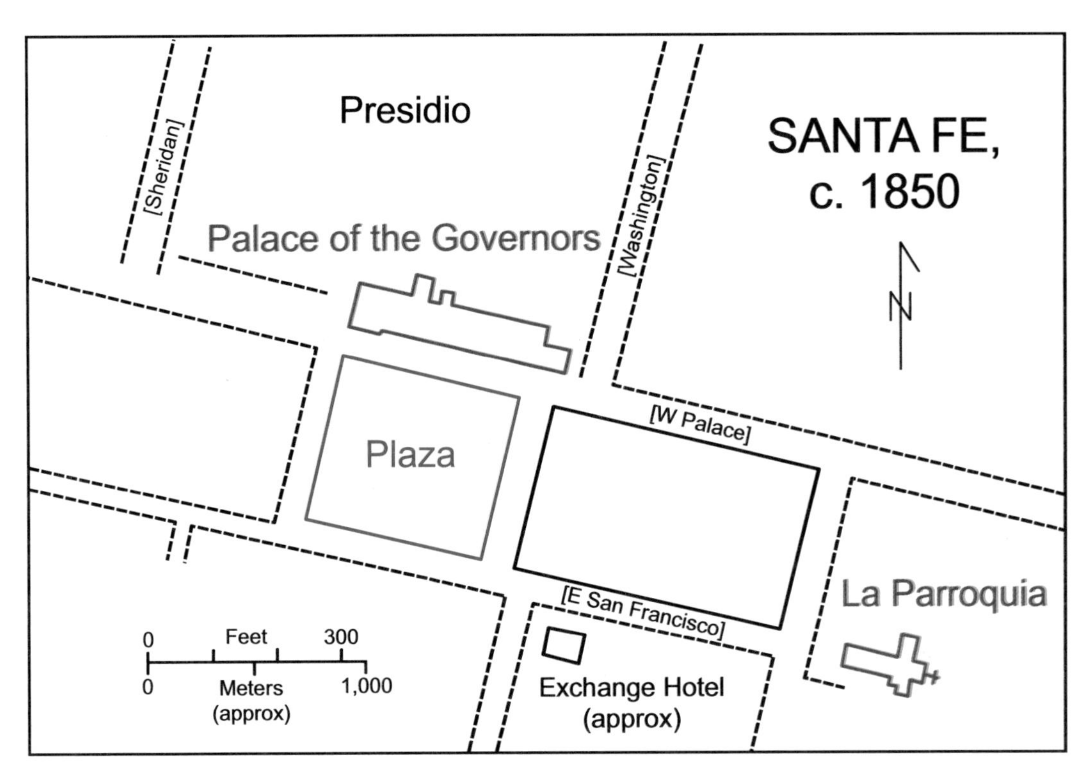

Map 15.5 Santa Fe, c. 1850.

Source: Lieutenant J. F. Gilmer's map of Santa Fe in 1847, reproduced in Moorhead, "Rebuilding the Presidio of Santa Fe, 1789–1791," 137 and 138. The map places the presidio approximately between Washington and Sheridan avenues.

The Santa Fe Trade

When a party of three would-be traders led by James Baird headed for Santa Fe in 1812, Spanish officials arrested them, confiscated their goods, and put them in prison. The message that New Mexico was off-limits was clear: What happened to Baird was similar to the fate of other prospective traders. In November 1821, however, William Becknell's party of six was treated quite differently. Becknell had left (Old) Franklin, Missouri, with a pack train of goods apparently intended for trade with nomadic Indians. Becknell reached New Mexico through Raton Pass (historian Janet Lecompte suggests that it was Trinchera Pass to the east), and just south of Las Vegas at Puertocito on Piedra Lumbre Creek, on November 13, he encountered Captain Pedro Ignacio Gallego. Neither apparently spoke the other's language, but Gallego made it clear to Becknell that his party could proceed to Santa Fe to trade. Becknell reached Santa Fe on November 16, spent the next several weeks trading his goods with local nuevomexicanos, and in early 1822 returned to Missouri using the shorter Cimarron River route. Word of Becknell's success in Santa Fe and of the profits he subsequently made in Missouri spread quickly. That same year saw the beginning of organized trade with Santa Fe; Becknell himself returned later that year with goods loaded in three Conestoga wagons.[11]

The Santa Fe Trail, which these wagons followed, linked Missouri and Santa Fe. St. Louis was the organizational headquarters for the largely American and French Canadian traders, and a fashionable hotel called the Planter's House was their focus. From St. Louis, traders would head by land or steamboat up the Missouri River to its first "Big Bend," where Independence was established in 1827 as an outfitting center and wagon head (see map 15.6). After 1833, Westport, now part of Kansas

Map 15.6 Santa Fe Trail, 1822–1880. The Trail was made a National Historic Trail in 1987.

City, Missouri, gave Independence much competition as a place to hire teamsters and acquire wagons, mules, or oxen (oxen could pull heavier loads and had less appeal to Indians). From the Big Bend, wagons went individually the one hundred miles west to Council Grove, the trail's jumping-off point on the Neosho River. Council Grove marked the end of hardwood trees needed for axle repairs. It was also the place where from perhaps twenty-five to one hundred wagons would be organized into caravans for protection, and where a wagon train captain was chosen. From Council Grove the trail went overland to the "Great Bend" of the Arkansas River, which traders followed to about the one-hundredth meridian (the future location of Dodge City). There they either crossed the Arkansas at one of three places where they could follow the trail's Cimarron Branch, or they continued along the north bank of the Arkansas to follow the trail's Mountain Branch.[12]

The Cimarron Branch was the Santa Fe Trail's principal route. It was the shorter and more direct branch. But, for some seventy miles, the Cimarron Branch was waterless, had less plentiful grasses for animals, and was more dangerous than the Mountain Branch. The Arkansas River (west of the one-hundredth meridian) was the boundary between the United States and Mexico until 1848. South of the Arkansas, there were no Army dragoons (mounted cavalry) offering protection from the raids of nomadic Indians. The Mountain Branch north of the Arkansas, on the other hand, was safer but longer. On this branch, William Bent built Bent's Fort, probably in 1834, according to historian David Lavender. Lavender notes that Bent chose to locate his private post on the American side of the river to avoid the arbitrary whims of Mexican tax officials. Surrounded by substantial adobe walls, the fort's compound offered a safe place for travelers to rest and refit. More than once, as those at the fort watched a wagon train ascend along Timpas Creek toward Raton Pass on the south side of the Arkansas in Mexico, rising dust told them that nomadic Indians were engaged in a raid and that help was needed. Crossing Raton Pass (7,834 feet) meant several days of slow and very difficult travel. The Mountain and Cimarron branches joined at La Junta (present-day Watrous), proceeded to Las Vegas (founded in 1835), and then forded the Pecos at San Miguel before skirting the southern end of the Sangre de Cristo Mountains to reach Santa Fe (map 15.6).[13]

Caravans of Conestoga wagons, or the smaller, lighter, and more maneuverable prairie schooners, left Council Grove in May or June. Progress was slow under heavy loads of textiles (calico or printed cotton cloth, hose, and silks), small hardware (knives, traps, guns, and ammunition), and liquor (rum and whiskey). Traveling at fifteen or twenty miles a day, caravans took a month and a half to two months to cross the 750 miles to Santa Fe. Gregg described how, in anticipation of seeing the señoritas and dancing at the fandangos, traders would "rub up" the morning before their wagons quickly descended into Santa Fe. Trading in Santa Fe lasted four or five weeks in July and August. The return trip to "the States" went faster: Goods were lighter, and if the wagons themselves had been sold, the traders were on horseback. Goods that returned included furs and skins, blankets and other woolens, saddles and other leather products, and Mexican silver pesos. For a time, the Mexican peso was the currency in Missouri, and for items worth less than a peso, the silver coins could be cut with an axe in half, in quarters, and even in eighths (thus the expression "two bits" to mean a quarter). In time, perhaps not surprisingly, the big traders squeezed out the little ones.[14]

The goods taken to Santa Fe did not stop there. Santa Fe became the portal through which American goods went down the Camino Real via El Paso and Chihuahua to Mexico City. So important did New Mexico become in Mexico as the source for American goods that in late 1847, during a truce in the Mexican War, Mexico agreed to cede Texas north of the Nueces as well as California, but not New Mexico. Santa Fe trade was in its heyday in the Mexican period, but it continued until about 1880, by which time the Santa Fe Railroad had built over Raton Pass to reach New Mexico. The importance of this first trail in the American West went beyond the two-way trade with international ramifications that it carried. A small but significant number of Anglos and French Canadians remained in New Mexico to marry Spanish women and become part of New Mexico's elite. Conversely, a number of the nuevomexicano elite got involved in the trade, and as early as 1831 their sons were being sent to schools in St. Louis. Cross-cultural borrowing was intriguingly underway.[15]

Spinning off from the Santa Fe trade was trade along the so-called Spanish Trail that led to California. In Santa Fe the demand by traders for horses and mules exceeded the supply. Thus in the winter of 1829 to 1830, an enterprising New Mexican by the name of Antonio Armijo led a party of some sixty nuevomexicanos to a source for horses and mules in Southern California. This trail led northwest into Utah, where Spaniards had once sought Ute Indians to be used as household servants, and from Utah it turned southwest across desert to reach the Los Angeles Lowland through Cajon Pass (see map 15.7). On the trail between November and April, when temperatures were cooler and grasses more plentiful, nuevomexicanos took pack animals loaded with woolen goods to be traded with californios for horses and mules in Los Angeles and on up the coast. Accounts of the Spanish Trail often end with the comment that no one knows what became of Antonio Armijo. Well, I found Armijo in California in the 1850 census: He was fifty-nine years old and living as a farmer in Solano County's Vacaville north of Suisun Bay, and besides one New Mexico–born son in his own household, the ages of three New Mexico–born and two California-born children in a second adjacent Armijo household suggest that Armijo had remained in California since the mid-1830s. Additional nuevomexicanos were recorded living in California in 1850, mainly at two villages (Agua Mansa and La Placita) known collectively as San Salvador on the upper Santa Ana River near Riverside. Trade over the Spanish Trail seems to have

Map 15.7 Spanish Trail, 1829–1848. The Old Spanish Trail was made a National Historic Trail in 2002.

Source: Hafen and Hafen, *Old Spanish Trail*, 317.

ended between 1847 and 1848, yet the fact that it took place is the only example of sustained contact between any of the four frontier outpost-clusters before the end of the Mexican period.[16]

Spanish Americans

What became of New Mexico's "Spanish Americans," to use their own self-referent of choice, is a most interesting chapter in America's changing geography. Between 1850 and 1900, the population of this subculture more than doubled from some 60,000 to 140,000. And as their numbers grew, they expanded quite dynamically beyond the upper basin of the Rio Grande. Their village-by-village spontaneous outward movement had slowly begun in the 1790s, accelerated after 1848 with the safer times that came from American control, saw its greatest areal gains in the 1860s, and ended in the 1890s. Several hundred new villages came to exist as sheepmen spread east to the panhandles of Texas and Oklahoma, north almost to the Arkansas Valley in Colorado, west to eastern Arizona, and south down the Rio Grande Valley in New Mexico (see map 15.8). Those pushing north into Colorado were largest in numbers. By 1900, a Spanish American region stretched into five states and covered an area the size of Utah. Areally, with minor exceptions, this was as large as the region would get.[17]

As the expansion took place, Anglos—to reiterate, an Anglo in the Southwest was anyone who was not Indian or Spanish/Mexican American—arrived in the region in relatively light numbers. Beginning about 1821, traders and trappers trickled in annually, married women of Spanish descent, learned their Spanish language, and

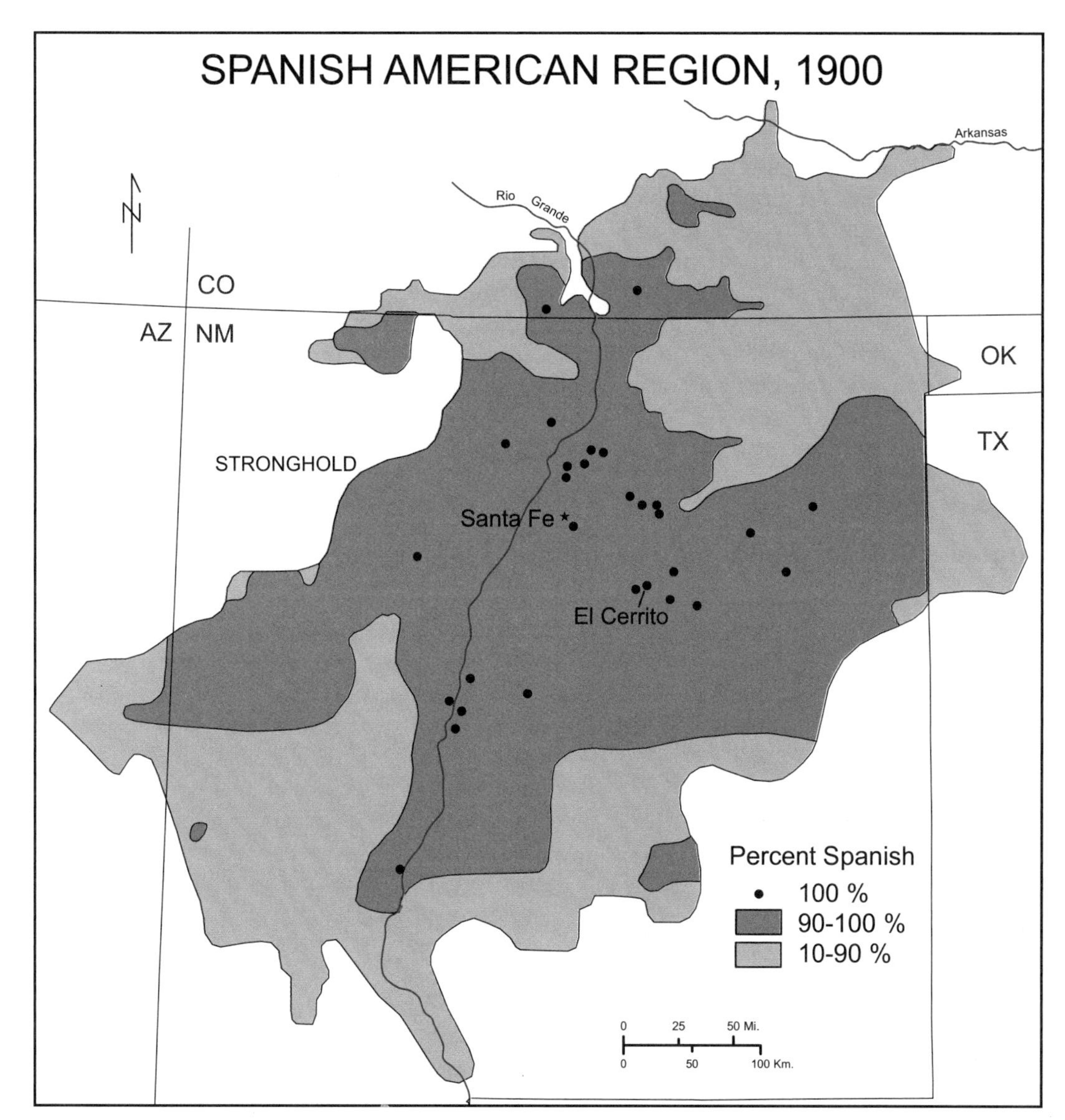

Map 15.8 Spanish American Region, 1900. This large region—85,000 square miles, or an area the size of Utah—contained 485 census precincts where Spanish people constituted a minimum of 10 percent of the population. The region's inner Stronghold—42,300 square miles, or an area the size of Tennessee—where Spanish people were a minimum of 90 percent of the population, contained 29 purely Spanish precincts. Omitted are thirteen nonregion inliers, thirteen region outliers, and twelve areas where Spanish people constituted less than 90 percent within the Stronghold.

Source: Population Schedules of the Twelfth Census [1900]. Reproduced with a new title from Nostrand, *El Cerrito, New Mexico*, 169. Courtesy of the University of Oklahoma Press.

named their children José and María. Adding to the intruders were especially Irish-born and German-born soldiers who came during the Mexican War and remained. By 1850 some 1,600 Anglos were recorded in New Mexico, half of them in Santa Fe. After 1850 the intruders included German-born Jewish merchants and French-born Roman Catholic priests. The 1860s seem to have been the pivotal decade in the transition that found the hispanicization of mainstream Anglo intruders shifting gradually to the anglicization of Spanish Americans. By 1900 the number of Anglos living in the region reached sixty-six thousand. Meanwhile, by 1900 the number of immigrants from Mexico who were drawn only to the southern part of the region reached ten thousand. Due to the influx of Anglos and Mexican immigrants to the region, the population of Spanish Americans was reduced to 64 percent. Map 15.8 shows, however, that Anglos (and the Mexican Americans) lived largely in the outer half of the Spanish American Region: With exceptions, its inner half was most amazingly almost everywhere greater than 90 percent Spanish American.[18]

Thus Spanish Americans were neither submerged by intruding Anglos nor were they absorbed into the arriving Mexican immigrant population—the fate of their californio and tejano brethren. What happened instead was that they preserved their distinctive subculture and they developed a new Spanish-centered identity. Nuevomexicanos had settled in the Borderlands more than a century before tejanos and more than a century and a half before californios, and they had arrived

Spotlight

L. Eugene "Cotton" Mather (1918–1999)

Cotton Mather shown in his broad-brimmed Stetson. Photograph courtesy of Cotton Mather, used in publicity for his colloquium held March 31, 1992, in the Department of Geography at the University of Oklahoma.

Cotton Mather, who was a direct descendant of colonial New England's Cotton Mather, loved New Mexico. Vivid childhood memories from a family road trip through Española and Taos in the 1920s prompted his many return visits to the state and, ultimately, his retirement in 1985 to a home he and his wife Julie built in an alfalfa field (later planted to pecans) in Mesilla. Born in West Branch, Iowa, on January 3, 1918, Mather attended the University of Iowa in the latter 1930s before earning AB (1940) and MS (1941) degrees at the University of Illinois and his PhD (1951) at the University of Wisconsin. Mather taught geography at the University of Georgia (1947–1956) before spending most of his career on the faculty at the University of Minnesota–Minneapolis (1957–1983). Most of his nearly fifty books, book chapters, and articles were coauthored, a dozen with John Fraser Hart and a second dozen with his good friend Pradyumna P. "Paul" Karan, of the University of Kentucky. Yet Mather is perhaps best known for his single-authored "The American Great Plains," which appeared in the *Annals of the Association of American Geographers* in 1972. Mather valued most being creative, traveling for adventure, and cultivating friendships. He inspired many, including me, through his storytelling, his upbeat attitude, and his many original insights. He was a strong advocate for field-based research, and I heard him say more than once that "the source of all geographical knowledge is in the field." And on one unrecorded occasion, he told me—correctly it would seem—that New Mexico is unique in America because its Spanish people shaped the intruding Anglos, and not the other way around. Mather died at eighty-one in Las Cruces on December 25, 1999.

more directly from Spain. Preserved in isolated New Mexico were certain archaic cultural traits not found elsewhere in the Borderlands: words and verb endings in their Spanish language; given names such as Esquipula, Belarmino, Secundido, and Onofre, as noted in Chapter 3; and examples of folklore handed down orally. Other distinctive cultural attributes included their identifying with a wooden statue called La Conquistadora, the carving from wood of religious icons called Santos, the heavy use of chili peppers in food, and uncommon Spanish surnames such as Abeyta, Barela, Maestas, and Tafoya. Besides their distinctive culture, they created a new Spanish American ethnic identity. Sorting out why this happened is complicated, but by being Spanish, one avoided the derisive use of Mexican, at least in the eyes of Anglos. "Spanish American" was in use by the 1870s, had become the strongly preferred self-referent by the 1920s, and began to decline only in the 1960s, when "Chicano" gained traction within the younger set. Roots planted deeply in the Spanish era seemed to justify this Spanish identity for nuevomexicanos.[19]

All this left a remarkable legacy in America's historical geography. First, because the small californio and tejano populations were both submerged by Anglos and absorbed into the Mexican immigrant populations, Spanish Americans became the only surviving Spanish colonial subculture. Second, because the inner half of the Spanish American Region—a "stronghold" with an area the size of Tennessee—was with exceptions everywhere greater than 90 percent Spanish American in 1900, nowhere in America (with the possible exception of the Mormon culture region) has so large an area been so purely ethnic so recently (map 15.8). Third, because this concentrated society was circular in shape with Santa Fe, its capital, located at the demographic center point, the Spanish American Region would have made an ideal political American state, giving its people a more secure hold on their territory, if political boundaries had been drawn to reflect cultural realities on the ground. Finally, as geographer Eugene "Cotton" Mather (see spotlight) observed, across America, as the advancing mainstream (Anglo) population overran relatively large concentrations of indigenous peoples, the mainstream group seems always to have transformed the indigenous groups. But, asserted Mather, this was not the case in New Mexico where, in a major way, Spanish people changed the intruding Anglos.[20]

Notes

1 The quote is from DeVoto in "Sea to Shining Sea," photographic section. In *The Shaping of America*, 3:167, Meinig rejects "simple surrogates," such as the twenty-inch isohyet and the one-hundredth meridian for defining the West, in favor of a transitional "band" that straddles the twenty-inch isohyet, as shown on his map, "The Edge of the West," 3:165. For the term "Middle West," see geographer Shortridge, "The Emergence of 'Middle West' as an American Regional Label." For the unabridged version, see Shortridge, *The Middle West*.

2 Allen, "Geographical Knowledge and American Images of the Louisiana Territory," 33–39. For Allen's unabridged version, see *Passage through the Garden*.

3 See DeVoto, ed., *The Journals of Lewis and Clark*; also, Lineback, "Celebrating Lewis and Clark," is a concise summary for geographers.

4 See geographer Lewis, "William Gilpin and the Concept of the Great Plains Region," 34–35, for Pike and Long and the "Great American Desert" label on Long's map, dated 1823.

5 Gregg, *Commerce of the Prairies*, 327 (agriculture) and 355 (grazing). For Bowden on the impact the desert image had on the settlement of the Great Plains, see his "The Perception of the Western Interior of the United States, 1800–1870"; "The Great American Desert and the American Frontier, 1800–1882"; and "Desert Wheat Belt, Plains Corn Belt" (where Bowden uses the term "literati").

6 Gregg, *Commerce of the Prairies*, 262, for the term *americanos*.

7 Ibid., 101, for the use of Rio del Norte; Nostrand, *The Hispano Homeland*, 214–17, for environmental adjustment in New Mexico.

8 Gregg, *Commerce of the Prairies*, 107, explains the contrast in cultural values concerning fences. See Nostrand, *The Hispano Homeland*, 217 and 220, for long lots and villages.

[9] See Nostrand, *El Cerrito, New Mexico.*

[10] For Pike's description, see *The Journals of Zebulon Montgomery Pike, with Letters and Related Documents*, 1:391. As early as 1697, houses were built in the east part of the plaza, according to Chavez, "Santa Fe Church and Convent Sites in the Seventeenth and Eighteenth Centuries," 92.

[11] For Trinchera Pass, Lecompte, "The Mountain Branch," 60. For the encounter between Becknell and Gallego, see Olsen and Myers, "The Diary of Pedro Ignacio Gallego," based on the two authors' discovery of the Gallego diary in the Mexican Archives of New Mexico.

[12] Simmons and Jackson, *Following the Santa Fe Trail.* Members of the Santa Fe Trail Association (founded in 1986) have actively mapped the trail and documented its landmarks.

[13] Lavender, "Bent's Fort," 15, for the year 1834; 14, for avoiding arbitrary taxes. Historian Lavender is an expert on the fort.

[14] For "rubbing up," see Gregg, *Commerce of the Prairies*, 78.

[15] For trade with Mexico, see Moorhead, *New Mexico's Royal Road.* About Mexico's unwillingness to give up New Mexico in late 1847, see Singletary, *The Mexican War*, 157–58. For nuevomexicanos engaged in the Santa Fe Trade, see Sandoval, "Who is Riding the Burro Now?" The date 1831 is from Sandoval as noted in Nostrand, *The Hispano Homeland*, 139.

[16] Hafen and Hafen, in *Old Spanish Trail*, give a comprehensive account. For contact between New Mexico and California, Nostrand, *The Hispano Homeland*, 132–39, esp. 134, for Antonio Armijo, who was found in Population Schedules of the Seventh Census [1850], Microcopy no. 432, 1964, roll 36: 14–15. Armijo's neighbor was wealthy, New Mexico–born Manuel Vaca [Baca], after whom Vacaville was named.

[17] See Nostrand, *The Hispano Homeland*, 70–97, for contiguous expansion; 96, for greatest gains in the 1860s; 97, for the size of Utah in 1900; 23, for 365,000 self-identifying in the 1980 census as "Spanish," "Spanish-American," or "Hispanic."

[18] Ibid., 98–130, for Anglo intruders; 129, for the 1860s as a pivotal decade; 20, for 1,600 Anglos in 1850; 22, for sixty-six thousand Anglos and ten thousand Mexican Americans in 1900.

[19] For discussions of cultural distinctiveness and ethnic identity, see ibid., 7–14 and 14–19, respectively.

[20] Ibid., 230–32, for points about the surviving Spanish colonial subculture and its uniqueness in 1900.

16 Oregon Country

To develop chronologically the story of European American settlement of the nineteenth-century West, we jump to far-removed Oregon Country on the Pacific coast. First, we see that "Four Nation-States Compete" for Oregon Country, and the outcome is a region split through the center, the northern half going to Great Britain and the southern half to the United States. The southern half that becomes American is also split environmentally, this time longitudinally along "The Cascade Divide." In the chapter's third section, we discuss "The Oregon Trail," the route taken by most Americans headed for Oregon Country and the most significant of all overland trails. And in what would soon be called "The Pacific Northwest," we relate how European American migrants created three distinct settlement areas, the Willamette Valley, the Puget Sound Lowland, and the Inland Empire, with Seattle by 1900 on its way to becoming the region's metropolis.

Four Nation-States Compete

In the first half of the nineteenth century, "Oregon Country" referred to the present-day northwestern corner of the United States and adjacent British Columbia (see map 16.1). The region stretched from the Rocky Mountains and the Continental Divide west to the Pacific Ocean, and from California (south of latitude 42° N) north to the southern tip of the panhandle of Alaska (at latitude 54°40' N). Four nation-states laid claim to this stretch of the Pacific coast. Spain and Russia based their interests on discovery, but their claims were relatively weak because neither made more than limited efforts to effectively occupy the area. Great Britain and the United States, on the other hand, had stronger claims to Oregon Country on the basis of both discovery and effective occupation.

Spain's mariner Juan Cabrillo had sailed up the Pacific coast to at least the forty-second parallel as early as 1542. In the 1780s Spain temporarily occupied Nootka Sound, on the west side of Vancouver Island. And from their explorations Spaniards left place names that have lasted: the Strait of Juan de Fuca, the San Juan Islands—and within these islands, Lopez Island—are examples. However, in the Adams-Onís Treaty of 1819, Spain gave up her rights to lands north of latitude 42° north, now California's northern border. Meanwhile, Russian traders in search of furs and especially sea otter skins were pushing east and south. They reached the Aleutian Islands by 1766, and their commercial interests found them advancing along coastal Alaska and south all the way to California. At Bodega Bay north of San Francisco, their Fort Ross (Rossiya) on the so-called Russian River existed from 1812 to 1841. But in 1824, in a treaty between the United States and Russia, Russia renounced her claims to land south of latitude 54°40' north, the southern limit of Alaska's coastal panhandle (map 16.1).

Thus the competition for who would control Oregon Country turned to Great Britain and the United States. In 1792 British captain George Vancouver sailed up the Pacific coast, chose not to investigate the discolored waters he saw at the mouth of the Columbia River, and on April 29 sailed into Puget Sound, which he claimed for Great Britain. Vancouver honored members of his crew and other naval friends with names for major landmarks: Peter Puget (Puget Sound) was the captain of Vancouver's second ship, *Chatham*; Joseph Whidbey (Whidbey Island) was the master of Vancouver's flagship, *Discovery*; and Peter Rainier (Mount Rainier)

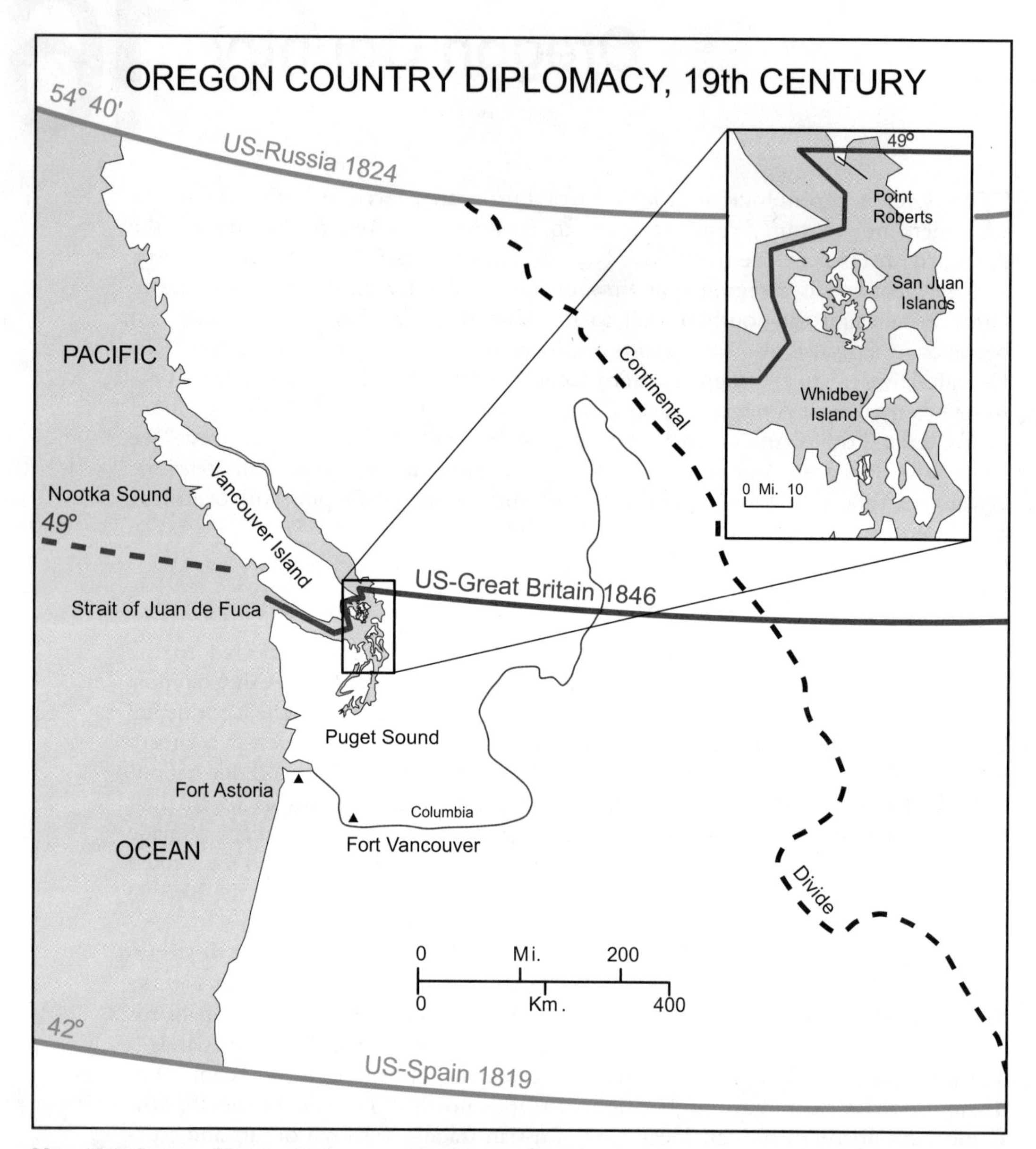

Map 16.1 Oregon Country Diplomacy, Nineteenth Century. The United States and Great Britain agreed on latitude 49° north for the international boundary in 1846.

was a Navy admiral not on the expedition. Following Vancouver came British fur traders who represented the North West Company and the Hudson's Bay Company (HBC); in 1821 the two companies merged, and Fort Vancouver (1824) became the headquarters for the HBC's regional Columbia Department. Meanwhile three events established the basis for American claims. In 1792 Robert Gray, representing mercantile interests in Boston, sailed up the Pacific coast and spotted the mouth of a river. But unlike Vancouver, on May 11, Gray found his way through the hazardous sandbars at the river's entrance, explored its estuary, and named the river after his ship, *Columbia*. Gray, who was after beaver pelts and sea otter skins, did not make formal claim to the huge Columbia River watershed (an area the size of France), but American diplomats would, nonetheless, make use of his discovery. They would also capitalize on the discoveries of Lewis and Clark, who wintered in 1805 at

Fort Clatsop at the mouth of the Columbia (see map 15.2). And in 1811, agents of John Jacob Astor's Pacific Fur Company set up headquarters at Fort Astoria at the mouth of the Columbia; threats posed by the War of 1812 prompted the sale of Fort Astoria to the British North West Company in 1813, and the fort was renamed Fort George.[1]

So both Great Britain and the United States had bases for claims to Oregon Country. In 1818 each agreed that for ten years Oregon Country would be "free and open" to both nation-states. And in 1827 they extended their "joint occupation" compact indefinitely. Given the HBC's fur trading and agricultural activities that were located notably at Fort Vancouver, the joint occupation agreement was, observes D. W. Meinig, a "diplomatic triumph" for the United States, because there had been no substantial American presence in Oregon Country since Gray, Lewis and Clark, and the Astorians. After 1827, British fur interests focused on the more productive northern parts of Oregon Country while Americans, who began to arrive in 1842 with intentions to farm, found the Willamette Valley attractive. In 1846, through diplomatic negotiation, both countries agreed to an international boundary at latitude 49° north (map 16.1). To include all of Vancouver Island in British America, the boundary was drawn south through the Strait of Georgia and out the midchannel of the Strait of Juan de Fuca to the Pacific. A miscalculation left in the United States the southern tip of Point Roberts, the site of a free-trade zone known to locals today for shopping bargains (see map 16.1 inset).[2]

The Cascade Divide

The Oregon Country that pulled European Americans west in the nineteenth century had a remarkably diverse environment. Two high mountain ranges (rising to ten thousand feet [three thousand meters]), the Rockies and the Cascades, formed its longitudinal structure (see map 16.2). The Olympic Mountains on northwestern Washington's Olympic Peninsula rose as a mountainous knot to some eight thousand feet (2,400 meters). Notable among the several low mountain ranges (of some three thousand feet [nine hundred meters]) were the Coast Range, located just inland from the Pacific, and the Blue Mountains, found largely in northeastern Oregon. Between the Rockies, the Cascades, and the Blue Mountains lay the large, roughly triangular Columbia Plain or Columbia Basin, a relatively level area covered with layers of basaltic lava flows topped with rich soils. And between the Cascade Range and the Coast Range were the Willamette Valley and the Puget Sound Lowland, a single structural depression where more than two-thirds of the region's population live today. The Columbia River and its main tributary, the Snake, drained fully two-thirds of Oregon Country to the Pacific; before the Snake joined the Columbia, it was deeply entrenched within Hells Canyon, along the Idaho-Oregon border.

Geologically speaking, the Columbia is an antecedent river. That is, the Columbia, most amazingly, preceded the orogeny that uplifted the Cascade Range. As the Cascade Mountains rose up, the Columbia continued to cut a roughly sea-level route—and a gorge—through the mountains. The Cascades, thus, are youthful mountains. Geologists tell us that only forty million years ago, they came to exist when a wayward slab of the Earth's crust called the Juan de Fuca plate began to subduct, or dive, eastward below the North American plate to create the volcanic Cascade Range (Baker, Rainier, Saint Helens, and Hood are notable among its volcanic peaks; map 16.2). Earthquakes occur when the two plates lock, and then lurch to unlock. A magnitude 9 earthquake is definitely major (San Francisco's 1906 earthquake registered 7.6), and in January 1700 such an earthquake shook "Cascadia," as the landform area is known. In 1986 geologist Brian Atwater of the United States

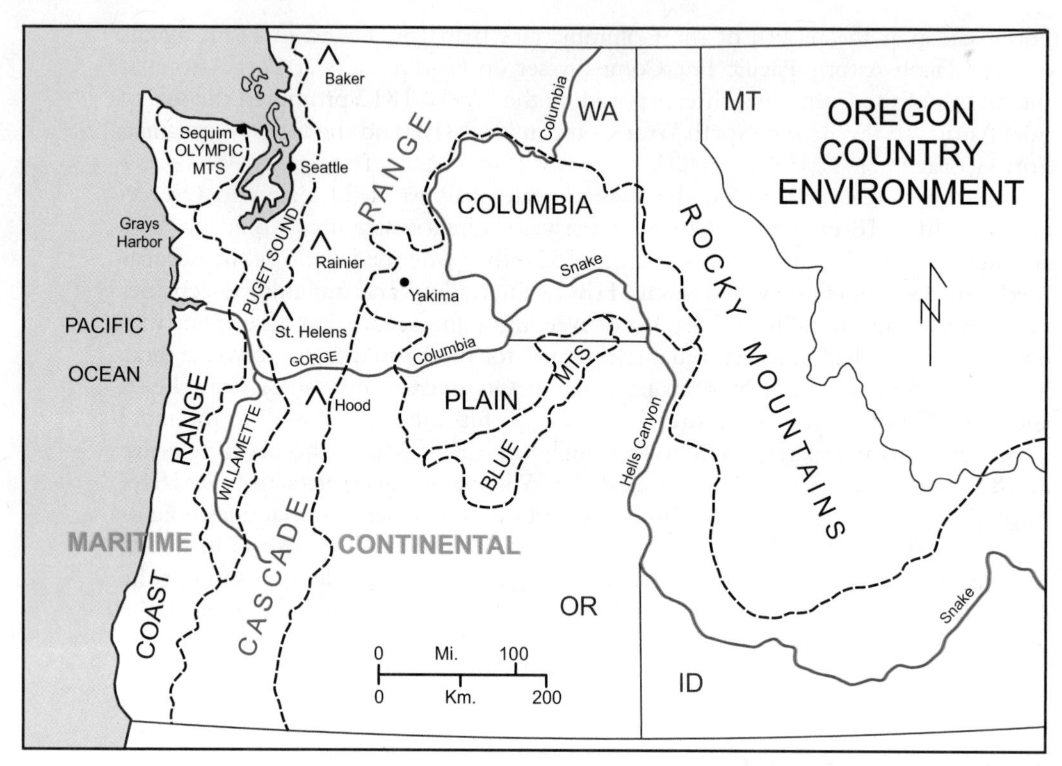

Map 16.2 Oregon Country Environment. The Cascade barrier creates two climatic regimes and environments, maritime and continental.

Geological Survey in Seattle pieced together the evidence for this major quake. It was the correlation of his discovery of coastal subsidence located in Washington's Grays Harbor area with tree-ring data that led to his arriving at 1700 as the approximate year of the magnitude 9 earthquake, and a devastating tsunami that hit the Japanese island of Honshu directly across the Pacific on January 27 and 28, 1700, corroborated his find. Volcanic explosions as at Mount Saint Helens in 1980 can be predicted, but not so earthquakes, and it is fair to say that the risks posed by major earthquakes to residents of youthful Cascadia are greatly underappreciated in the region today.[3]

Because the Cascades are the first high continuous upland encountered inland from the Pacific, they divide Oregon Country into two environments: maritime to the west and continental to the east. West of the Cascades, year-round temperatures are mild, due to the prevailing westerlies that bring marine-moderated air onshore. For example, snows at low elevations, as in Seattle in winter, occur only occasionally and do not last. West of the Cascades, conditions are also humid. The windward side of the Cascades at about six thousand feet (1,800 meters) receives some fifty to one hundred inches of precipitation annually, because moisture-laden air is cooled as it is forced to rise (the orographic effect). And vegetation is luxuriant. Douglas fir, Western hemlock, and Western red cedar (the three main lumber trees) cover the area in forest, with an undergrowth of brackens and nettles and other plants. Because the knot of the high Olympic Mountains rises up so close to the Pacific, its windward slopes receive in excess of 140 inches of precipitation annually. However, on the northeast or lee side of the Olympics, there exists a "rain shadow" where precipitation amounts are low. On forty-seven-mile-long Whidbey Island, for example, precipitation drops from roughly thirty-two inches at the southern end to twenty-one inches at the northern end as the rain shadow effect increases, and at Sequim, located on the Olympic Peninsula at the maximum point of rain shadow effect, only sixteen inches of precipitation are measured annually.

Continental climatic conditions prevail east of the Cascades. Like the Olympic Mountains, the Cascade Range creates a rain shadow on its eastern or lee side. The mountains are also a barrier to the moderating effects of the Pacific. Thus temperatures east of the Cascades are extreme, with hot summers and cold winters with some snow often covering the ground. East of the Cascades average annual precipitation amounts are low—generally between ten and twenty inches (semiarid), although the western Columbia Plain in the Yakima area receives less than ten inches (arid), and the higher elevations as in the Blue Mountains and the Rockies receive more than twenty inches (subhumid). Reflecting these precipitation amounts are three vegetation covers: Grasses exist where it is semiarid, sagebrush where it is arid, and ponderosa and other pine where it is subhumid.

When Oregonians began their westward migration in 1842, they headed to the western humid side of the Cascades—specifically, the Willamette Valley. Nearly all of the immigrants originated in the humid East. Careful analysis of the original schedules of the 1850 census by geographer William A. Bowen reveals that fully half of those mainly farm families who reached the Willamette Valley in the 1840s were from the five states of Missouri, Illinois, Ohio, Kentucky, and Indiana. Bowen believes that a combination of disease and sickness was the leading push factor for the migration, and in the Ohio and Missouri valley lowlands of these five states, diseases, including tuberculosis (consumption), malaria, and cholera, were widespread. The leading pull factor behind the arrival of nearly twelve thousand pioneers to the Willamette Valley by 1850 was the familiar desire of pioneers to better themselves economically, in this case coupled with the strong hope (realized in the Donation Land Law of 1850) for free land. Because the Willamette Valley was humid environmentally, newcomers who arrived on the Oregon Trail found little to adjust to when practicing their agriculture. The important exception was growing coarse-grained corn used for livestock feed. Summers in the Willamette were too cool and, given the winter-maximum precipitation regime, too dry for this corn to do well.[4]

The Oregon Trail

The Oregon Trail, known to migrants as the "Emigrant Trail," was very much a nineteenth-century phenomenon. In 1812 fur traders returning from Fort Astoria used parts of the Oregon Trail, including South Pass. Mountain Men like Jedediah Smith, when crossing the Rockies in the 1820s, went through South Pass. So did Marcus and Narcissa Whitman, Presbyterian missionaries on their way in 1836 to establish the Whitman Mission—an event carefully chronicled by Narcissa in letters to relatives in New England. The year 1842 marked the first major use of the trail by migrants moving to Oregon. Beginning in 1847 Mormons followed the first half of the trail before cutting off to Utah on the Mormon Trail. Soon, traffic got even heavier on the first half of the trail when the "forty-niners" joined others before cutting off on branches to the California Trail and the gold fields. Peak use of the Oregon Trail came in 1852. The completion of the Pacific Railroad in 1869 diverted traffic going west away from the trail and its Oregon destination, yet some families heading to Oregon Country in covered wagons made use of the trail down to the 1890s. In 1906 old Ezra Meeker, himself an immigrant to Oregon Country in 1852, lamenting how few really remembered the trail, took it upon himself to commemorate the trail by erecting stone monuments at points of significance as he traveled the trail from west to east in a trip that ended with a big welcome at the White House (see figure 16.1).

Figure 16.1 Ezra Meeker's Monument at South Pass, Wyoming, 1977. The "great migration" to Oregon Country began in 1843, but Meeker's use of 1857 is less clear.

Photograph by the author, June 8, 1977.

The Oregon Trail had two main beginning points, and both were on the Missouri River (see map 16.3). One was Independence, also the outfitting point for the Santa Fe Trail; the other was Council Bluffs-Omaha. From either, wagons converged at Fort Kearny, an Army post founded in 1848 on the Platte River near Kearney (note spelling), Nebraska. The Platte, which meant "flat" in French, was then followed to beyond where it forked; in its shallow, braided channel flowed water popularly spoken of as "too thick to drink, too thin to plow." Beyond where it forked, the North Platte was followed past Chimney Rock and other anticipated erosional monuments to Fort Laramie. Built in 1849 at the site of earlier trading posts, Fort Laramie resupplied emigrants and became the major Army post in the northern Great Plains. The North Platte then led to present-day Casper, Wyoming, site of private Fort Caspar (note spelling), founded late (1858) and of minor importance except as a fording point and later toll bridge across the North Platte. Several days of travel in country that lacked drinkable water brought migrants to Independence Rock, so named by someone who hoped to reach it by the Fourth of July. From Independence Rock wagons followed the small and well-named Sweetwater River to South Pass (elevation 7,550 feet [2,250 kilometers]) at the continental divide. Named

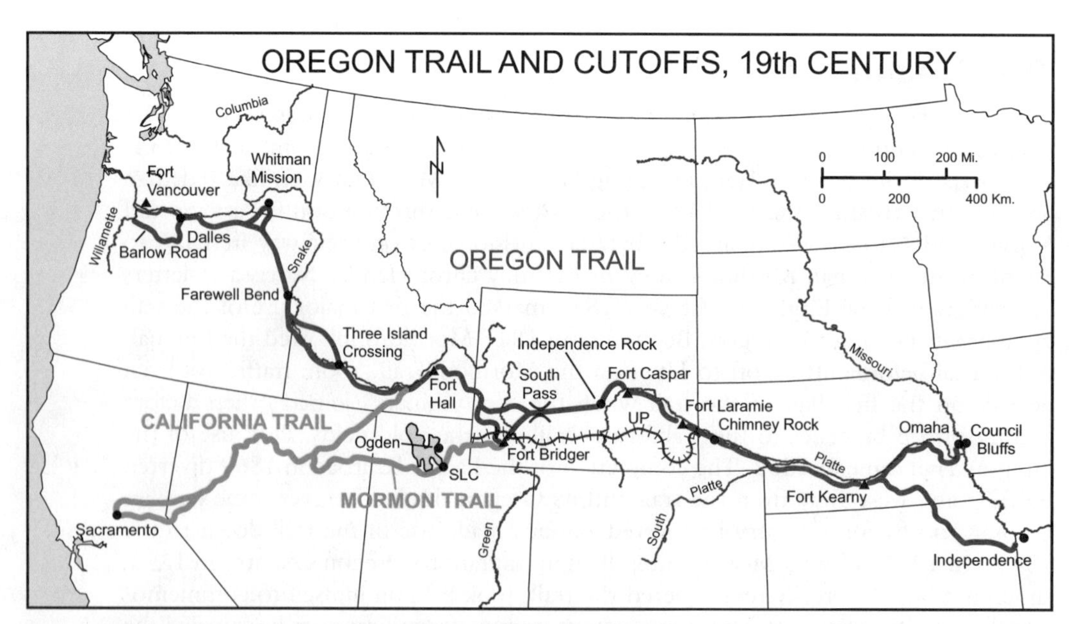

Map 16.3 Oregon Trail and Cutoffs, Nineteenth Century. In *Lectures on the Historical Geography of the United States as Given in 1933*, Barrows explained that immediately after leaving Council Bluffs-Omaha, travelers followed the Platte River's south bank (192).

because of its relative location south of passes used by Lewis and Clark, South Pass was a twenty-plus-mile-wide (thirty-two-kilometer-wide) saddle in the central Rockies. The grade leading to its summit was so gentle that John C. Frémont likened going through South Pass to driving up Pennsylvania Avenue from the White House to Capitol Hill.[5]

South Pass marked the approximate halfway point in the two-thousand-mile-long (3,200-kilometer-long) trail. Lacking water and grass, the fifty miles (eighty kilometers) between South Pass and the Green River was the trail's most difficult stretch. In this segment some wagons dipped south to private Fort Bridger (1842), while others crossed high, open country that led more directly to Fort Hall near the Snake River. Fort Hall, initially an American private post dating from 1834, represented to emigrants the entrance to Oregon Country. Beyond Fort Hall, wagons stayed on the south bank of the Snake; if water in the river was sufficiently low, Three Island Crossing was a fording point to the north bank and its better travel conditions. Wagons left the Snake at Farewell Bend; not to do so would bring them to Hells Canyon. The Blue Mountains (summit elevation 4,193 feet or 1,220 meters) were actually a welcome sight, for the more rugged terrain meant water and grass in valleys for overnight camping, and passable open forest on ridges when traveling in daylight. Beyond the Blue Mountains many early caravans looped north to the Whitman Mission, but this ended in 1847 when a band of the Cayuse Indians, upset by Whitman's failure to cure an epidemic of measles, killed all fourteen at the mission, including the Whitmans themselves. The south bank of the Columbia River led to the Dalles (French for "trough"), beyond which emigrants were either forced to build rafts and float their wagons through the treacherous gorge or hire ferrymen to do the job. Barlow's toll road, opened in 1846 through heavy forest around the south flank of Mount Hood, was a difficult and expensive alternative for the exhausted travelers. Happily, John McLoughlin, chief factor at Fort Vancouver, welcomed Americans, to whom he sold provisions on credit.

It took four or five months between April and September for migrants to reach the Willamette Valley. Saying good-bye to relatives at embarkation points was indeed sad, for this was a one-way trail and those who took it did not expect to return. Wagons (prairie schooners, not the heavier Conestogas) were often organized into trains, yet unlike on the Santa Fe Trail, protection from Indian raids was a lesser issue, and only a few wagons could travel safely. Each wagon contained clothes and bedding, a plow and tools, medicine, guns and ammunition, cooking utensils, and food: "pilot" bread (hardtack), flour, dried beans, corn meal, bacon, coffee, salt, and sugar (see figure 16.2). To reduce weight, the jettisoning of prized furniture, trunks, books, and unessential items began, usually after Fort Laramie. The elderly and the sick or injured rode, as did small children, but adults walked, and this meant from ten to twenty miles a day, repeated some 150 times. People died of diseases, especially measles and cholera; from drownings at river fords; from being crushed, in the case of children, under heavy wagon wheels; and from accidental firearm discharges—but only rarely from conflicts with Indians who, if anything, assisted emigrants at fords and other points. It is estimated that the Oregon Trail and its cutoffs were used by possibly five hundred thousand people, making it the most heavily traveled of all overland routes. For the nineteenth century, personal accounts written about the Oregon Trail are apparently exceeded only by those written about the Civil War.

After 1900 the Oregon Trail pretty much disappears, and the modern traveler wishing to follow its route must seek out the interpretive centers at restored forts and other points between Independence and Oregon City. My much esteemed mentor, Howard H. Martin (see spotlight), pointed out to those of us in his course, Historical Geography of the United States, that the Oregon Trail violated the principle of

Figure 16.2 A Pioneer Family on the Oregon Trail.

This remarkable photograph (n.d.), according to the Denver Public Library, was found in a family album belonging to Alice Stewart Hill. When used by the *National Geographic* on the cover of its August 1986 issue, a request was made of readers to volunteer information about the image (see Gibbons, "The Itch to Move West," 149). Photograph courtesy of Denver Public Library, Western History Collection, X-11929.

"locational succession." Often, Martin noted, a one-time Indian trace would become a pioneer trail that would become a paved highway and perhaps also a route for a railroad. But not so the Oregon Trail. Geographer James E. Vance Jr. explained in much detail why the Union Pacific (built between 1865 and 1869) abandoned the route of the Oregon Trail through South Pass in favor of one along the southern tier of Wyoming, along which the railroad founded Cheyenne, Laramie, Rawlins, and Rock Springs (the modern-day route of I-80—map 16.3). The Union Pacific, Vance explained, was *economic* transportation, while the Oregon Trail was *noneconomic* transportation. Determining the route of the Union Pacific were directness between Omaha and Ogden, avoiding areas of heavy snowfall in winter, and terrain where grades would not exceed ninety feet per mile. Determining the route of the Oregon Trail were the presence of grasses and water for animals, timber for fuel, and the easiest possible grades. Oregonians sought the noneconomic goal of reaching their destination, which necessarily took place during the warm months, while the railroad men of the Union Pacific sought the economic goal of carrying people and goods for profit over the entire year. With a new purpose came a new route.[6]

The Pacific Northwest

The Willamette Valley developed quickly into Oregon Country's first important subregion. Prairie covered much of the valley in the 1840s. Indians had created these grasslands with their annual fall burnings in order to confine deer beyond the burned areas for hunting. The first to settle the prairies were French Canadian HBC retirees from Fort Vancouver who, with their Indian wives and children, became farmers in the early 1840s at "French Prairie," south of Champoeg (see map 16.4A). Then came American migrants, who overwhelmingly selected land claims in prairie areas

Spotlight

Howard H. Martin (1892–1966)

Howard H. Martin. This photograph appeared as a tribute to Martin in the *Yearbook of the Association of Pacific Coast Geographers* 31 (1969): 2. Courtesy of James Craine, Yearbook editor.

Howard Hanna Martin was my undergraduate mentor at the University of Washington (UW). In 1960 I took his Geography 325, Historical Geography of the United States, the course that changed my life. And when I went off to UCLA for graduate work, Martin made sure that original copies of his lecture notes caught up with me. A native of Little York located in the Corn Belt of northwestern Illinois, Martin attended nearby Monmouth College, then George Washington University, as an undergraduate. Following service in World War I, he returned to George Washington University to earn both his master's and doctoral degrees. Martin taught briefly at the University of Cincinnati and at Columbia University before joining the faculty at UW in 1930, where he served as geography's first chair (1935–1950) and implemented its doctoral degree. He retired in 1962. Martin was primarily an economic geographer, but a story he told me underscores his strong interest in historical geography. While a graduate student, Martin knew of University of Chicago geography professor Harlan H. Barrows and his renowned course, Historical Geography of the United States. One summer quarter Martin moved to Chicago to be able to audit Barrows's lectures. Martin's eyes were overly sensitive to light (I often saw him wearing a green visor). One day as he sat by a window in Barrows's class, direct sunlight flooded his face, forcing him to close his eyes. A somewhat impetuous Barrows thought Martin was napping and threw a blackboard eraser to wake him. An embarrassed Martin sought out Barrows after class to explain why he had closed his eyes and to say that he was in Chicago expressly to audit that class.

---

that bordered on forests; the prairie would not need to be cleared of trees, and the forest supplied logs for log cabins. Geographer Bowen explained the settlement process: Most migrants who arrived in 1842, 1843, and 1844 spent the winter and early spring clustered near Fort Vancouver or at Oregon City (founded in 1842) at the falls of the Willamette, seeking lodging that was in very short supply while the men often took jobs, importantly at McLoughlin's sawmill or on his timberlands. In the spring, these migrants headed south up the valley, often in kinship clans, to preempt a farm claim and quickly put in a crop of wheat. Newcomers after 1845 were better able to find a relative or friend to winter with before repeating the process of heading south up the valley to preempt a farm. Passage of the Donation Land Law in 1850 accomplished what these migrants had hoped for—namely, those who were eighteen and over who had arrived by December 1, 1850, were entitled to 640 acres (for married couples) and 320 acres (for individuals) of free land, provided they make improvements over four years.[7]

Oregon City became the Willamette Valley's first urban center beyond Fort Vancouver. Located at the river's falls 11.5 miles (19 kilometers) from the mouth

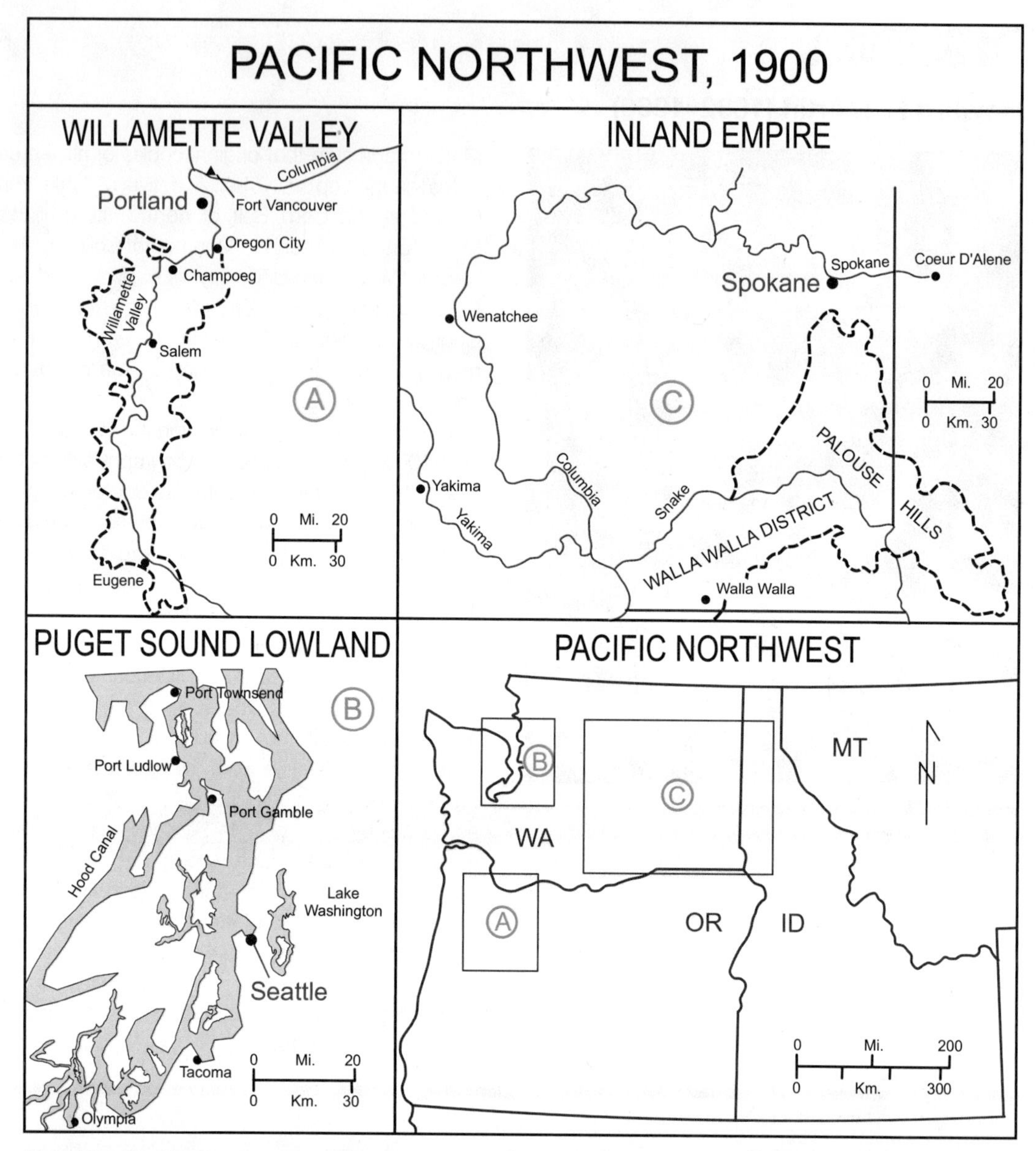

Map 16.4 Pacific Northwest, 1900. Shown are the larger region's three important subregions.

of the Willamette River, Oregon City could be reached from the Columbia by river steamers. McLoughlin's water-driven sawmill was the place where the migrants of 1842 clustered for the winter. Indeed, McLoughlin himself moved permanently to Oregon City when political control passed to the United States in 1846. By 1848, Oregon City had some seven hundred residents and was expected to become the regional metropolis. But Portland, founded in 1845 near the mouth of the Willamette at the head of navigation for oceangoing ships, would soon surpass Oregon City. When gold was discovered in California in 1848, Portland, in a better position to export the valley's two major products of wheat flour and timber, grew quickly to become Oregon's largest community. Oregon moved rapidly through the political stages of territorial status (1848) and statehood (1859). In 1853 the more centrally located Salem wrested Oregon's territorial capital from Oregon City, and with statehood officials made Salem the permanent capital; Salem had developed around a Methodist mission founded in 1834 as Oregon Country's first permanent American settlement.

The Willamette Valley became the major source area for settlers headed to the Puget Sound Lowland, a second notable Oregon Country subregion. About 1850, stands of Douglas fir around Puget Sound's indented coastline attracted lumber entrepreneurs. Early lumber-oriented communities came to exist at Port Townsend, Seattle, and Olympia; small company-owned sawmill towns were formed at locations such as Port Ludlow and Port Gamble (map 16.4B). Olympia, located at the head of Puget Sound, led in size, and was chosen in 1853 to be the capital of a new Washington Territory. Seattle's leader, Arthur Denny, an overland emigrant from Illinois who had found the Willamette Valley too crowded, in November 1851, with twenty-two people, settled at Seattle's Alki Point. Southwesterly winds on the point made anchorage for sailing vessels difficult while loading timber, and in February 1852, the community moved around the point to a better-protected Elliott Bay. That fall Henry Yesler was given a stretch of Elliott Bay waterfront, well-chosen between high bluffs to the north and tide flats to the south, and on it he built the only steam-powered sawmill then operating on Puget Sound. Logs were rolled or dragged down First Hill along (the original) "Skid Road," now Yesler Way, to be milled for export. Denny helped secure for Seattle the territorial university, and he gave ten acres at the top of First Hill for its first campus. An ample harbor, Yesler's mill, and good leadership would help Seattle rise in regional prominence.[8]

The dry eastern side of the Cascade divide attracted settlers more slowly. In the late 1860s farmers began to find that the moisture-retentive loessal soils found especially in the Walla Walla District and the Palouse Hills located in the eastern Columbia Plain were excellent for dry-farmed wheat (see map 16.4C). These farmers discovered that "wheat will grow and mature wherever the 'bunch grass' grows."[9] The Willamette Valley again contributed the larger part of settlers going to this new frontier. Local railroads tied this interior country initially to Portland. Urban centers developed at Walla Walla, Yakima, Wenatchee, and most notably at Spokane. Besides wheat, timber (pine) harvested in the Blue Mountains and the Rockies and minerals mined in Idaho became the area's exports, and these were channeled down the Columbia through Portland and, by the end of the century, increasingly by rail to Seattle. What emerged about 1900 was an imprecisely defined area called by promoters the "Inland Empire," a third notable Oregon Country subregion.

Meanwhile, Seattle grew steadily, and between 1900 and 1910 it overtook Portland in population (in 1910: Seattle 237,194; Portland 207,214) to become the region's leading metropolis. At the turn of the century the label "Oregon Country" was also being replaced by "Pacific Northwest" as the region's name. Rail connections and the Alaska gold rush seem to explain Seattle's successful rise in regional prominence. The Union Pacific had built north from Portland to Seattle, and in 1893 the Great Northern built over Stevens Pass to Seattle; terminals of each line stood side by side as symbols of Seattle's land connections. With the discovery of gold in Alaska in July 1897, the harbor of this northernmost city became the shipping gateway to Alaska. Seattle also made some bold decisions about environmental issues that, arguably, gave it a boost. Environmental historian Matthew Klingle recounts how R. H. Thompson, Seattle's powerful city engineer between 1891 and 1911, masterminded Seattle's acquisition of the Cedar River watershed for a clean municipal water supply, and then used some of this water to level three hills (the Denny, Jackson, and Dearborn regrades) by sluicing hydraulically muddy glacial materials downhill to fill in adjacent Elliott Bay tideflats. Klingle emphasizes that acquiring the Cascade foothill watershed and reshaping central Seattle's topography did not all go smoothly, and it saddled Seattlites with heavy social and economic costs.[10]

Notes

[1] Drawn on here is Johansen and Gates, *Empire of the Columbia*, 53 and 69–71, where author Johansen explains how, on April 29, 1792, Vancouver (represented by his officers) and Gray actually met near the entrance of the Strait of Juan de Fuca, just before Vancouver entered Puget Sound and Gray sailed south to explore the Columbia River.

[2] Meinig, *The Shaping of America*, 2:75, for "diplomatic triumph"; 2:118, for the boundary settlement in 1846 as "a major geopolitical victory" for the United States.

[3] See Hunt, *Physiography of the United States*, 54 and 383, for the Columbia as an antecedent river; see Gore, "Cascadia," 14, 15, 20, 25, for the Juan de Fuca plate; 18, for Brian Atwater.

[4] In *The Willamette Valley*, 17–21, Bowen argues that sickness and disease were likely the leading push factors for the migration to Oregon in the 1840s. See ibid., 25, where his table on population origins by state in 1850 shows Missouri, Illinois, Ohio, Kentucky, and Indiana contributing 5,431 (46 percent) of the 11,873 total. If the Oregon-born element (2,191), largely the children of these immigrants, is deducted from the total, the people from the five states represent 56 percent of the total.

[5] Much has been written about the Oregon Trail. The "too thick to drink" quote is from Gibbons, "The Itch to Move West," 158. For Frémont and the gradual ascent to South Pass, see Franzwa, *The Oregon Trail Revisited*, 277.

[6] Vance, "The Oregon Trail and Union Pacific," 357–79, esp. 357, for "locational succession," although as a label for the concept, he seemed to prefer "continuous use"; and 374, for the grade limit of ninety feet per mile. Vance's article appeared in December 1961; Martin's cited course was given in the spring quarter of 1960.

[7] Bowen, *The Willamette Valley*, 60, for maps of areas of prairie in 1850; 9, for the French Prairie; 61, 62, 71, for how most settlers selected claims on the prairie-woodland margins; 50, 52, 53, for migrant clustering in kinship clans; 12 and 63, for a succinct presentation of the settlement process that began in 1842.

[8] Historian Murray Morgan, *Skid Road*, 29, for Yesler's steam-powered sawmill being unique on Puget Sound; 9, for Yesler Way as the original "skid road." Also see English professor Roger Sale, *Seattle*, 8, for the problem of winds at Alki Point; 21, for the qualities of the site given to Yesler for his mill; 24, for the naming of the first building completed in 1895 on the new University of Washington campus as Denny Hall, in Arthur Denny's honor.

[9] Meinig, *The Great Columbia Plain*, 264.

[10] The census figures for Portland are 90,426 (1900) and 207,214 (1910); for Seattle they are 80,000 (1900) and 237,194 (1910). For Seattle's rise, see Sale, *Seattle*, 50–93. For R. H. Thompson's engineering feats, see Klingle, *Emerald City*, 86–118.

Mormons and the Great Basin 17

Mormons have made the Great Basin a truly distinctive American culture region. And how members of the Mormon religion and subculture went about creating the region's unique features is a significant chapter in American history. In this chapter's first section, "Great Basin Setting," we establish that the region is both rugged and dry. Under "LDS Origin and Zions," we then follow the Mormon migration to this relatively uninviting part of the American West, a migration that gradually led from upstate New York, the birthplace of the new faith, to Salt Lake City, the religion's headquarters and regional focus. Under "Mormon Colonization," we find a centralized church hierarchy in Salt Lake City directing Mormons in the founding of agricultural villages in the Great Basin and beyond. Finally, under "The Mormon Impress," we note those qualities of the cultural landscape and the Mormon society itself that make the Great Basin different.

Great Basin Setting

Between the Wasatch Mountains and the Sierra Nevada is an area that John C. Frémont named the "Great Basin." The name is most appropriate because the area is, indeed, one giant interior-drainage basin (see map 17.1). Both the Wasatch Mountains and the Sierra Nevada (Spanish for "snowy mountains"; to say "Sierra Nevada Mountains" is redundant) are high fault-block mountains. They represent steep-sided blocks of Earth's crust that have been pushed upward along "thrust" faults. Between them are fully one hundred lesser north-south aligned fault-block mountains with their own interior-drainage basins. These rugged landforms of the Great Basin constitute "basin and range" topography; the Great Basin itself comprises the western half of the larger Basin and Range Province that stretches east to Arizona, New Mexico, and Texas.[1]

The Great Basin is as dry as it is rugged. Nearly all of it receives ten inches or less average annual precipitation, which classifies it as arid or a desert. When it rains, which is often in the form of downpours of short duration, watercourses come to life. The materials carried by the intermittent streams are deposited as alluvial fans along the sides of the steep-sided mountains. If watercourses reach the bottom points of basins, they deposit fine particles in the form of brown mud or white salt—from an airplane, the small circles of brown ("playas") or of white ("salars") at the lowest points of basins are quite conspicuous in an otherwise uniformly brown-gray landscape. Snowpacks in the high Wasatch Mountains and the Sierra Nevada feed watercourses that are perennial. Such streams replenish the Great Salt Lake, which would otherwise evaporate away: The Great Salt Lake is four to six times saltier than oceans, and swimmers find that it readily holds them afloat. A source of water is critical for survival in a desert, and it was to the perennial streams at the foot of the Wasatch Mountains that the Mormons were headed in 1847.[2]

LDS Origin and Zions

The founder of the Mormon religion was Joseph Smith Jr. Born in Sharon, Vermont, in 1805, he became, with his family in 1815 and 1816, part of the Yankee exodus to upstate New York, where they settled on a farm near Palmyra (see map 17.2).

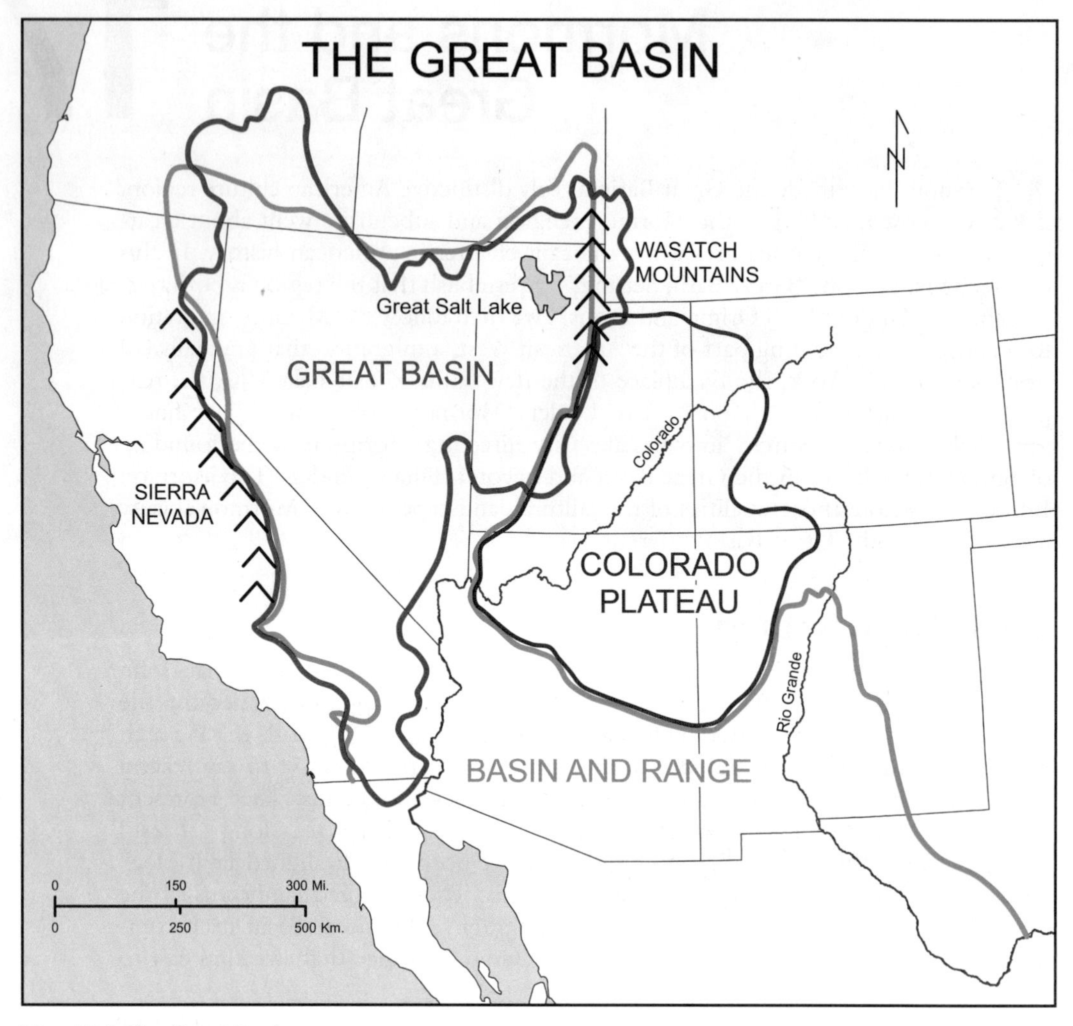

Map 17.1 The Great Basin.

Sources: The boundary in red distinguishes interior from exterior drainage for the huge Great Basin and is based on maps drawn by geographer Guy-Harold Smith in Fenneman, *Physiography of Western United States*, 226, 327, 368. Boundaries for the Basin and Range Province and the Colorado Plateau come from Hunt, *Physiography of the United States*, 278, 310–11.

According to Mormon belief, near Palmyra in 1823, Smith unearthed some inscribed plates from which he "translated" the Book of Mormon. This book purports to be the record of some of the early inhabitants of the New World—the descendants of a lost tribe of Israel known to Mormons as Nephites and Lamanites. The emphasis placed on the Lamanites (who eliminated the Nephites) explains why Mormons over the decades have made special efforts to convert Native Americans. In 1830 in Fayette, New York, Smith founded what has since 1838 been officially called The Church of Jesus Christ of the Latter-Day Saints. Better known as Mormons, members refer to themselves as "LDS" or "saints" and to non-Mormons as "gentiles."

In 1831 Smith and some seventy converts moved to Kirtland, Ohio (map 17.2), located in the Western Reserve east of Cleveland. That same year, Smith had a revelation that the Mormon Zion—or the final gathering place of the true believers—lay in western Missouri, specifically Jackson County, which contains Independence. On August 3, 1831, on a hill in the new community of Independence, Smith designated a site (Temple Lot) for the building of a future Mormon temple. Thus the 1830s

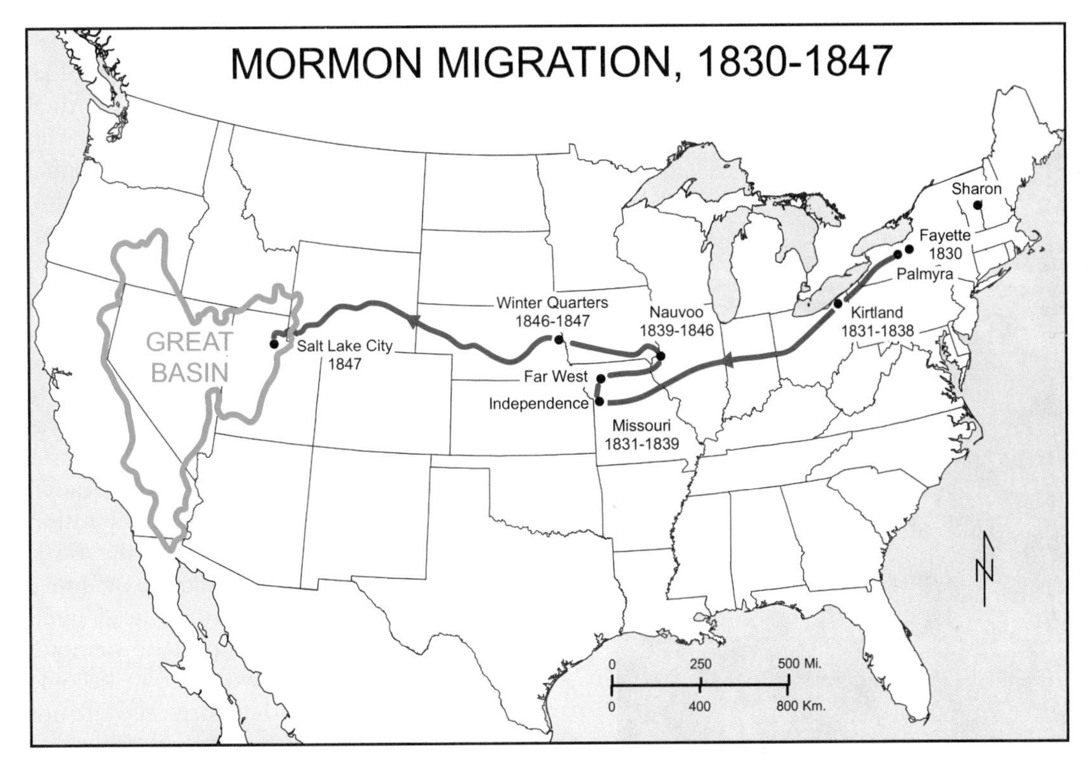

Map 17.2 Mormon Migration, 1830–1847.

found Mormons—who were growing in number under Smith's charismatic leadership and aggressive church missionary policies—in both Missouri and Ohio. Smith stayed in Ohio, and in Kirtland he oversaw the construction of the Mormons' first temple, built from local sandstone, and dedicated in 1836 (see figure 17.1). Meanwhile, tensions between Mormons and non-Mormons in Independence forced the Mormons to flee that community late in 1833. Those Mormons settled first in Clay County, north of Independence, and thirty months later they moved to Caldwell County, where they founded Far West. In 1838 Smith and many of the Ohio saints joined the Missouri Mormons at Far West, following internal conflicts as well as problems with non-Mormons in Ohio.[3]

In Missouri, trouble between the Mormons and their neighbors continued. Part of the problem was that Missouri was a border state, and southerners suspected the Mormons of siding with northern abolitionists. The Mormon attitude that western Missouri had been given to them by God offended some non-Mormons. And that Mormons were clannish and dressed and behaved in ways peculiar to non-Mormons made them generally suspect. Indeed, feelings got so heated that the governor of Missouri issued an "extermination order" that authorized officials to use the state's militia to force Mormons from Missouri. Thus in 1838 and 1839, the Mormons began backtracking to Illinois where, on a bluff above the Mississippi River, they established the town of Nauvoo (map 17.2).

By 1846 Nauvoo had some eleven thousand inhabitants, which made this mostly Mormon community even larger than Chicago. But the problems for Mormons continued. Some non-Mormons resented the rumored (and secret) Mormon practice of polygamy, or having multiple wives, which began in 1841 or 1842. And whether justified or not, non-Mormons accused Mormons of cattle rustling. Incensed by such accusations, Smith ordered church officials to destroy the press of an anti-Mormon newspaper. Smith was thrown into jail in the county-seat community of Carthage,

Figure 17.1 The Kirtland Temple in Kirtland, Ohio, 1996. Completed in 1836, this first temple is owned by the Reorganized Latter Day Saints (RLDS) who, unlike the LDS, open their temples to non-Mormons. A weather vane is mounted on top of the steeple; crosses are not displayed on any Mormon church.

Photograph by the author, June 3, 1996.

Illinois. An angry mob of non-Mormons stormed the jail, and on June 27, 1844, Smith and his older brother, Hyrum, were shot and killed in their jail cell. The Mormons were unprepared for this disaster and the untimely death of their prophet and church president. In the struggle that followed over who would lead, the Mormons splintered into at least sixteen bodies. Brigham Young (1801–1877) emerged as the leader of the major body, the LDS, while a son of Joseph Smith Jr., Joseph Smith III, would become the head of the reorganized LDS. In 1846 a second temple was dedicated in Nauvoo; eighteen months later it was largely destroyed by fire.

Young turned out to be a superb leader. From 1846 to 1847, he planned the relocation of Mormons to the Great Basin. At the time, the Great Basin was part of Mexico, so a move to the Great Basin meant leaving the United States. However, the general expectation among Mormons was that the area would soon become part of the United States, so the Mormons were not trying to leave the country as much as they were trying to separate themselves from non-Mormons. En route to the Great Basin, the Mormons spent the 1846–1847 winter in Florence, Nebraska (present-day North Omaha), which they called Winter Quarters. Despite Young's planning for food and shelter, a severe winter took its toll, and the cemetery at Florence is full of Mormon graves. In the spring of 1847 the Mormons followed the Oregon Trail west, but they stayed on the north side of the Platte River to minimize contact with non-Mormons, especially Missourians, and for better forage for their animals. The first migrant party led by Young arrived at the site of Salt Lake City on July 24, 1847. There, on a terrace above Salt Lake City at the foot of the Wasatch Mountains—the location of the "This is the Place" monument today—Young, who was sick at the time, propped himself up in his wagon, looked over the valley below, and reportedly declared, "This is the right place—drive on" (map 17.2).[4]

When identifying the Great Basin as the right place in 1847, Young in effect sanctioned the Great Basin as a *new* Mormon Zion. But to this day western Missouri is still viewed as *the* Mormon Zion, because Smith proclaimed it so. Today, the top of the hill in Independence that was visited by Smith in 1831 symbolizes in microgeography the lasting tensions within the larger LDS community. On the block of the grid of Independence that contains Temple Lot stands only the small frame church of the Hedrickites, one of the splinter groups that formed after Smith's assassination. The relatively small congregation of Hedrickites has been unable to raise the money to build a temple. Facing Temple Lot on a second block is a large stone "auditorium" built by the RLDS. In 1867 the RLDS, now known as the Community of Christ, returned to Missouri, although the order banning Mormons from the state was not repealed until 1976. Next to the auditorium, the RLDS in 1992 completed a spiral-shaped temple that, like the temple they own in Kirtland, is open to the public. And on the corner of a third contiguous block of the Independence grid, the LDS (Mormons) operate a large visitor center. Meanwhile, in Nauvoo, the LDS have rebuilt the temple destroyed in 1847, and on 1,250 acres they have restored Nauvoo's village-farm setting. The RLDS, however, are the owners of the forty-four acres that contain Smith's Nauvoo home and grave.

Mormon Colonization

In 1847 Brigham Young knew that his Great Basin destination would be a desert. Some of his followers arrived in the Great Basin with a knowledge of how Spanish people went about irrigating in New Mexico and California. Arriving in late July meant that crops would have to be planted quickly, so under Young's direction Mormons immediately began to build dams and dig irrigation ditches to divert water by gravity into fields from what they called City Creek. Meanwhile, Salt Lake City was surveyed into a grid with spacious ten-acre blocks, and streets that were wide enough to allow a wagon pulled by a team of oxen to make a U-turn. By the winter of 1847 some 1,600 Mormons had entered the Salt Lake Valley, and within three years there would be more than eleven thousand Mormons living in thirty-five agricultural communities located where patches of pasture and arable land coincided with a perennial spring or watercourse along the Wasatch front range.[5]

To maintain their relative freedom and identity, Mormons quickly sought political control over the Great Basin and a large part of the interior West. In 1849 Young petitioned Congress to create the state of Deseret. Taken from the Book of Mormon, "Deseret" is a term for honeybee; the beehive is a symbol of industriousness, and in the "Beehive State" of Utah this symbol is found on state road markers and elsewhere. Deseret was to include the vast area stretching from the Continental Divide west to the Sierra Nevada, and from the Columbia River watershed south to the Gila River—which marked the border with Mexico until the Gadsden Purchase of 1853 (see map 17.3). In their grand scheme Mormons even allowed for a window on the Pacific that would facilitate the arrival of an ever-growing number of converts especially from Great Britain and the Scandinavian countries. Congress refused to create Deseret and in 1850 established instead Utah Territory, with Young appointed as territorial governor. But Mormons acted as if Deseret were a reality. In 1851 they founded, around Deseret's outer margins, outposts that would contain one thousand or more Mormons, at San Bernardino (on a purchased Mexican ranch) and at Mission Station in Carson Valley. Smaller outposts included Fort Supply (1853), located near Fort Bridger, and a mission to Indians at Las Vegas (1855).[6]

During this period of expansion in the 1850s Mormons were rumored to be rebelling. In 1857 some three thousand federal troops were sent to Utah to quell the supposed rebellion. Although there was little bloodshed in the following so-called Utah War (1857–1858), the event found Mormon leaders changing their strategy. The outposts of Deseret were abandoned (Mormons had already been recalled from San Bernardino), and renewed energy now focused on the establishment of contiguous agricultural villages along the Wasatch and ranges farther south and into the higher tiers of valleys to the east. Villages came to exist from Brigham City north of Salt Lake City to St. George and the Virgin River Valley of "Dixie"; the arc-shaped pattern followed the base of the escarpments of the Colorado Plateau (map 17.3). Meanwhile, during the 1860s, a hostile federal government managed to whittle down the territory of those suspect Mormons as follows: On the west the new state of Nevada was founded (1861), which would need enlarging by one degree of longitude (1864) and, later, a second degree of longitude (1866). And on the east Congress created Colorado Territory (1861), then to the north Idaho Territory (1863), and finally Wyoming Territory (1868; map 17.3).[7]

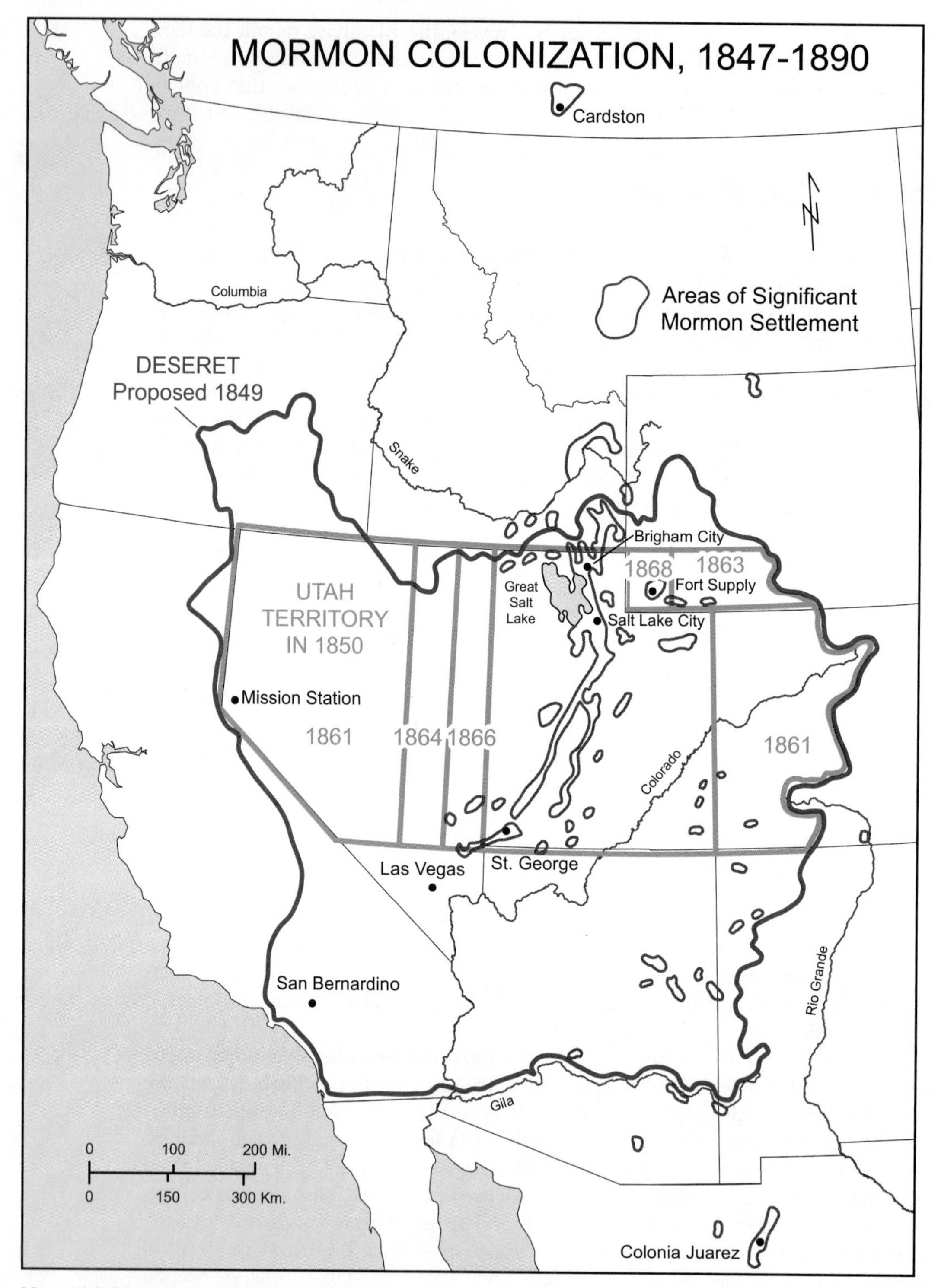

Map 17.3 Mormon Colonization, 1847–1890.

Sources: Meinig, *The Shaping of America*, 3:97, for Deseret and Utah Territory; and his "The Mormon Culture Region," 205, for areas of significant Mormon settlement.

Brigham Young directed Mormon colonization until his death in 1877. Under his thirty years of leadership in the Great Basin, 350 villages were founded, mainly along the west sides of the Wasatch and other ranges. Young knew that through irrigation, Mormon farmers could solve the problem of aridity. But his initial perception that by moving south and not north Mormons would encounter warmer temperatures and longer growing seasons was somewhat flawed. Because temperatures in a general way decrease to the north, Young was persuaded that Mormons should not colonize in that direction; certainly they should not exceed Utah Territory's northern border at latitude 42°. Young advocated instead that Mormons settle to the south where warmer temperatures, notably in Utah's Virgin River district—"Dixie"—would allow Mormons to grow subtropical crops (cotton, grapes, and figs) and further the goal of self-sufficiency. But the problem was that killing frosts on either end of a growing season are more a function of *altitude* than of *latitude*, and to the south, elevations increased, especially when Mormons followed Young's directives to settle the Colorado Plateau. Not until the 1870s did many Mormons begin moving north into the lower-elevation Snake River Valley in southeastern Idaho. In Idaho the growing season was longer, but because non-Mormons had preceded them, Mormons there were forced to compete for land.[8]

After 1877, Mormon expansion continued, but in more limited ways and with mixed success. Indeed, in the 1880s the issue of polygamy drove some Mormons to move beyond the boundaries of the United States. Americans were overwhelmingly opposed to polygamy. Responding to this, Congress in 1882 and 1887 passed the Edmunds and the Edmunds-Tucker acts that, aimed at Utah, strongly prohibited polygamy in United States territories. In manifestos issued in 1890 and 1904, the Mormon Church itself responded by ending the practice. Polygamous families, reports geographer Lowell C. "Ben" Bennion (see spotlight), represented an average of 25 to 30 percent of the LDS population as of 1870 (see figure 17.2). When polygamy was made illegal, husbands had to abandon given wives and the children of these wives, causing much hardship and even bitterness by some toward the church for complying with federal law and church manifestos. Anticipating that polygamists would be imprisoned, some Mormons decided to leave the United States. After 1885 eight Mormon villages or *colonias* were established in Mexico, within 120 miles (200 kilometers) of the US border; six were in northwestern Chihuahua and two were in northeastern Sonora. Colonia Juárez became the principal colony (map 17.3). And beginning in 1887, to preserve their plural families, some Mormons settled in the Cardston area of southwestern Alberta. In Mexico, the raiding of villages that came with the outbreak of the Mexican Revolution of 1910 prompted most Mormons to return to the United States in 1912; yet some remained in Mexico, others drifted back, and a Mormon presence in northern Mexico continues, as symbolized by a late-twentieth-century temple in Colonia Juárez.[9]

By 1890, at the time Mormon expansion basically ended, some 170,000 Mormons then living in five hundred villages had become the dominant people in the Great Basin and adjacent areas (see map 17.4). Mormons were different in several ways from others in the American West. First, especially where mining and ranching prevailed, men far outnumbered women in the West during the second half of the nineteenth century. But in Mormon society—also in the Spanish American Region and in Oregon Country—men and women were about equal in number. Second, Mormons were set apart demographically by their percentage of foreign-born settlers. Owing to aggressive missionary policies that began overseas in England in 1837, a steady flow of converts, many with LDS travel support, reached the new Mormon Zion. In 1870 Utah's foreign-born percentage was 35.4, compared with 14.4 for the rest of the nation. Third, and perhaps most distinguishing, Mormons

Spotlight

Lowell C. "Ben" Bennion

Lowell C. "Ben" Bennion. Photograph taken in 2012, courtesy of Bennion.

Ben Bennion was born in Salt Lake City and earned his bachelor's degree in geography (1960) at the University of Utah. His geography education continued at Syracuse University, where he earned both a master's degree (1965) and a doctorate (1971) under D. W. Meinig. His dissertation focused on patterns of German emigration to North America, Prussia, Russia, and the Hapsburg Empire in the eighteenth century. Bennion taught at Indiana University before moving in 1970 to Humboldt State University in Arcata, California. Shortly after 1970, Meinig introduced Bennion to LDS church historian Leonard J. Arrington, who offered Bennion two summer fellowships for research in the church archives. These fellowships redirected Bennion's focus to Mormons, notably their changing geography and polygamy's place in early Mormon society. Many publications have followed, including a guidebook coauthored with Gary B. Peterson, *Sanpete Scenes: A Guide to Utah's Heart* (1987, 2003), for which the authors received geography's prestigious J. B. Jackson Award. Now retired from Humboldt State (1999) and living in Salt Lake City, Bennion's steady stream of publications continues, interspersed with frequent trips to various world destinations with his wife, Sherilyn.

Figure 17.2 The Watkins-Coleman-Tatge House, Midway, Utah, 1977. This Gothic Revival house was built symmetrically to accommodate the polygamous builder's second and third wives equally. Located in a village founded by German-Swiss emigrants in 1859 above Salt Lake City in the Wasatch Range, the house is constructed of hand-pressed adobe brick.

Photograph by the author, April 26, 1977.

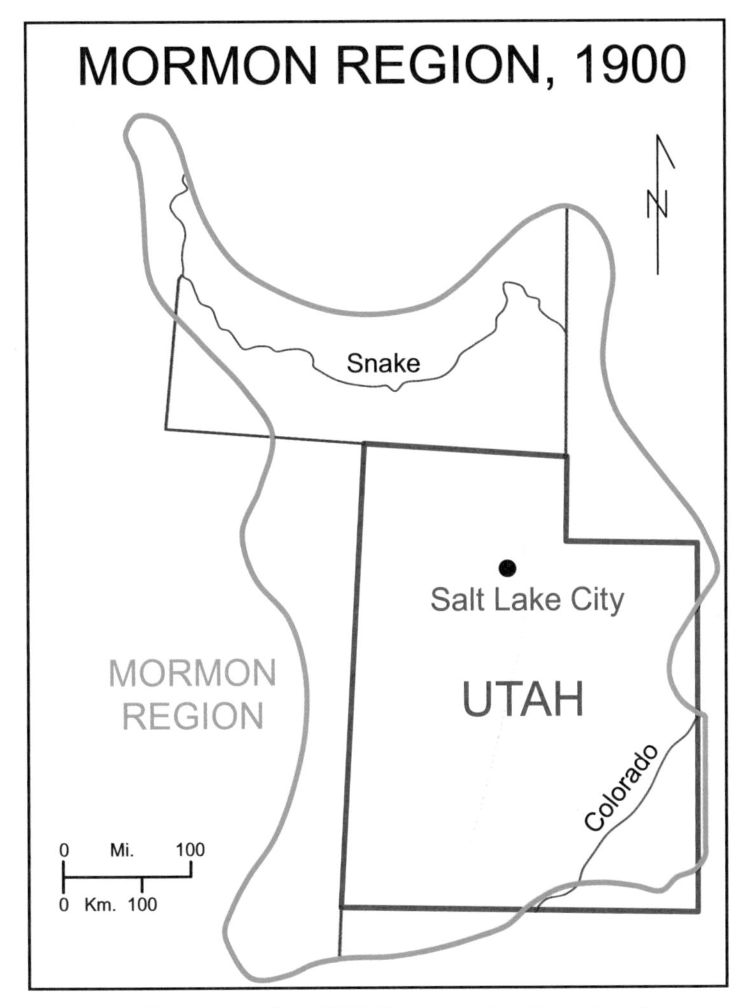

Map 17.4 Mormon Region, 1900. The generalized boundary shows the region to be focused on Salt Lake City and Utah (granted statehood in 1896).

were bound by religion into a strong communal society. A central church hierarchy directed LDS efforts to colonize, and this hierarchy carefully planned how to go about it. Believing in the divine origin of church directives, Mormons participated as congregations. The outcome was success for a people bent on taming a forbidding environment.[10]

The Mormon Impress

In one of his several studies about Mormon landscapes, geographer Richard V. Francaviglia lists ten characteristics or elements that offer clues to identifying Mormon settlements. He notes that before 1900 these clues were more obvious; searching for examples in the landscape today, he could have added, requires diligence. The context for three of Francaviglia's clues is that Mormons, having had roots in community-minded New England and upstate New York, settled in villages. Francaviglia notes (1) *wide streets*, (2) *barns on village house lots*, and (3) *open field landscapes around a village* as examples. Living in the Middle West explains another clue, the significant use by Mormons of the (4) *I-style*, or in Francaviglia's words,

Figure 17.3 I-Style House, South of Brigham City, Utah, 1974. Built of stone, this classic example of an I-house—tall and shallow, with two rooms over two rooms and a central door—lies near an arm of the Great Salt Lake.

Photograph by the author, August 25, 1974.

Figure 17.4 Three Mile Creek Meeting House, Perry, Utah. Built in 1889, this ward church located south of Brigham City was labeled a "Meeting House"—suggestive of Mormon roots in New England—in the circular plaque above the door.

Photograph by the author, August 24, 1974.

"Nauvoo-style," *house type* (see figure 17.3). Adjusting to a dry and treeless environment is the context for two additional clues: (5) *village roadside irrigation ditches* and the use of (6) *brick or stone building materials* for houses. (Geographer Richard H. Jackson notes that, lacking wood and brick, adobe was the "dominant building material" for Mormons before 1900.) Rounding out Francaviglia's list of ten clues are (7) *unpainted barns and farm buildings*, (8) *Mormon hay derricks*, (9) *Mormon fences*, and (10) *Mormon ward chapels*—always, he notes, without crosses.[11]

Mormon ward chapels bring us to the ecclesiastical landscape. Mormons organize their churches into three levels. A "ward" church represents the village or neighborhood level. Some three to five ward churches are then clustered into "stakes" that represent the middle level. Stake centers today were known as tabernacles in the nineteenth century. And stakes are then gathered geographically into "temple" areas. Mormons use temples for performing ordinances such as marriages and baptisms for deceased persons. Mormons, then, have organized their religious landscape into geographical units, with the outcome that where one lives within a Mormon society determines where that person worships. Having geography determine where one worships

Figure 17.5 The Box Elder Tabernacle and Brigham City, Utah, 1974. From I-15 looking east, Brigham City can be seen within the horizontal strip of dark green vegetation at the foot of the Wasatch Mountains. Above the vegetation the horizontal bench cut by Glacial Lake Bonneville (which shrank to become the Great Salt Lake) is visible on the Wasatch front range. This much-photographed tabernacle was built in 1876.

Photographs by the author, August 24 (Brigham City) and August 25 (tabernacle), 1974.

Figure 17.6 The Mormon Temple, Salt Lake City, Utah, 1974. The temple is so large that, if photographed from within the walls of Temple Square, only half can be taken at a time. It was built between 1853-1893 of granite quarried in the Wasatch Mountains.

Photograph by the author, August 25, 1974.

would appear to make Mormons unique among religions. Figures 17.4, 17.5, and 17.6 show examples from the nineteenth century of churches representative of the three levels; geographer Paul F. Starrs points out that Mormon church architecture was modernized and became standardized in the twentieth century.[12]

That Francaviglia's examples of the Mormon imprint have increasingly become relic features reinforces how much has changed since 1900. Utah in 2000 had two million residents, of whom some 70 percent were Mormon, making the state one of America's most religious in all ways that religiosity might be measured. Salt Lake City is now the headquarters of what has become a world religion that claims some fifteen million members. But less than half of all Mormons now live in North America. Beyond Utah, Mormons in North America live especially in urban areas along the Pacific coast, urban areas that became major destinations for much Mormon expansion after 1900. Because of its tithing policy, the Mormon Church is quite wealthy, and church policy makes it a priority to fund missionary activity. When Mitt Romney, a Mormon, ran for president of the United States in 2008 and again in 2012, reports about Mormons in the press characterized this subpopulation as a "mainstream force" in American society. Because of their family values, their emphasis on education, and their self-sufficiency, Mormons are a much-respected American subculture today. The generally held image of Mormons as a "peculiar people" in the nineteenth century has evolved into that of "model Americans" today.[13]

Notes

1 Fenneman, *Physiography of Western United States*, 326–95, for Basin and Range Province; esp. 348, for Frémont having named the Great Basin. Also see Hunt, *Physiography of the United States*, 308–47, esp. 308, where Hunt reports more than two hundred ranges in the province as a whole.

2 Fenneman, *Physiography of Western United States*, 351, for the Great Salt Lake as about four times saltier than oceans. E. C. Pirkle and W. H. Yoho, *Natural Landscapes of the United States*, 306, claim the Great Salt Lake is six times saltier than oceans.

3 Many facts and dates used here are drawn from Lowell C. Bennion's richly informed essay, "Mormondom's Deseret Homeland," 184–209.

4 Bennion notes how Mormons followed the north bank of the Platte in ibid., 192–93. Billboards that welcome motorists to Utah today advertise, "Still the right place."

5 Meinig's *The Shaping of America*, 3:89–113, for "Zion, Deseret, and Utah," a review of Mormons in the nineteenth century; esp. 96, for thirty-five settlements with eleven thousand Mormons.

6 For Deseret and its "portal" on the Pacific, see Meinig, "The Mormon Culture Region," 197–201. Also see Brightman, "The Boundaries of Utah," 87–95, esp. 90, for detail on outposts.

[7] Meinig, *The Shaping of America*, 2:171, records the number of troops as three thousand. In "The Boundaries of Utah," Brightman reviews Utah's areal reduction in the 1860s.

[8] Geographer Richard H. Jackson, in "Mormon Perception and Settlement," esp. 317, 321, 326, emphasized that for three decades, Young directed Mormon settlement to the south, believing that north of latitude 42° would be too cold for crops. See also ibid., 334, for Jackson's research drawing upon the diaries of both the elite leaders and the common folk (among other original church sources).

[9] See Bennion, "Plural Marriage, 1841–1904," 122–25. For an overview of colonias in Mexico, see Young, "Brief Sanctuary."

[10] For the figures 170,000 Mormons and 500 villages in 1890, see Meinig, *The Shaping of America* 3:104. See geographers Kay and Brown, "Mormon Beliefs about Land and Natural Resources, 1847–1877," 253, for figures with regard to the foreign-born in 1870; 264, for the notion that Mormons complied with church directives because of their "implicit belief in the divine origin of their church's teachings."

[11] See Francaviglia, "The Mormon Landscape: Definition of an Image in the American West," 59–61; and Jackson, "The Use of Adobe in the Mormon Cultural Region," 84 and 85.

[12] Starrs notes that the term "meetinghouse" is commonly used to refer to ward chapels and stake centers, and that stake centers are comprised of three to five ward chapels, in "Meetinghouses in the Mormon Mind," 352.

[13] For "peculiar people" and "model Americans," see Meinig, *The Shaping of America*, 3:113.

California 18

To have state boundaries define a culture region is, admittedly, a bit of a stretch. But for California, like Texas, state boundaries define an especially distinctive people and culture. While one of the most diverse states environmentally and demographically, California is generally perceived as one entity in part because of its general allure; its lifestyle, which has developed around automobiles and suburbs; and its romanticism based on Hollywood. We begin by noting California's pronounced environmental differences in "Three Longitudinal Bands." We then focus on California's all-important Hispanic legacy in "Mexican California." The discovery of gold brought a sudden surge of people—and a parallel rise in notoriety—to California, as seen in "Gold and the Mines." And after the gold rush, two parts of California develop quite differently, as noted in "Northern versus Southern California."

Three Longitudinal Bands

The "grain" of California runs north-south largely in three longitudinal bands. The high Sierra Nevada on the east is one of these bands (see map 18.1). The Sierra Nevada is a fault-block range whose more gentle slopes face the Pacific. This four-hundred-mile-long (650-kilometer-long) north-south landform stretches from about Lassen Peak south to where it curves west to become the lower Tehachapi Mountains. Mount Whitney (14,495 feet [4,280 meters]) is its tallest peak, but ten other peaks rise above fourteen thousand feet. The Giant Sequoia (*Sequoia gigantea*), trees that are among the largest and oldest (some over three thousand years old) on Earth, grow only on the Sierra Nevada's western slopes. Their cousins, the Coastal Redwoods (*Sequoia sempervirens*), are less massive and live less long (two thousand years), but they are taller (at 367.8 feet [112 meters], one California redwood is the world's tallest tree).

Coastal mountains along the Pacific Ocean form a second longitudinal band (map 18.1). The longest stretch, known as the Coast Ranges, has only one sea-level break: the mile-wide Golden Gate that leads into spacious San Francisco Bay. These mountains end at the Transverse Ranges, so named because they run counter to the grain of the state. And continuing south are the Peninsular Ranges, which become the backbone of the Baja California Peninsula. Along the entire eight hundred miles (1,290 kilometers) of California's coast is found only one major lowland that faces the Pacific. This is the Los Angeles Lowland, an area hemmed in by the local Santa Monica, San Gabriel, and San Bernardino mountains. Cajon Pass, which is located near the eastern edge of the Los Angeles Lowland, separates the San Gabriel from the San Bernardino mountains, and is the transportation corridor (followed by Interstate 15) that links the Los Angeles Lowland with the Mojave Desert and the Basin and Range Province.

Between the Sierra Nevada and the coastal mountains lies the final longitudinal band, the more than four-hundred-mile-long Central Valley (map 18.1). The northern third, the Sacramento Valley, is drained to San Francisco Bay by the Sacramento River. Its middle third, the San Joaquin Valley, is also drained to San Francisco Bay by a much smaller San Joaquin River. These two rivers converge in the so-called Delta, a flat area underlain by soft muds. The southernmost third of the Central Valley, the Tulare Basin, is one large inter-drainage basin. Successful efforts in the twentieth century to divert water from north to south to irrigate the San Joaquin Valley and

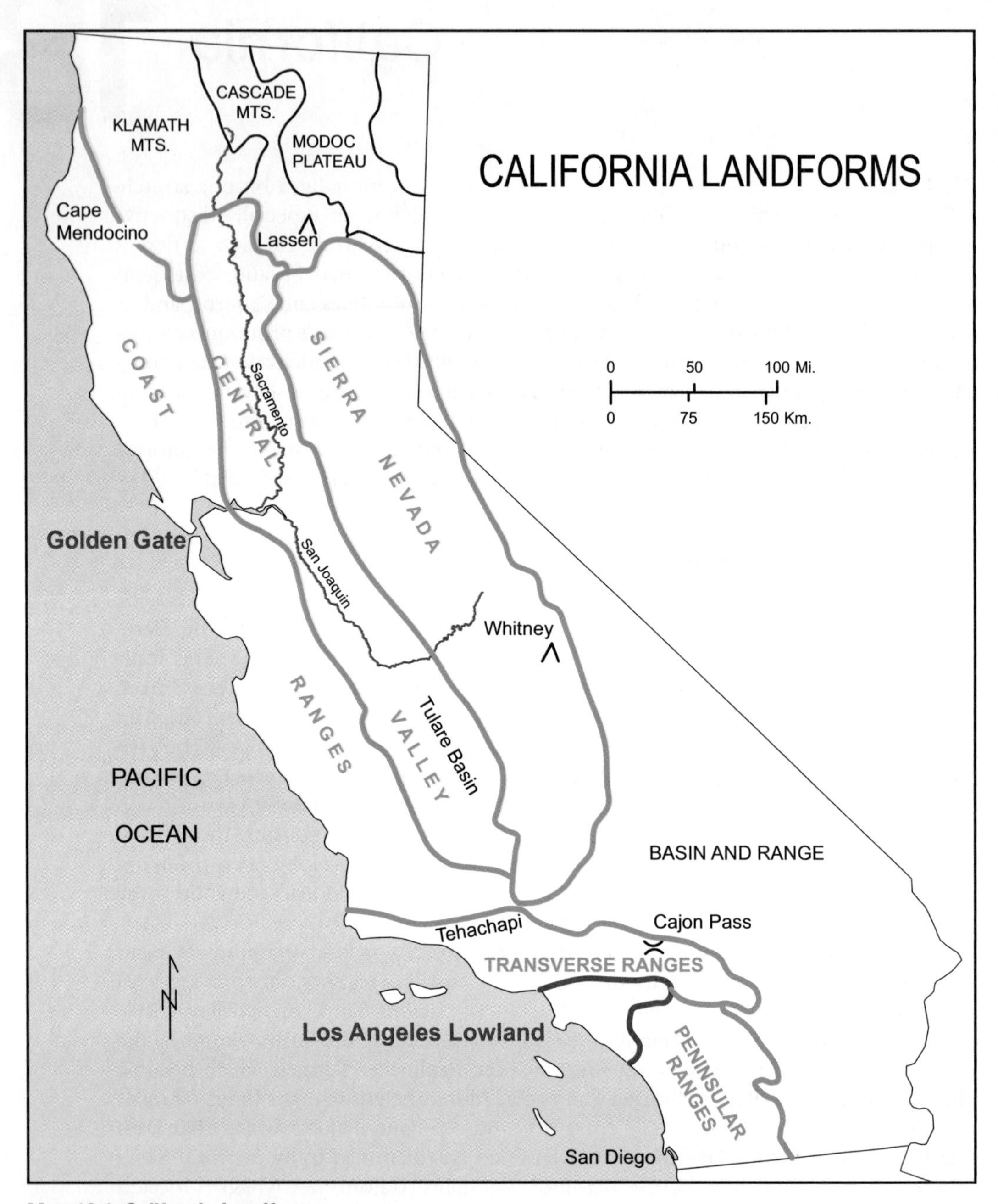

Map 18.1 California Landforms.

Source: Modified from "Geomorphic Provinces," in Beck and Haase, *Historical Atlas of California*, plate 3.

Tulare Basin segments of the Central Valley have created one of the world's most productive agricultural areas.

A Mediterranean climate, defined by its two seasons—a hot, dry summer and a mild, moist winter—stretches along California's coast from Cape Mendocino south almost to San Diego. Unique because it is the only humid climactic type with a pronounced summer drought, the Mediterranean climate is considered by many to be the most delightful of all climates—which is part of California's allure. This climate supports oak-parkland vegetation. The coastal hills and mountains are covered with grass and studded with oak that, near the coast, are evergreen (*Quercus agrifolia*), and inland in the dry season are deciduous (*Quercus lobata*). This coastal environment during the Mexican period (1821–1848) became the setting for what a number of Americans in the East found quite fascinating.

Mexican California

Two important developments made Mexican California different from pre-1821 Spanish California (see Chapter 2). The first was a flourishing hide and tallow trade. Franciscan missionaries used their captive Indian labor to raise longhorn cattle on their expansive mission lands. The major products were cowhides to be made into leather goods, and tallow (solid fat) used for candles and soap. The trade between the missionaries and sailing vessels plying California's coast began before 1800, yet before 1821 Spanish regulations inhibited the commerce. Without such regulations after 1821, the trade peaked in the Mexican period. Ships coming mainly from Boston, if they followed the rules, would put in at Monterey, the capital, to register an invoice of the ship's cargo and pay customs. Then they would skip down the coast, trading at important collection points that included San Francisco Bay (for missions San Jose and Santa Clara), Monterey, San Luis Obispo, Santa Barbara, San Pedro (the port for Los Angeles), and San Diego (see map 18.2). On the beach at San Diego, sailors would scrape and flatten the hides and store them in guarded "hide houses," while the crew made additional runs from north to south down the coast in order to fill the ship's hold. Vessels could carry between thirty thousand and forty thousand hides, each of which weighed about twenty-five pounds. Once loaded, ships would sail south, stopping perhaps at Lima, where tallow was in perennial demand, and then around Cape Horn and on to Boston. So important was Boston in the endeavor that californios knew of the place where Americans lived simply as "Boston."[1]

Information about the hide and tallow trade was beautifully chronicled by Richard Henry Dana Jr. in *Two Years before the Mast*, published in 1840. Dana (1815–1882) contracted the measles while a junior at Harvard. To regain his disease-weakened eyesight and for some adventure, he and two companions signed on a ship as common sailors in Boston harbor. Their ship, the *Pilgrim*, took them to California in 1834, and Dana returned on a company sister ship, the *Alert*, in 1836. In California for sixteen months, Dana jotted down his thoughts in a pocket notebook while on shore, and he expanded his thoughts into a journal when back on ship. So eager was Dana to see his family in Cambridge when he returned to Boston that he had his trunk, which contained his journal, sent to the family home. The trunk never arrived. In 1838 and early 1839, while attending law school at Harvard, Dana wrote his classic account of California and the hide and tallow trade, relying only on his pocket notebook, his memory, and perhaps letters that he had sent home. Interest in California was high, and the book was an immediate success. Unfortunately, in 1840 Dana sold the rights to his book to his publisher and did not reap the financial rewards from its many subsequent editions until 1868, when the copyright reverted to him. Dana took away from the experience a deep concern for the possible cruelties that resulted from the autocratic behavior of a ship's captain, and with his law degree (awarded in 1840) he focused on reforming the law of the sea.[2]

In the mid-1830s, Dana witnessed but said little about mission secularization, the second development to make Mexican California different. California's Franciscans, as clergymen who represented a cloistered order, were one day to be replaced by "secular" clergy who lived in dioceses within the community. Politicians in Mexico in the 1820s and 1830s were annoyed that the Franciscans in California were prospering from the labor of Indian recruits, who were not being integrated into California's

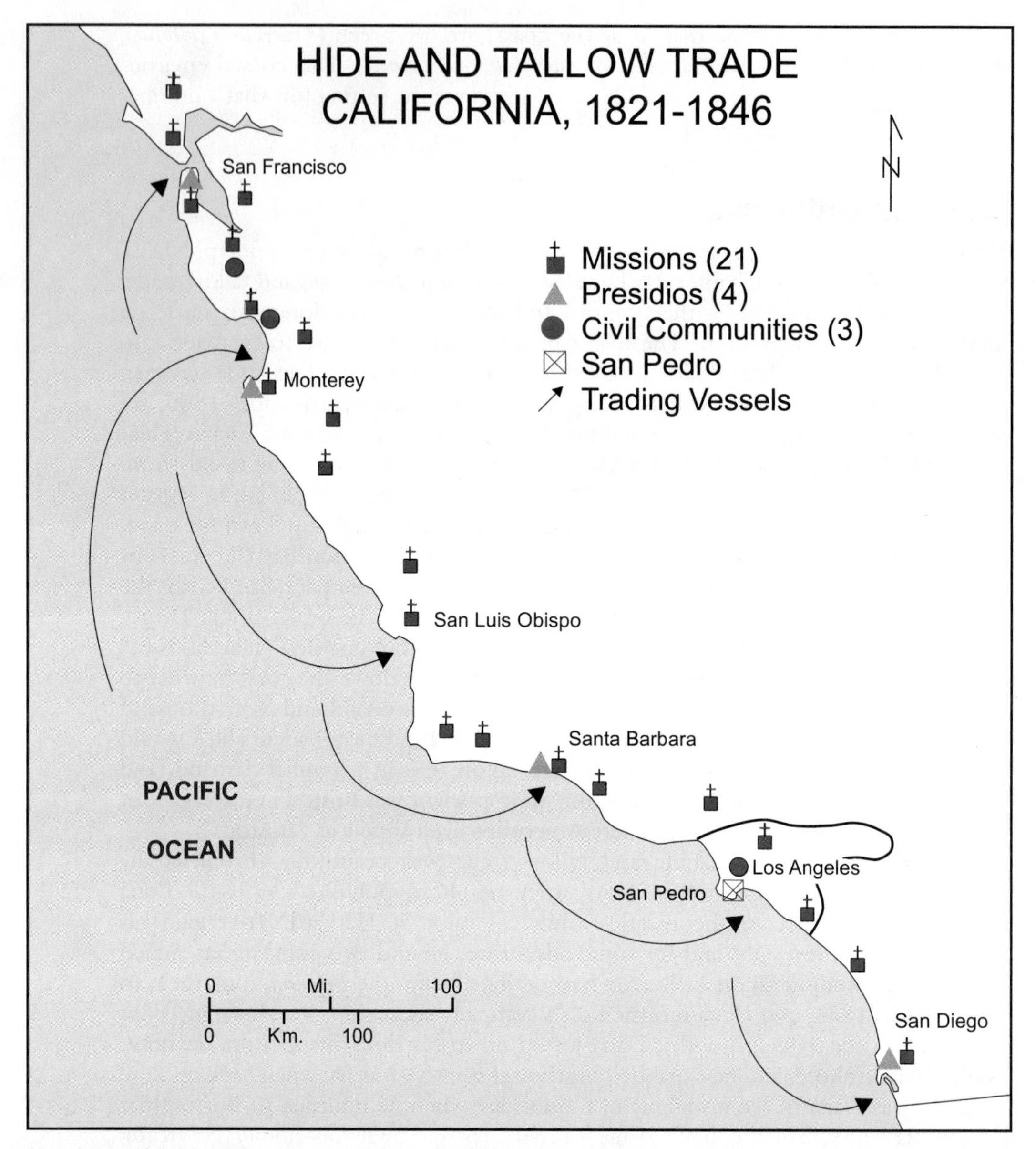

Map 18.2 Hide and Tallow Trade, California, 1821–1846. The trade began before 1800; the outbreak of war in 1846 and the gold rush brought about its demise.

society. Persuaded that the missionary order was stifling California's economic development and population growth, they passed several laws, notably one in 1833, that brought about the transferring of California's missions to secular clergy. Despite the strong protests of the Franciscans, local authorities were empowered to strip the order of its landholdings, including the mission compounds themselves (compounds that were not restored to the Roman Catholic Church until the 1860s). During the 1830s and 1840s, California's governors awarded the mission holdings as private ranchos to well-connected citizens. For starter herds, missionaries were obligated to "lend" some of their cattle to the new rancheros. And many of the mission Indians—those who neither returned to the places from which they had been recruited nor otherwise drifted away—became the laborers of the new rancheros. These developments are discussed by historian Robert Jackson.[3]

The Mexican period, then, became one of rancho granting. Nearly eight hundred ranchos were awarded between 1822 and 1846, compared to about thirty in the Spanish era. David Hornbeck, geography's expert on the subject, gives the figure 809 for total claims filed in the American period (almost all claims were for ranchos), of which 604 were patented. The Franciscans themselves, as noted by Jackson, documented their own demise in landholdings, livestock, and crops produced. Besides californios, a few of those favored with rancho grants were recent intruders to California. For example, in 1842 one Nicholas Den, who had arrived in California from Ireland in the 1830s, was awarded the Dos Pueblos Rancho in the Goleta area west of Santa Barbara. Den was among the new elite class who, until the end of the Mexican period, profited from the hide and tallow trade. By 1846 the number of californios living along California's coast had grown to at least ten thousand, and among them lived perhaps seven hundred intruders like Den, largely American and British men who typically became Mexican citizens and married señoritas from the californio upper class.[4]

In the process of confirming California's ranches in the American period, boundaries that were often vague and imprecise had to be surveyed and new ones established. An interesting study of the fifty-five ranchos awarded in the Los Angeles Lowland shows what impact these boundaries had on the landscape. In 1962 or 1963, UCLA geography professor Howard J. Nelson (see spotlight) and a class of his graduate students went to the field to determine what was lasting about ranchos. They found no example where a ranch house and its adjacent open-spring water source became the focal point for a future urban community. But they did find significant coincidences between rancho boundaries and a number of other features. As measured in linear miles, rancho boundaries persisted as roadways (173), political (county and city) lines (88.6), street disruptions (66), railroad rights-of-way (12), power line easements (about 12), and drainage ditches (limited miles). The class concluded that a relatively small number of californios, through surveys undertaken in American times, had indeed etched their property lines in Southern California's permanent landscape.[5]

If we revisit the four frontier outpost-clusters one final time, we can fit California into the larger Borderlands picture. The evolution is shown diagrammatically in figure 18.1. In the Spanish period the four frontier outpost-clusters—California, Pimería Alta (Arizona), New Mexico, and Texas—were isolated from one another and were only tenuously tied to a distant Mexico City (figure 18.1A). North of these outpost-clusters, the Adams-Onís boundary was drawn in 1819. In the Mexican period, Anglos intruded upon three of the four outpost-clusters (figure 18.1B). They engulfed the tejanos (see Chapter 12), they intruded lightly upon nuevomexicanos (see Chapter 15), and now in California they intruded lightly as hide and tallow traders. And during the Mexican era, it was over the Spanish Trail that the first sustained contact between two outpost-clusters developed, as noted in Chapter 15. To initiate the American period shown (figure 18.1C), the United States–Mexico border was realigned with the annexation of Texas in 1845, the Mexican Cession in 1848, and the Gadsden Purchase in 1853. The discovery of gold in California in 1848 brought a sudden and huge influx of Anglos to that state. Like the tejanos, californios (in stages) were engulfed. Thus only the New Mexico and Pimería Alta outpost-clusters were spared from heavy Anglo intrusion—until the twentieth century.

Spotlight

Howard J. Nelson (1919–2009)

Howard J. Nelson. Photograph courtesy of the Department of Geography, UCLA.

Howard Nelson was an urban historical geographer who spent his entire career at one institution, UCLA, from 1949 to 1986. Born in 1919 near Gowrie, Iowa, where he attended high school, Nelson earned his Bachelor of Arts degree (with high honors) at Iowa State Teachers College in 1942. Following service in the Army between 1943 and 1946—he was in charge of the Map Department at Fort Leavenworth, Kansas—Nelson enrolled at the University of Chicago where he earned both a master's (1947) and a PhD (1949) in geography. At UCLA, Nelson's publication, "A Service Classification of American Cities" (1955) brought him early recognition. In his many publications to follow, Nelson focused mainly on the evolution of urban morphologies in American cities. A gem of an example is his "The Spread of an Artificial Landscape over Southern California" (1959). A second gem, "Townscapes of Mexico" (1963), found him making several research trips beyond American borders, which was rare for him. From my UCLA days I remember Nelson best for two reasons: When chair of the department (1966–1971), Nelson had a couch in his office. When visitors like me would enter his open door, he would immediately get up from behind his desk and motion the visitor to sit opposite him at the end of that couch. Nelson championed equality, and his genuine desire to help matched his accessibility. And in the lunchroom Nelson would bring out his lunch in a brown bag. He promptly ripped this brown bag down the middle to make himself a placemat. This seemed quite extravagant to me, a graduate student whose tight budget required reusing paper sacks. Nelson retired from UCLA in 1986 at the age of sixty-seven and died at ninety in Carmichael, California.

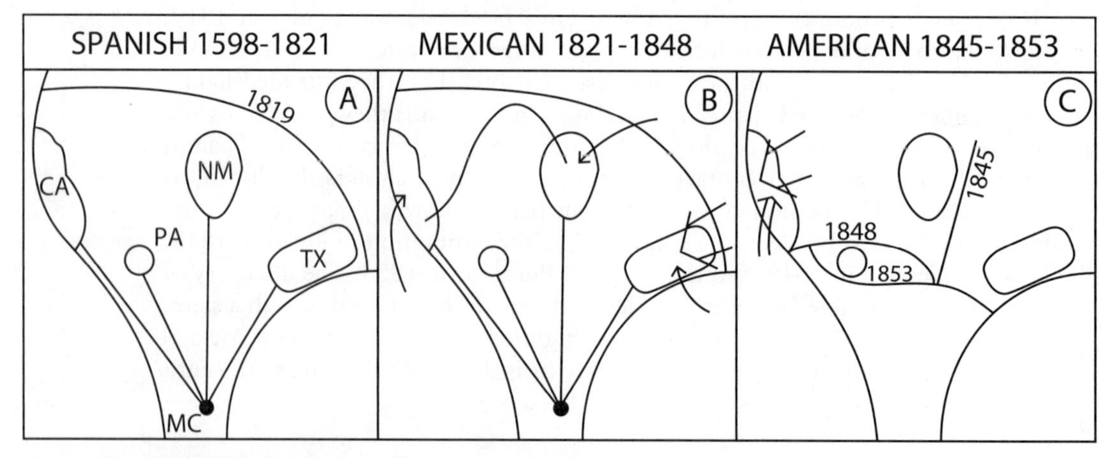

Figure 18.1 Borderlands Outpost-Clusters, 1598–1853.

Source: This figure was inspired by Meinig, *Southwest*, diagrams on 24 and 122.

Gold and the Mines

On January 24, 1848, James Marshall, a carpenter employed by John Sutter, discovered gold in the millrace at Sutter's sawmill. Sutter, a onetime Santa Fe trader (1835–1837), had moved to California in 1839, and in 1848 he operated a small sawmill on the south fork of the American River. The site is now in the town of Coloma, located in the Sierra Nevada foothills fifty miles east of Sacramento (see map 18.3). Marshall's discovery, made only nine days before the signing of the Treaty of Guadalupe Hidalgo (February 2) which ended the Mexican period, precipitated one of the largest migrations in the nineteenth century. Word soon spread to the coastal californio population and beyond, to Mexico, and in 1848 perhaps ten thousand Sonorans, as Mexicans

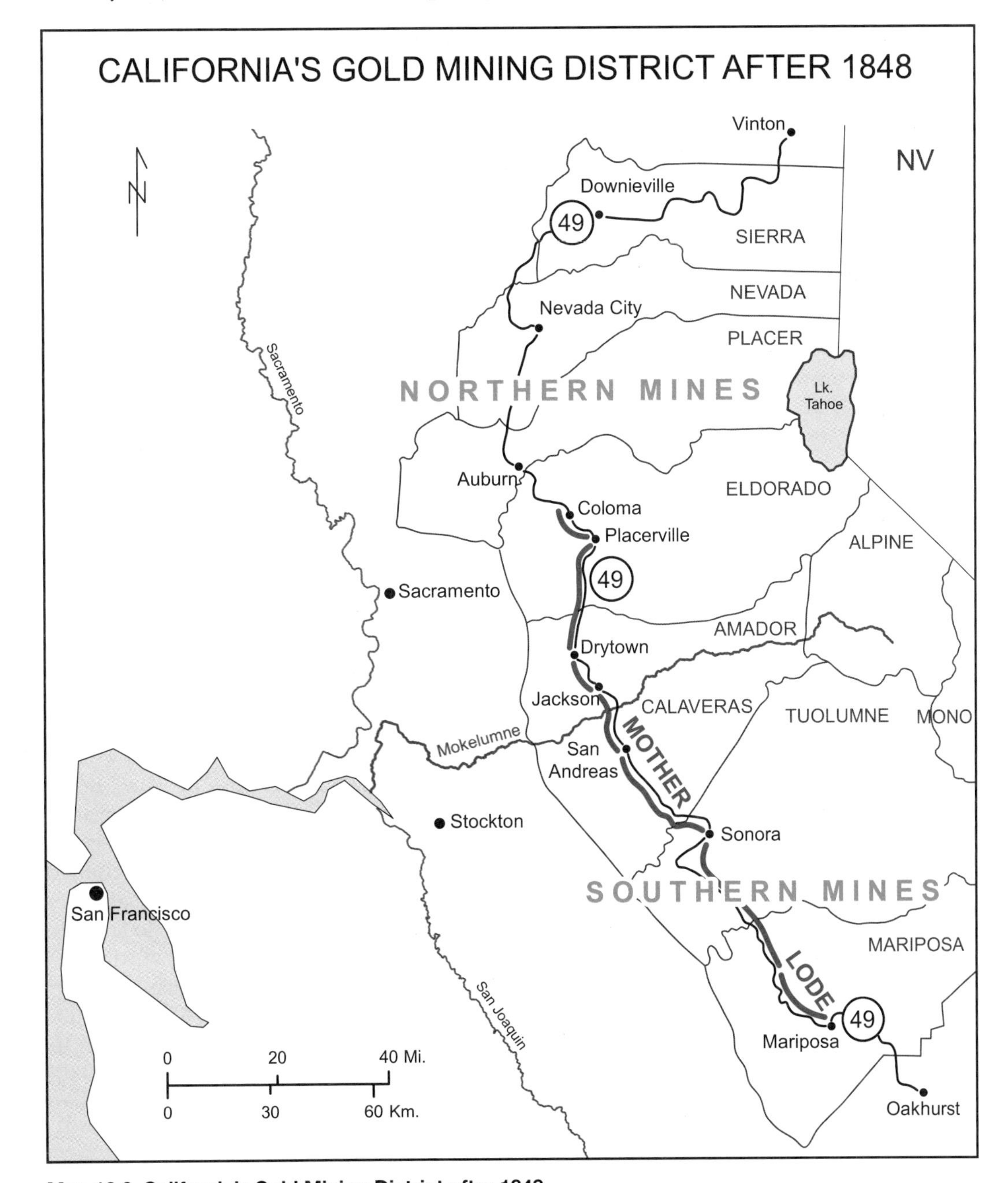

Map 18.3 California's Gold Mining District after 1848.

Source: Beck and Haase, *Historical Atlas of California*, plate 50, regarding the mother lode.

were called because so many came from the northwestern state of Sonora, in addition to men from Chile, Peru, and elsewhere in Latin America, came to mine. Word reached the New York newspapers in August, too late for anyone in that readership to travel to California for the mining season of 1848. Instead, that onslaught came as the "forty-niners." By 1850 California's population was enumerated at 92,597 (definitely an undercount), and by 1852 it exceeded 200,000.[6]

Sonorans who arrived in 1848 came overland up the coast; their migration was seasonal, except for some who stayed over the winter months largely in Los Angeles. The forty-niners and those who followed used several routes. Those from New York and the New England states, both major contributors, favored traveling by sea, either the five- to six-month trip around Cape Horn, or by crossing over Panama, Nicaragua, or Mexico. The most popular route for most Americans, however, was the Oregon Trail and its cutoffs. Ironically, the Mormons, who sought to be isolated in 1847, found themselves on a heavily traveled route (from which they did profit by selling supplies). A number traveled via the Santa Fe Trail, beyond which many used a new route that followed the Gila River. Like Americans on the East Coast, European gold-seekers who came, notably from Ireland and Germany, arrived by ship. San Francisco became an instant metropolis and the gateway for those arriving by sea.[7]

The mines themselves stretched for about 150 miles (240 kilometers) along the foothills of the Sierra Nevada from roughly Downieville south to Mariposa (map 18.3). Activity basically focused on the eight counties shown whose boundaries had come to exist by 1860 (Alpine and Mono counties would be created out of four of these counties between 1860 and 1870). Quartz rock impregnated with gold, known as the mother lode, outcropped sporadically between Coloma and Mariposa, and north of Coloma was more gold-bearing bedrock. Over the millennia streams had eroded these gold-bearing deposits and had deposited the metal along riverbanks and in stream beds to the west. Miners distinguished between what they called the Northern Mines and the Southern Mines, which were separated at the Mokelumne River. American and European miners worked the Northern Mines, whose streams flowed to the Sacramento River, while Latin Americans (who had arrived first) worked the San Joaquin–watershed tributaries of the Southern Mines. The Latin American miners were resented and treated quite brutally, and by 1851 many had been driven permanently from the gold diggings. Sacramento and Stockton were the outfitting centers for the Northern and Southern mines, respectively; San Francisco was the ultimate supplier.[8]

Mining went through three phases. In the first, "placer" mining, miners worked the surface sands and gravels along stream beds using a shallow pan or *batea*. Being heavier, the small particles or nuggets of gold would sink, as foreign material washed over the rim of the swishing pan. By 1852, stream beds had been pretty much worked over and mining turned to two additional phases, both of which required capital to acquire machinery and timbers for shafts. One of these was "quartz" or "lode" mining, where miners went after veins of gold in the mother lode itself. Quartz mining lasted well into the twentieth century. In a third phase, "hydraulic" mining, water under pressure in hoses was used to blast away at riverbanks and overburden, from which the gold was captured in sluices. First used in 1852, hydraulic mining destroyed topsoil and filled stream channels with sediment, resulting in its being outlawed in 1884. Geographer L. Allan James made a case study in the 1980s of the destructive results of hydraulic mining in the Bear River Valley, a tributary to the Feather River some thirty miles (fifty kilometers) north of Sacramento.[9]

The high point in the amount of gold extracted seems to have been reached early—in 1852. Thereafter, placer mining declined, quartz and hydraulic mining picked up, and miners came to be concentrated in fewer centers. Meanwhile, the number of miners decreased as some returned home, others drifted to new mining districts like Comstock Ledge in Nevada, and still others turned to farming at lower elevations or to lumbering. Mining had unfortunate consequences for the environment. Streams today are choked with sediment, especially from hydraulic mining; officials must control floods, manage reservoir sedimentation, and monitor pollution and aquatic life. The damage is also obvious aesthetically. But on the bright side, the economies of the eight mining counties benefit today from tourism. Highway 49 (an appropriate number) connecting Vinton and Oakhurst passes through more than a dozen onetime mother lode and other mining towns. False-front buildings that line narrow streets re-create the atmosphere: Hotels, dry goods stores, assay offices, blacksmith shops, and the ubiquitous saloons (Drytown had twenty-six of them) are found alongside museums and ice cream parlors that attract a Bermuda-shorts-wearing public.[10]

Northern versus Southern California

Because of the gold rush, Northern and Southern California developed quite differently. The influx of people to Northern California was so large that by 1852, of the state's two hundred thousand total, some fourteen of every fifteen Californians lived north of Monterey. The consequences of this population disparity for the old californio population were major. In 1852, some twelve thousand to fifteen thousand californios, together with more than seven hundred mainly American and British expatriates, lived along the coast between Sonoma and San Diego. Californio proportions in the 1850 census were especially high in two counties: Santa Barbara (93 percent) and San Luis Obispo (87 percent). The larger californio populations in Los Angeles County (2,700) and Monterey County (1,400) had attracted newcomers, but still californio percentages in those two counties were 76 and 74, respectively. But north of Monterey, by 1852, county californio percentages (which in 1850 had ranged between 7 and 35) had plummeted. Californios there found themselves besieged by waves of settlers and squatters who appropriated their cattle and crops and even cut down their orchard trees for lumber. Historian Leonard Pitt records the fate of prominent Bay Area families like the Peraltas and Vallejoses who, under the onslaught, quickly sold their ranchos and literally fled in fear for their lives.[11]

Hornbeck outlines the contrast in rancho adjudication between Northern and Southern California. In Northern California, grants were adjudicated more quickly; judging from surnames on grants, more Americans received grants than did californios, and grants were subdivided more quickly into parcels for farming. In distant Southern California, ranch grants were adjudicated less quickly, generally after 1870; californios received the majority of the patents; and grants remained intact and in the hands of californios longer—some beyond 1900. In the 1850s rancheros in Southern California profited by selling cattle to the streams of migrants coming through, or by driving cattle to the mines themselves. Unsuccessful miners did arrive early in Southern California, but only in small numbers. For example, in 1856, several onetime Argonauts seeking to farm squatted temporarily in a corridor of government land sandwiched between two ranchos in the Santa Ynez Valley of central Santa Barbara County—in the very heart of the californio population stronghold. A severe drought in Southern California from 1862 through 1864 decimated the livestock of californios, and this accelerated the selling of ranchos and subdivisions before it might otherwise have happened. Even so, down to the early 1880s, Southern California was known as the Cow Counties because development had lagged so behind Northern California.[12]

The "Boom of the Eighties" is what launched Southern California's growth in a major way. According to historian Glenn Dumke, real-estate interests that focused on developing especially Los Angeles but also Santa Barbara and San Diego, were behind the boom. Advertising Southern California as a paradise for health seekers and as a subtropical location for agriculture, promoters subdivided urban tracts for houses and sold undeveloped land to people intent on creating vineyards and orange groves. Facilitating the arrival of Americans from the East were two railroads, the Southern Pacific, which had built south to Los Angeles through Tehachapi Pass in 1876, and then on to New Orleans via San Gorgonio Pass, and the Santa Fe Railroad, which had arrived from the East via Cajon Pass in 1885 (see map 18.4). The frenzy of the real-estate boom peaked in 1887, when competition between the two railroad carriers grew so fierce that the price for a ticket from Kansas City to Los Angeles fell

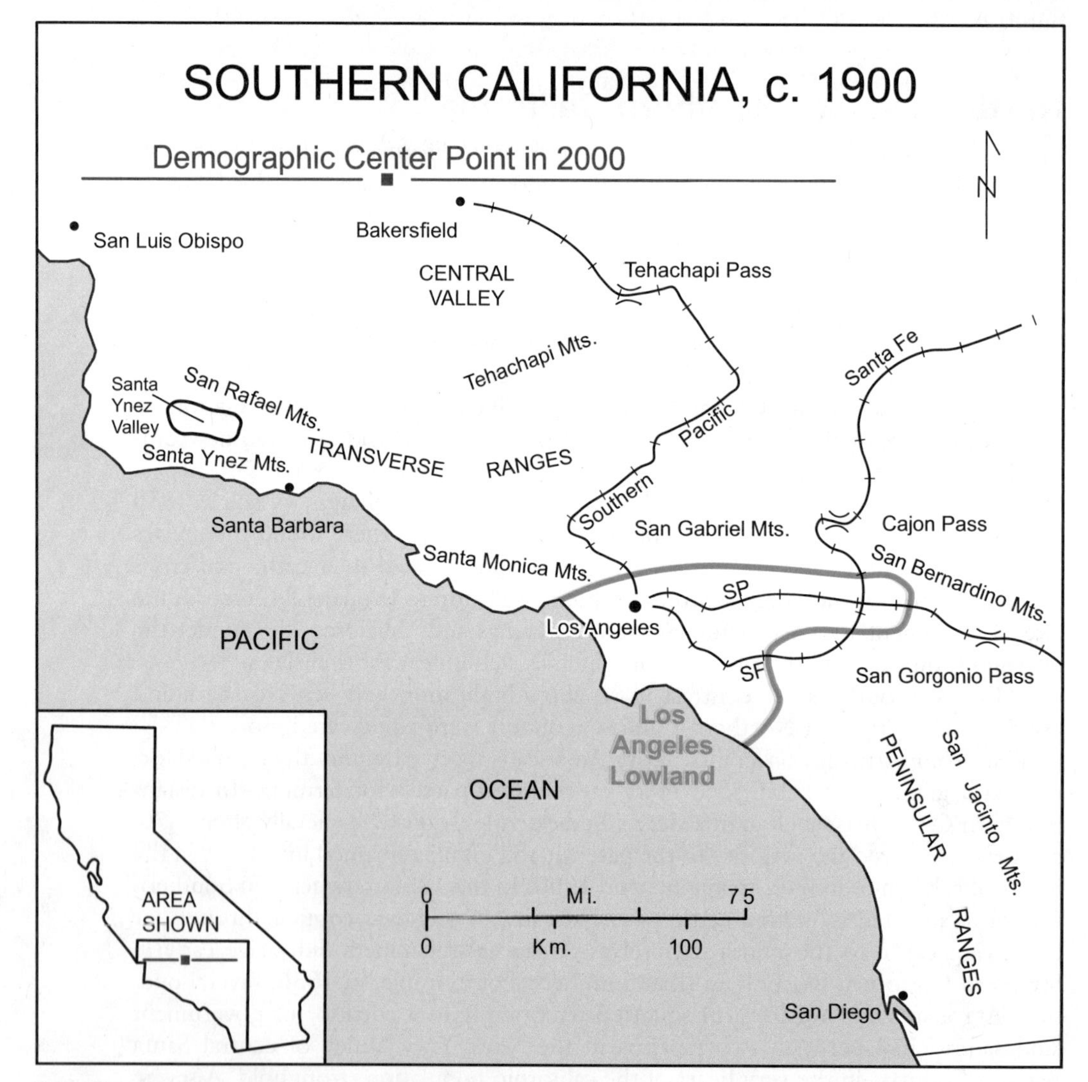

Map 18.4 Southern California, c. 1900. California's demographic center point (half of California's population above and half below) in 2000 was located approximately thirty miles west of Bakersfield.

Source: Meinig, *The Shaping of America*, 3:58, with regard to railroads.

briefly on March 6 to one dollar. Southern California's remarkable growth in population would continue, and between 1910 and 1920 Los Angeles would overtake San Francisco as California's largest urban center.[13]

Few who have not lived in California seem to realize how small, areally, Southern California is thought to be in the minds of residents. For Californians, Southern California is only that stretch of coast—that corner of land—that lies between the mountains and the sea (map 18.4). It lies south of the Tehachapi (Transverse) Range—the extension of the Sierra Nevada that closes off the southern end of the Central Valley—and it lies west of the San Jacinto and San Bernardino (Peninsular) ranges. The Los Angeles Lowland is the focus of this coast that is 275 miles (440 kilometers) long, stretching between Santa Barbara and San Diego. In his insightful book, *Southern California Country*, Carey McWilliams used Helen Hunt Jackson's apt phrase, "An Island on the Land," for his book's subtitle. This island and California more generally would become the destination for so many people after 1900 that in 1964, California overtook New York to become the most populous state, and it reached nearly thirty-four million in the year 2000. California's demographic center point illustrates the importance of Southern California. Most amazingly, in 2000, the point where half of California's population lived to the north and half lived to the south was located only fifty miles north of the Tehachapi Range, west of Bakersfield (map 18.4).[14]

Notes

1 For registering an invoice of the ship's cargo at Monterey, see Ogden, "Hides and Tallow," 257. And for the fact that the three-hundred- to four-hundred-ton vessels could hold thirty thousand to forty thousand cowhides weighing some twenty-five pounds each, see Ogden, "New England Traders in Spanish and Mexican California," 404–5, 407.

2 See Dana, *Two Years before the Mast*. For details about Dana's writing of his book and its copyright, see Kemble, "The West through Salt Spray," 65–75.

3 Jackson, "The Impact of Liberal Policy on Mexico's Northern Frontier," 195–225.

4 For 809 claims and 604 patents, see Hornbeck, "Land Tenure and Rancho Expansion in Alta California, 1784–1846," 373. For Nicholas Den, see Tompkins, *Santa Barbara's Royal Rancho*. The estimate of 10,000 californios in 1846 is based on 9,178 californios identified mainly in the 1850 original census schedules. See Nostrand, "Mexican Americans Circa 1850," 388. And see Pitt, *The Decline of the Californios*, 43, for reports that in 1849 californios "did not exceed 13,000"; 53, for his use of the estimate of fifteen thousand in 1850.

5 Nelson et al., "Remnants of the Ranchos in the Urban Pattern of the Los Angeles Area," 1–9. Nelson had five coauthors, all my graduate student contemporaries at UCLA; for reasons I don't remember, I was not part of the team.

6 In 1850, 291 californios (4.4 percent of the population), nearly all young male "miners," were in the gold diggings, 178 of them in Calaveras County. They were intermingled with the Sonorans, of whom 5,456 were enumerated in 1850, according to Nostrand, "Mexican Americans circa 1850," 384 and 388. In other examples of the undercount, census enumerators in 1850 would report a number of miners whom they could not reach on the opposite bank of a stream, and dozens more unrecorded Chinese in the lobby of a San Francisco hotel. Richard M. MacKinnon, in his "The Sonoran Miners," 26–27, explains that to escape the Foreign Miner's Tax, many Sonorans left the Southern Mines and returned to Mexico in the summer of 1850, before the census was taken.

7 For trails that led through Santa Fe and across Mexico, see Bieber, *Southern Trails to California in 1849*.

8 William Robert Kenny summarizes the ill treatment of Sonorans in the Southern Mines in "Mexican-American Conflict on the Mining Frontier, 1848–1852," 582–92.

9 James, "Sustained Storage and Transport of Hydraulic Gold Mining Sediment in the Bear River, California," 570–92.

10 Dilsaver discusses demographic and occupational change in "After the Gold Rush," 1–18.

[11] See Nostrand, "Mexican Americans Circa 1850," 388–89, for a reconstruction of californio county numbers and percentages. For besieged californios in the Bay Area, see Pitt, *The Decline of the Californios*, 95–100, 118.

[12] See Hornbeck, "The Patenting of California's Private Land Claims, 1851–1885," especially 445 and 448. The unsuccessful miners in 1856 had identified government land in the Alamo Pintado corridor in the Santa Ynez Valley, where two small communities, Ballard and Los Olivos, are located today. See Nostrand, "A Settlement Geography of the Santa Ynez Valley, California," 72–74.

[13] Dumke, *The Boom of the Eighties in Southern California*, 24–25, for the railroad rate war.

[14] See McWilliams, *Southern California Country*, 1–8, esp. 7, for Helen Hunt Jackson's phrase. For California's center of population in 2000, see Turner, "Buttonwillow, Heart of Cotton Country," 4. The precise location was between Shafter (13,700) and Buttonwillow (1,300).

The Great Plains 19

In physical terms, the Great Plains is largely flat, semiarid, and grass-covered. In human terms, it is a region where inhabitants are still not in adjustment with nature. In examining this huge part of the United States—an area that occupies about one-fifth of the Lower 48—we look first at "Webb's Thesis" concerning environmental adjustment. Under "The Cattle Kingdom," we see cowboys and cattle diffuse northward into the region—a time when adjustment to the environment was initially compatible. But when farmers arrive to fence in the open range and put it under the plow, human-environment problems begin, as noted in "Farmers and Ranchers." And as settlers exploit "The Ogallala Aquifer," new challenges confront their adjusting to the environment.

Webb's Thesis

For geographers, Walter Prescott Webb has a rightful place alongside geographical historians Frederick Jackson Turner and Herbert Eugene Bolton. The geography profession made this clear when it invited Webb to present in a plenary session at its 1960 annual conference in Dallas. In Texas, moreover, Webb has become an academic folk hero. Born in the East Texas community of Panola in 1888, Webb grew up in a sharecropper-schoolteacher family in more centrally located Ranger, Texas. In 1918 he was hired to teach history at the institution where he had previously earned a bachelor's degree, the University of Texas–Austin. Two years later, Webb earned a master's degree at UT with a thesis about the Texas Rangers. His department chair, Eugene C. Barker, then urged Webb to earn a doctorate, and for this, Webb enrolled at the University of Chicago in 1922. At Chicago, Webb focused more than he should have on Texas and the Great Plains, and for lack of preparation, he failed his comprehensive exams, returning to UT after the 1922–1923 academic year without a PhD. Back in Austin, Webb immersed himself in his regional Great Plains interest. In 1931 he published *The Great Plains*, and the book was so acclaimed by Barker that, in 1932, based on this book, UT awarded Webb a PhD. Webb's book quickly became a classic, and in recognition of it and of Webb's other accomplishments, in 1958 Webb was elected president of the American Historical Association, the same year the University of Chicago awarded him an honorary doctorate. Webb had persevered in his own way, and Texans loved him for it. He died at age seventy-four in an automobile accident near Austin in 1963.[1]

In *The Great Plains*, Webb emphasized that the region has three salient environmental attributes. First, the region is *flat*. Indeed, the Llano Estacado or Staked Plains section in West Texas and eastern New Mexico is one of the flattest places on Earth (see map 19.1). But there are major exceptions to flatness, for within the Great Plains in South Dakota, the Black Hills rise to four thousand feet (1,220 meters), and in South and North Dakota the Badlands are anything but flat. Second, the Great Plains is *treeless*. The region is covered with short "steppe" grass that is measured only in inches. Again, there are exceptions, for cottonwoods and willows are found along rivers like the Platte and the Arkansas, and forests grow at higher elevations, as in the Black Hills. Third, the Great Plains is *semiarid*. Average annual precipitation amounts range between only ten and twenty inches. That precipitation is scant and unreliable is what most challenges settlers environmentally.[2]

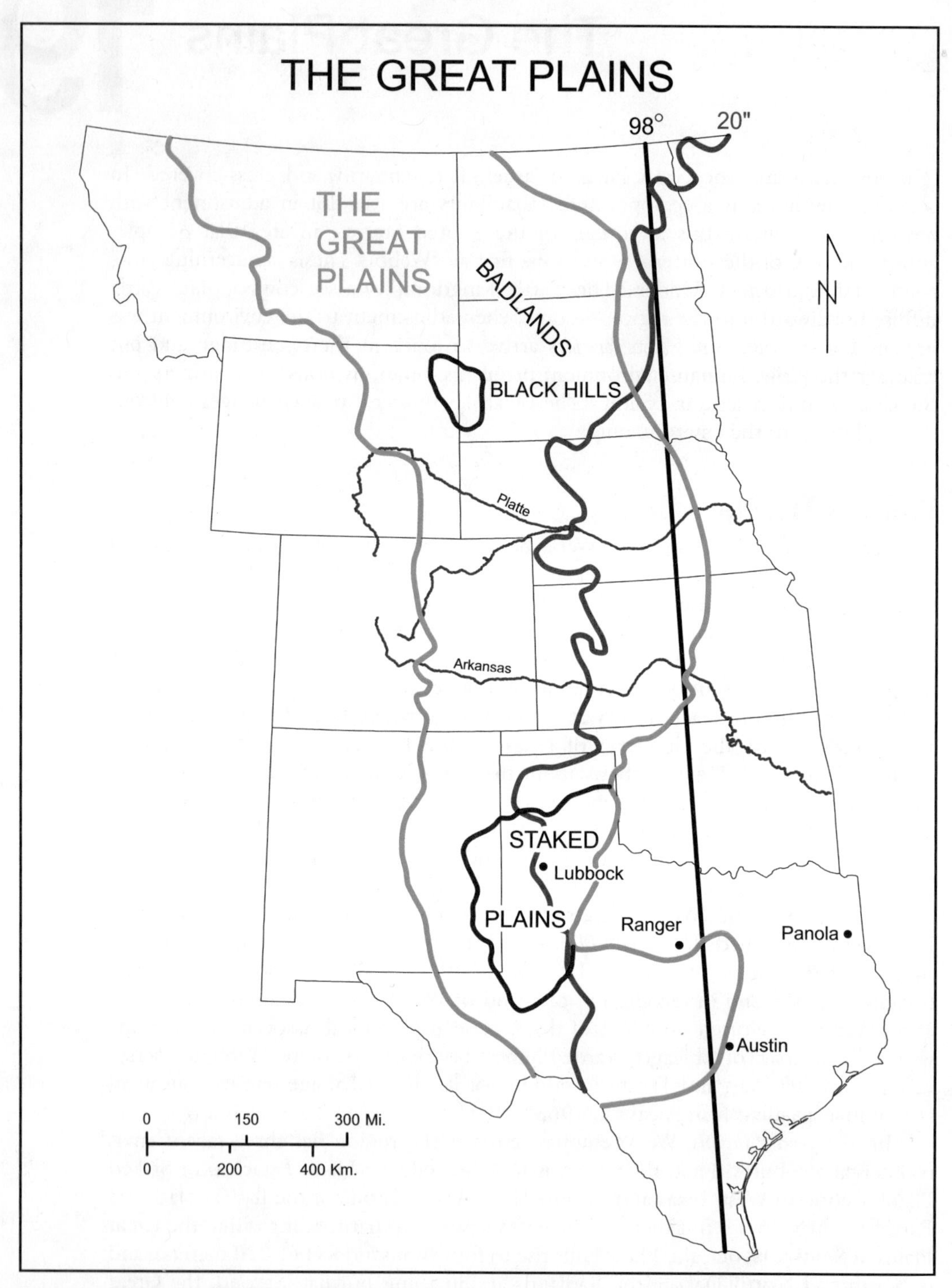

Map 19.1 The Great Plains. The plains rise in elevation from east to west from a discontinuous escarpment at 1,500–2,000 feet (450–600 meters) to the Rocky Mountain front range at 3,000–5,000 feet (900–1,500 meters).

Sources: The boundary shown is from Mather, "The American Great Plains," 240; Mather drew on Fenneman, *Physiography of Western United States*.

Developed around these three Great Plains attributes, Webb's thesis was original and powerful. What Webb did was draw a line down the middle of the United States. The line he used was the ninety-eighth meridian, essentially the twenty-inch rainfall line. He then argued that east of that line American civilization stood on water, timber, and land—and the land, he noted, was often hilly to mountainous. West of that line, he continued, water and timber were withdrawn, and the land that remained was flat. America's institutions and its methods of pioneering, Webb asserted, were worked out in the humid East. But when taken west of the ninety-eighth meridian onto the Great Plains, these institutions broke down and had to be modified. In short, Webb's thesis held that because Americans were not equipped to proceed beyond the ninety-eighth meridian, these westward-moving people stalled at the eastern edge of the Great Plains. In *The Great Plains*, Webb then marshaled considerable evidence to explain just how colonists modified their institutions in adjusting to semiaridity, treelessness, and flatness.[3]

In discussing adjustment to semiaridity, Webb observed that to preserve soil moisture, farmers perfected ***dry-farming techniques***: They planted crops in widely spaced rows, left fields fallow in alternate years, and, before a rain, plowed deeply and then sealed in the moisture within furrows with a harrow. Both farmers and ranchers drilled ***wells*** to tap the groundwater, and ***windmills*** (mass produced after 1873) brought this water to the surface. Webb wrote about how both ***land and water laws*** were modified to reflect dry conditions. In adjusting to treelessness, farmers built ***sod houses***—the Great Plains have been called the sod house frontier. And lacking timber for rail fences, pioneers turned to ***barbed wire***, invented by Joseph Glidden in DeKalb, Illinois, in 1873. Barbed wire had the advantage of being durable, cheap, and most importantly, available. And innovations were made in ***agricultural machinery*** to adjust to flatness, notably large combines used for harvesting wheat. ***Railroads***, described as the great space conquerors, after the 1850s and 1860s were built rapidly across this flat area to bring people and timber west and to haul cattle and wheat east.[4]

To his book and its thesis, Webb had his detractors. Historian George Wolfskill, an appreciative student of Webb's, points out that much of the criticism resulted from Webb's unorthodox research method "of first developing a hypothesis and then gathering data to support it." Webb gave easy answers to complex questions, and he was prone to overgeneralize, as exemplified in the paper he read before geographers in 1960. Webb was also overly supportive of the view that the environment determined human behavior in the Great Plains, a theme of special sensitivity to geographers. For one example of the environment as a causative factor, Webb wrote, "The plains put men on horseback and taught them to work in that way." Furthermore, geographer David J. Wishart (see spotlight) analyzed carefully the westward movement of pioneers into the central and northern Great Plains (Kansas, Nebraska, South Dakota, and North Dakota) decade by decade between 1860 and 1890. Using the criteria of two to six people per square mile and 5 to 15 percent of improved land, Wishart found not a stalled frontier but one that advanced steadily in each of the four decades analyzed. The frontier of settlement did not stall at the ninety-eighth meridian, after all. Webb himself seems to have largely ignored the critics after his book's publication in 1931. When geographer Terry Jordan took Webb's Great Plains course as a graduate student at UT in the 1960–1961 academic year, Jordan reported disappointedly that Webb was still teaching his course "by the book."[5]

Spotlight

David J. Wishart (1946–)

David Wishart at the University of Nebraska–Lincoln, January 2015. Photograph courtesy of David Wishart.

I first met David John Wishart in 1971 in Tucson, Arizona. We shared an office briefly while we both taught summer school at the University of Arizona. Our conversations suggested to me that we looked forward with optimism to careers in teaching our common interest, historical geography. Born in County Durham, England, in 1946, Wishart earned his BA degree at the University of Sheffield in 1967. He came directly to America to study under Leslie Hewes at the University of Nebraska, where he earned both MA and PhD degrees in 1968 and 1971, respectively. Following two years teaching at Beloit College (1972–1974), Wishart returned to Nebraska, where he continues to put in a long career. Wishart excels at teaching and was awarded the University of Nebraska's Distinguished Teaching Award in 1978. He has also published dozens of articles and five books on his focused research agenda—the Great Plains, particularly Native Americans and white colonization in the northern Great Plains. His *An Unspeakable Sadness: The Dispossession of the Nebraska Indians* (1994) was awarded geography's J. B. Jackson Prize. Wishart oversaw and was the editor of the *Encyclopedia of the Great Plains* (2004), a ten-year project funded by the National Endowment for the Humanities. At Nebraska Wishart has been the major professor for twenty doctoral students, nine of whom, to honor him, read papers in two appreciative sessions held at the 2011 Association of American Geographers convention in Seattle.

The Cattle Kingdom

In *The Great Plains*, Webb wrote that the "cradle" of the industry he called the "Cattle Kingdom" lay in a diamond-shaped area south of San Antonio (see map 19.2). To this part of South Texas, beginning in the 1750s, Spaniards had introduced longhorn cattle—those light-bodied, long-muzzled, hardy animals whose meat was tough and stringy. Webb's diamond and its Nueces Valley focus had mild subtropical winters and flat expanses covered with scrub brush and grass—conditions ideal for longhorns to multiply in. Between about 1845 and 1860 in this diamond, Webb argued, Anglo Texans and Mexican ranchers overlapped, and from the Mexicans, Anglos learned about roping, roundups, and horsemanship. Here, said Webb, lay the origins of the Great Plains cattle industry as well as the source area for those long drives to get Texas cattle north to railhead towns for shipment east to markets.[6]

Not so, wrote Terry Jordan in his most definitive *North American Cattle-Ranching Frontiers*. Jordan points out that the antecedents of the Great Plains cattle industry were far more complex and came to exist at earlier times. According to Jordan, three cattle-ranching systems reached Texas and the Great Plains more generally. In the first, Spanish longhorns and the British cattle of Lowland Southerners from South Carolina had mixed in the humid prairies of Louisiana before moving into coastal Texas about 1820. Jordan labels this mixture the "Texas System." In the second, British cattle accompanied Upland Southerners as they settled the Blackland Prairie and Crosstimbers of northeastern Texas in 1815. And reaching "North Central Texas" came a third

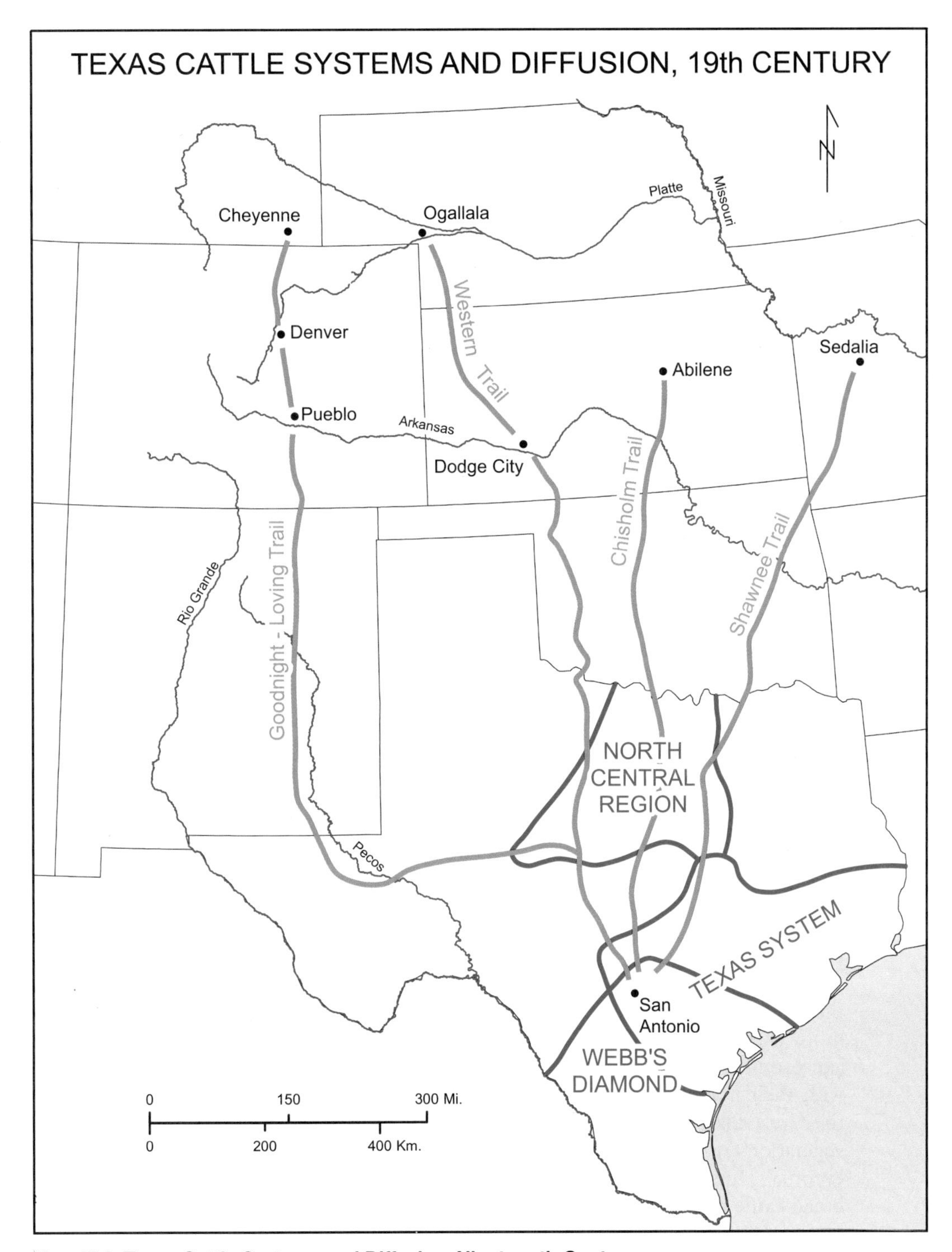

Map 19.2 Texas Cattle Systems and Diffusion, Nineteenth Century.

Sources: Webb's diamond is from *The Great Plains*, map between 224 and 225. Jordan's Texas System and North Central Region are from *North American Cattle-Ranching Frontiers*, maps 176, 209 (inc. cattle trails), 224, 229, 268.

system, the "Anglo-Celtic" stream of British cattle brought by Midwesterners coming especially from Missouri, Illinois, and Iowa. Systems two and three came to mix in the North Central Texas Region, which became the "funnel" for diffusion north. The Midwestern (third) stream, Jordan argues, ultimately prevailed in the Great Plains, for Midwesterners fed and took better care of their cattle in winter months than did Texans who usually neglected their longhorns. Indeed, Jordan's conclusion is that, compared to the Midwestern system, the importance of the other two Anglo Texas ranching systems have been "greatly overstated" in the evolution of western America's cattle ranching.[7]

Texas cattle ranching, then, emerged from the convergence of three cattle ranching traditions, but the cattle themselves that left Texas were basically longhorns. Before the Civil War, long drives to get these animals to markets were sporadic. New Orleans, communities in the Middle West, and even California, were destinations. Between about 1866 and 1885, however, long drives north to railheads in the Great Plains became major annual events. Apparently leading the way in 1866 was the Goodnight-Loving Trail, which tracked west to the Pecos Valley and north to Colorado (map 19.2). The three trails and railheads that generated much of the lore of the western cowboy, fiction and otherwise, were the Shawnee Trail to Sedalia, Missouri, starting in 1866; the Chisholm Trail to Abilene, Kansas, founded in 1867; and the Western Trail to Dodge City, Kansas, founded in 1876. Besides heading to eastern markets by rail, Texas cattle were sold at western Army posts, Indian agencies, and mining camps. In a remarkably short twenty years, longhorn cattle had transformed Great Plains economic activity.[8]

Jordan put the number of Texas cattle driven north after the Civil War at five million. Others have estimated twice this number. However many cattle notwithstanding, this was, in Jordan's words, "the largest short-term geographical shift of domestic herd animals in the history of the world." A prolonged wet cycle in the climatic regime facilitated this expansion. The Cattle Kingdom gradually came to an end before a final calamity, the disastrously cold winter of 1886–1887, in different areas killed 60, 80, and even 90 percent of all cattle between Montana and West Texas. Despite their hardiness, longhorns were, quite simply, ill-adapted to cold temperatures and drought. Beyond problems of weather, the encroachment of farmers with their fenced-in fields also contributed to the demise of the Cattle Kingdom by reducing the availability of free grass in an open public domain.[9]

The Texas cattle industry had the potential to be in balance, ecologically, with the Great Plains environment. Terry Jordan points out that *moderate* grazing yields more grass than does complete protection. By eating grass, animals stimulate herbage production, they create contoured terraces with their trails, and they fertilize with their manure. But the herds of the Texans exceeded the carrying capacity of the Great Plains grasslands, and this overgrazing reduced the volume of the native vegetation cover and led to erosion. Texans, wrote Jordan, had laid the basis for the eventual Dust Bowl. Wishart points out that it made good sense, ecologically, to breed cattle in South Texas, where winters were mild and calving rates high, and fatten these cattle on the more lush grasses of the northern plains. But Texas longhorns, unless crossbred with other cattle, were not favored for breeding in the northern plains. An ecologically sound symbiotic possibility became a lost opportunity.[10]

Farmers and Ranchers

A summary sketch of the settlement of the central and northern Great Plains by farmers and ranchers might go something like this: Farmers entered the plains gradually but steadily after 1865. The progress between 1870 and 1900 is shown in map 19.3.

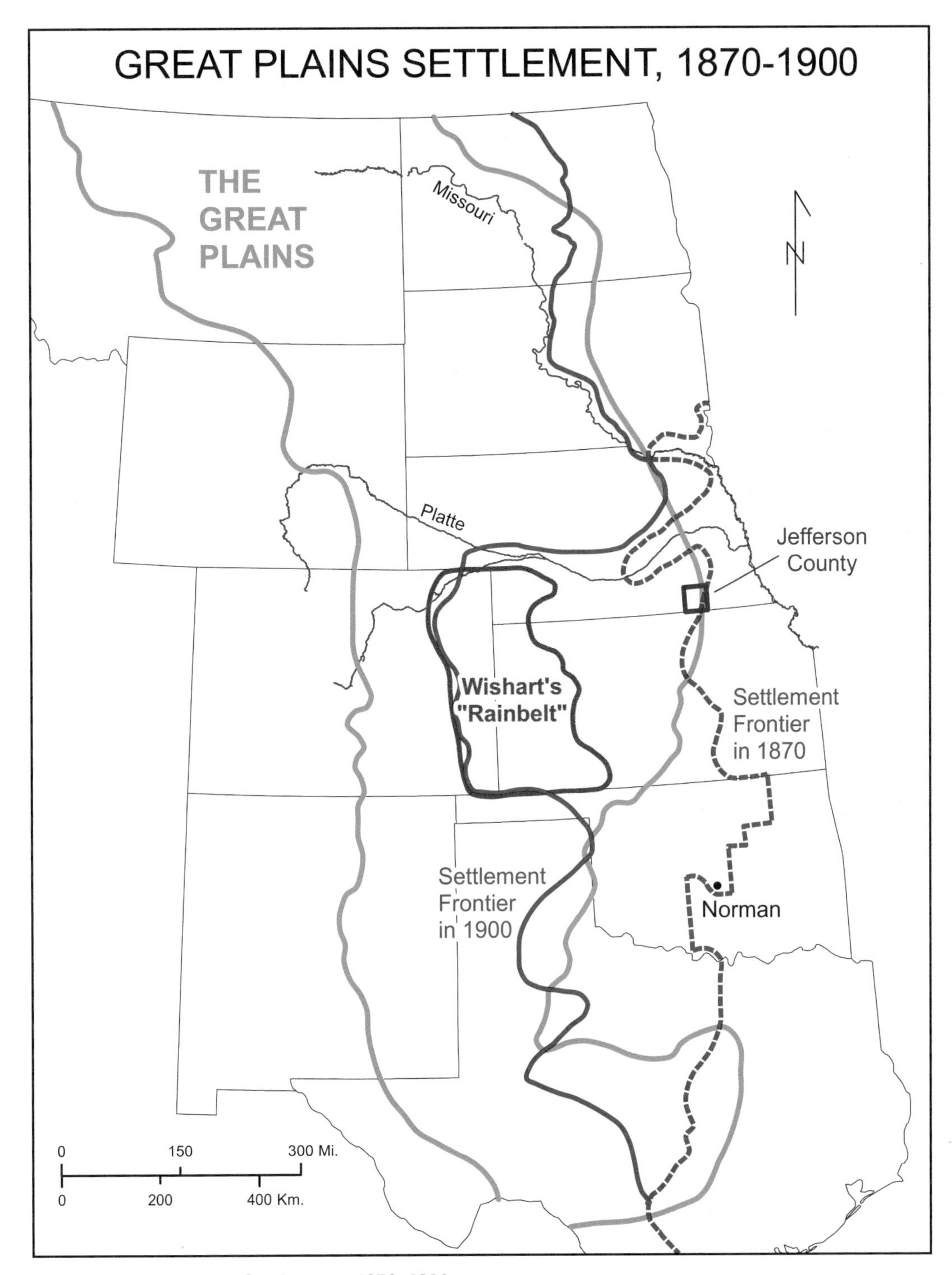

Map 19.3 Great Plains Settlement, 1870–1900.

Sources: Great Plains boundary is from Mather, "The American Great Plains," 240. Settlement frontiers in 1870 and 1900 are from Meinig, *The Shaping of America*, 3:165.

The 1870 and 1900 settlement frontier lines represent the western limits of areas inhabited by two or more people per square mile, which, the Bureau of the Census suggested, represented a beginning farming population. Martyn Bowden's case study of Jefferson County, Nebraska, found farmers in the 1860s growing corn and wheat in the county's shallow river valleys. In the 1870s farmers were also found on the flat uplands above the valleys, plowing the grasslands to grow wheat. To break the virgin sod of the uplands, ideally a farmer needed several yoke of oxen and a steel plow. Facilitating the arrival of farmers were the westward-building railroads. The evenly spaced towns railroads founded along their tracks became the delivery points for the agricultural produce of farmers. D. W. Meinig points out that where railroads crossed the plains, farmers actually purchased more farmland from railroad companies than they acquired through homesteading. And as fields fenced with barbed wire gradually enclosed the open range, ranchers with livestock were forced to reinvent themselves by acquiring land with a source of water and becoming sedentary.[11]

Wet cycles in the precipitation regime encouraged the arrival of farmers. The Great Plains is a region where "wet" cycles and "dry" cycles occur in something approximating a decade-by-decade rhythm. For most of the nineteenth century, data are lacking for a precise picture of these cycles, but a wet cycle between about 1876 and 1888 was followed by a dry cycle in the 1890s, notably from 1893 to 1895. In wet-cycle times, farmers took up new land, plowed under the native grasses, and exposed the soil as they planted crops. In the dry cycle times, crops failed, wind eroded the soil, and farmers often had to abandon the land. In the twentieth century, the disastrous dry cycle, or Dust Bowl, of the 1930s was followed by dry cycles in the 1950s and 1970s. Wishart points out that the 1893–1895 drought in western Kansas and eastern Colorado, an area that would be part of the 1930s Dust Bowl, was shorter but no less severe than the latter. But in the 1890s, he noted, there was no Dorothea Lange to photograph the victims, no John Steinbeck to write a novel about what happened, and no Works Progress Administration to provide jobs for those affected.[12]

Wishart has written an insightful case study about what happened to farmers and town builders who moved onto the high plains of western Kansas, southwestern Nebraska, and eastern Colorado—an area labeled the Rainbelt on map 19.3. Wishart drew heavily on interviews conducted in 1933 and 1934 of the elderly pioneer survivors of the 1890s drought. Told in *The Last Days of the Rainbelt*, the essence of Wishart's story is this: Between 1885 and 1889, Midwesterners from especially Iowa, Illinois, Indiana, and Ohio rapidly settled this flat, treeless, and semiarid (less than fifteen inches, average annual precipitation) area. Typically, these settlers arrived by train to railhead towns; in 1885 trains had reached only the eastern edge of this Rainbelt area. The new arrivals were relatively young people, often couples with several children, and they were relatively poor. At a land office in the railhead town, the new arrivals filed for a preemption or homestead claim, and if one was available, they took out a 160-acre tree claim (which everyone understood would be held in speculation of rising land values, not for the intended planting of trees). Before leaving town and heading for their claim, these settlers, provided they could afford them, purchased a wagon and a team of horses, a plow, and provisions that included flour, cornmeal, and a slab of bacon. At the claim, a makeshift dugout provided shelter while a sod house, made from blocks cut from the steppe grass, was built. A well had to be hand-dug or drilled by professionals, and the first crop of wheat or corn produced in the virgin soil was invariably "bountiful" (see figure 19.1).[13]

The availability of free land in this last frontier of the central Great Plains had lured these farmers, and by 1888 the area's population had reached the two people or more per square mile threshold that indicated a farming population. What

Figure 19.1 Deep-Well Drilling on the High Plains, c. 1900.

Source: Photograph by Willard D. Johnson, of the United States Geological Survey, from his "The High Plains and Their Utilization," 734.

also lured these pioneers was the firm belief, confirmed in agricultural journals and the publications of railroad companies and state immigration societies, that if the grasslands were plowed and trees were planted, rainfall would move west. "Rainfall follows the plow" was the thinking; the "Rainbelt" would shift west to include them. Encouraging this settlement was a wet cycle that had two dry year setbacks, in 1887 and 1890. But what ended the process was a disastrous dry cycle that lasted from 1893 into 1895. This severe drought caused farmland to revert to rangeland and settlements to become ghost towns. In 1899, when agricultural agent J. E. Payne toured eastern Colorado, he reported that very few farmers and townspeople were left. Nonetheless, some settlers, including those interviewed, survived, and after 1900, wet cycles brought in optimistic newcomers. Indeed, the rhythm of settlement followed by abandonment continues to this day in Wishart's study area. Unlike in the 1890s, however, farmers are now supported by federal programs, and many also use irrigation.[14]

In settling the Great Plains, the role played by railroads and the towns they founded was profoundly important. Railroads brought in the settlers, and railroad towns were the purveyors of goods and services. Of special interest to geographers is how railroad companies laid out, and in time reconfigured, their towns. In his *Plains Country Towns*, John Hudson discusses three town morphologies—symmetric, orthogonal, and T-town—used by railroads between the 1850s and the 1920s. Early towns were symmetric—that is, business blocks with their grain elevators, lumberyards, fuel dealers, and rows of false-front business buildings faced both sides of the tracks beyond the railroad rights-of-way. When trains arrived, streets facing the rights-of-way and the tracks became congested with people, animals, and wagons, and there

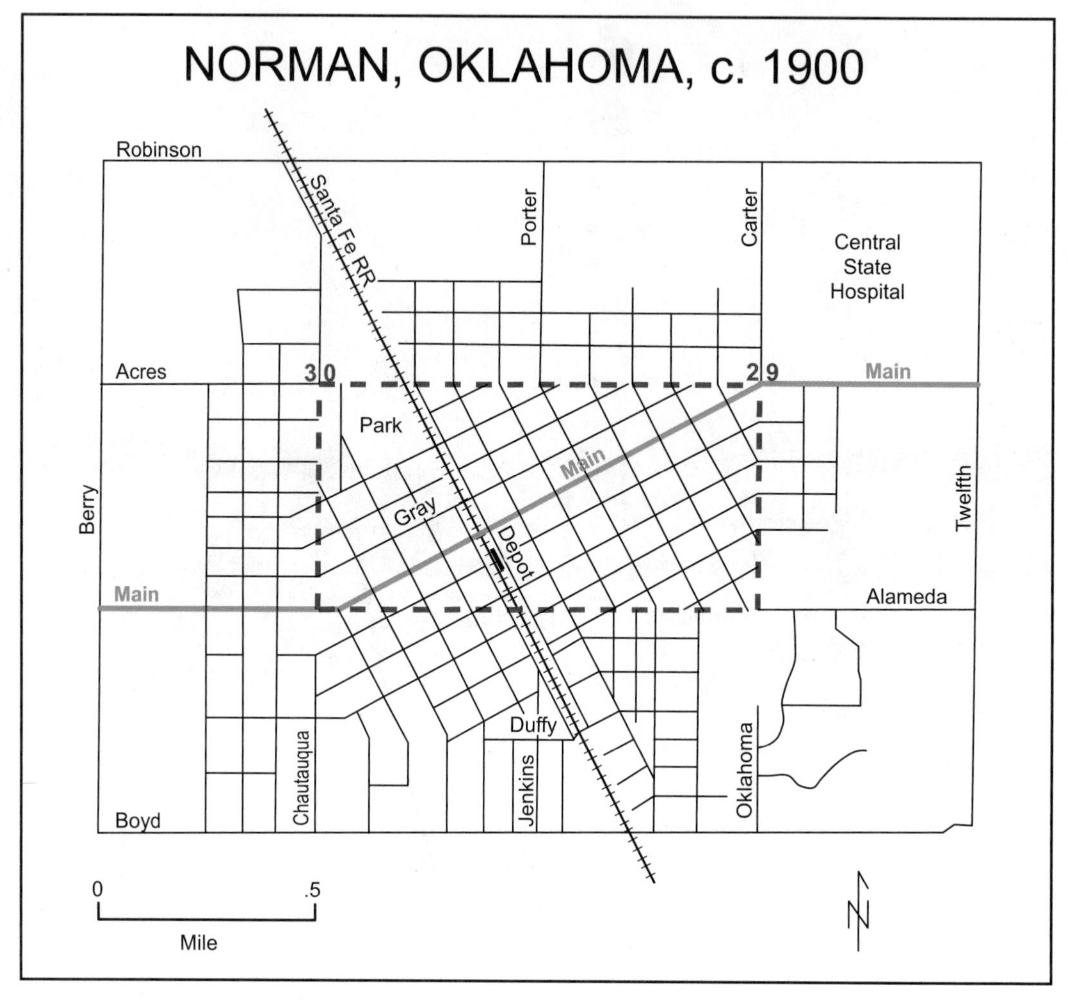

Map 19.4 Norman, Oklahoma, c. 1900. Norman follows the orthogonal layout where the community grid straddles the railroad tracks in alignment with Main Street, which is perpendicular to the tracks. Laid out by the Santa Fe Railway Company in 1889, Norman's original townsite contained 320 acres (dashed line—SW ¼ Sec 29; SE ¼ Sec 30; T.9N [Township 9 North], R.2W [Range 2 West]). As Norman grew, with the exception to the south, the grid was realigned at the townsite boundaries to conform with township and range cardinal directions (surveyed in 1873). A strip of land (two hundred by three thousand feet) designated for station grounds that stretched south from Gray to Duffy streets may have helped "pull" the railroad grid orientation south of the original townsite.

were accidents. About 1890, to alleviate congestion, railroad companies introduced two new layout designs. The orthogonal plan put Main Street and its businesses at right angles to the tracks, which eliminated much of the congestion but, nonetheless, left a busy intersection where Main Street crossed the tracks. Norman, Oklahoma, was such a town (see map 19.4; see figure 19.2). The T-town layout also put Main Street perpendicular to the tracks, but in T-towns the community was entirely on one side of the railroad. Hudson cites examples of all three morphologies in his book's North Dakota study area.[15]

The Ogallala Aquifer

Aquifers are water-bearing layers of underground sands, gravels, and permeable rock. If tapped by wells, these layers can yield important quantities of water. Underlying the Great Plains is the huge Ogallala Aquifer (see map 19.5). Areally the size of

Figure 19.2 Norman, Oklahoma, in 1889 and 1911. The Santa Fe Railway Company founded Norman on April 22, 1889, the day the unassigned lands were opened to white settlement. Major business activity came to focus on Main Street east of the railroad tracks. The first photograph, taken June 29, 1889, shows the false-front wooden business buildings on Main Street, looking east from the tracks. The second photograph, taken in 1911, shows what had become more substantial brick buildings taken from about two blocks east of the tracks, looking west.

Photographs courtesy of the Western History Collections, University of Oklahoma.

California, the Ogallala has strata saturated with water that measure one-thousand-feet thick or more under Nebraska, but to the south the strata become thinner until under West Texas they measure one-hundred-feet thick or less. Aquifers replenish as precipitation percolates downward by gravity. If the amount of water pumped out of an aquifer exceeds its natural recharge, it may take many centuries for present levels of saturation to be regained.

In the twentieth century the Ogallala Aquifer offered new hope to settlers in the precipitation-challenged Great Plains. Farmers first tapped the Ogallala for irrigation water in the 1930s, and after World War II heavy pumping began to supply new center-pivot sprinkler irrigation systems. By the 1950s annual overdrafts from pumping in places began to exceed recharge, and deeper wells had to be drilled. Overdrafts became especially critical in West Texas where the aquifer is thinnest. Compounding the thinness of the Ogallala under Texas is Texas groundwater law: Of the eight states situated over the Ogallala Aquifer, only Texas retains the

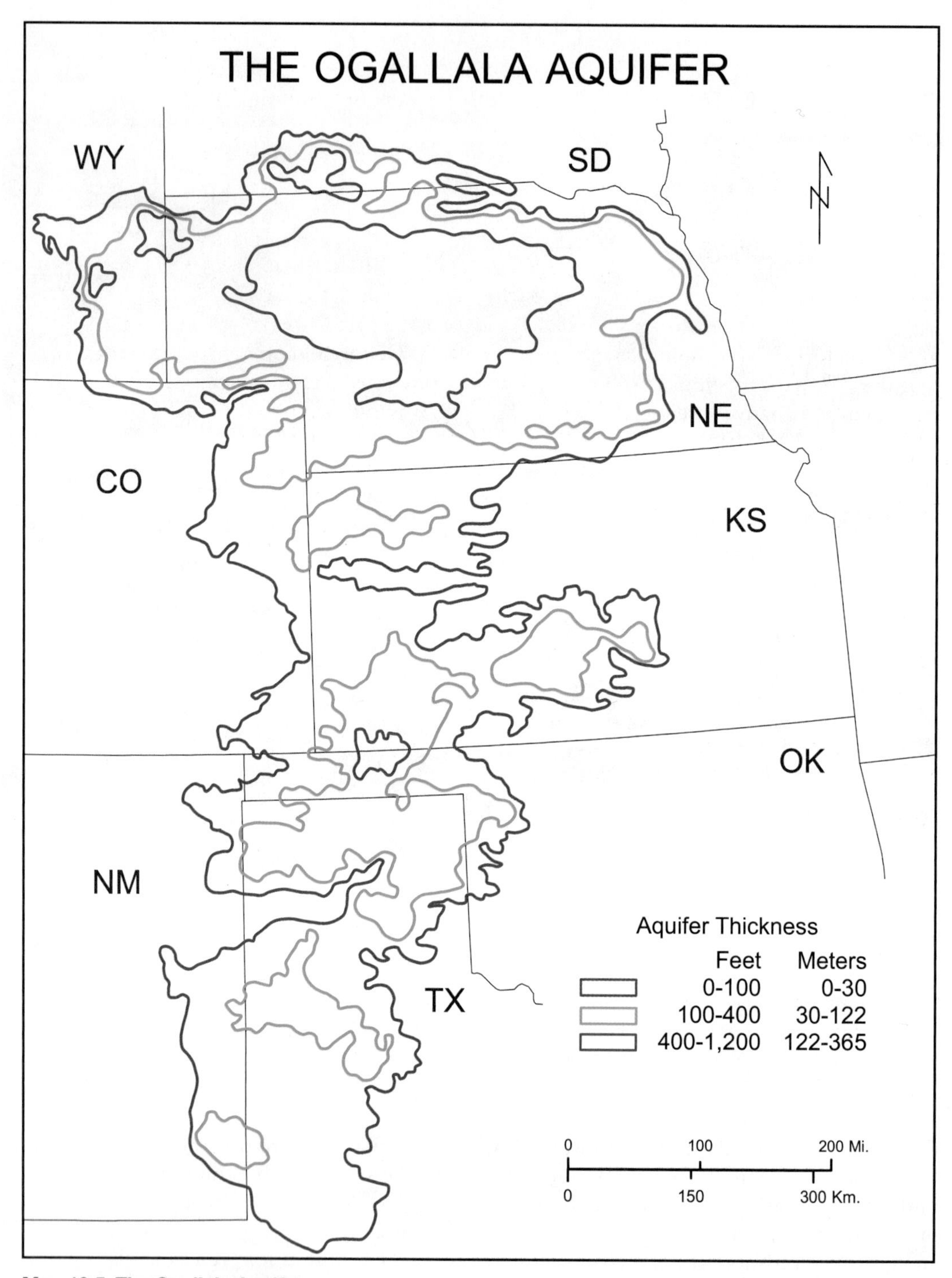

Map 19.5 The Ogallala Aquifer.

Sources: Kromm and White, "Interstate Groundwater Management Preference Differences," 5, with data from the 1980 US Geological Survey.

common-law rule that gives landowners "the right to capture and use percolating groundwater beneath the land." Aided legally by basically regulation-free exploitation, the Ogallala under Texas by the year 2000 had been significantly depleted, and something like half of the once-irrigated lands in West Texas had by then reverted to ranching. Annual overdrafts are also depleting the Ogallala north of Texas, where long-term predictions warn of the exhaustion of this finite groundwater resource if pumping is not curtailed.[16]

And so the struggle to get humans in ecological adjustment with their Great Plains environment continues. Carving the area into 160-acre farms was a flawed policy because such farms were too small for ranching and too large for irrigated farming. Plowing under the grasslands in wet cycles subjected soils to wind erosion in dry cycles. Some, including geographers Frank and Deborah Popper, have argued that the Great Plains should become a "Buffalo Commons": The region should be replanted with native grasses and repopulated with herds of buffalo. Buffalo are easier on the land than cattle, and their thicker hides and longer hair allow them to better withstand cold winters. Indeed, buffalo are presently on the rise numerically, due to increased consumption of buffalo meat. Although remaking the Great Plains for buffalo could happen in selected places, proposing a Buffalo Commons as a regional development policy, while good for the environment and the buffalo, seems to be unrealistic.[17]

Notes

1. For Webb's invited Plenary Session paper, Webb, "Geographical-Historical Concepts in American History," 85–93. Necah Stewart Furman supplies much detail in her biography, *Walter Prescott Webb: His Life and Impact*, 81–84, for the University of Chicago exam failure; 88, for Barker's use of Webb's book for a doctorate; 166 and 180, for Webb's death on I-35 near Buda, south of Austin.
2. See Webb, *The Great Plains*, 3ff., for flatness, treelessness, and semiaridity.
3. Webb's thesis is spelled out in ibid., 8–9. See also page 206, for his statement that "the agricultural frontier stood at ease" at the ninety-eighth meridian.
4. Ibid., 366ff., for dry farming; 339, for windmills mass produced after 1873; Chapter 9, for land and water laws; 298, for barbed wire invented, "with some qualification," by Glidden.
5. For Wolfskill's quote about Webb's unorthodox research method, see his "Introduction," 4. "The plains put men on horseback" quote is in Webb, *The Great Plains*, 226. For Wishart's analysis, see "The Changing Position of the Frontier of Settlement on the Eastern Margins of the Central and Northern Great Plains, 1854–1890," 153–57. That Jordan found Webb disappointing is in "Reviews of *The Making of a History*," 31.
6. See Webb, *The Great Plains*, 205–69, "The Cattle Kingdom," esp. 208, for "cradle" and "diamond."
7. Jordan, *North American Cattle-Ranching Frontiers*, 170–89, 224, esp. 176 (map), and 224 (map), for Carolina's coastal expansion; 189–207, 209, esp. 190 (map), and 209 (map), for Upland Southern expansion; 267–75, esp. 268 (map), for the Anglo-Celtic Midwestern system; 224, for "funnel" for diffusion; 210, for "greatly overstated."
8. For the diffusion of Texas cattle, see ibid., 221–27, and 270, which states that longhorns were "more numerous" than shorthorn British stock trailed north out of Texas.
9. See ibid., 236–40, for the demise of the Cattle Kingdom; 222, for five million head and for the "geographical shift" quote; 223, for long wet cycle; 238, for the winter of 1886–1887.
10. Ibid., 10, for moderate grazing as a stimulus to herbage production; 239, for how overgrazing brought deterioration and the eventual Dust Bowl; 274, for longhorns usually not bred, 274. See Wishart, "Settling the Great Plains, 1850–1930," 250, for the complementary relationship of breeding cattle in Texas and fattening them in the northern plains.
11. For the advance of farmers after 1865, Wishart, "The Changing Position of the Frontier of Settlement on the Eastern Margins of the Central and Northern Great Plains, 1854–1890," 153–57. For Jefferson County, see Bowden, "Desert Wheat Belt, Plains Corn Belt," 193–94. See Trewartha, "Climate and Settlement of the Subhumid Lands," 173, 175, for several yoke of oxen and a steel plow. Meinig makes the point that more land was purchased than homesteaded, in *The Shaping of America*, 3:166.
12. In 1971 Borchert wrote in "The Dust Bowl in the 1970s," 2, that beginning with the period of instrumental weather records, the midpoints for four major periods of drought were 1892, 1912, 1934, and 1953. For "no John Steinbeck," etc., in the 1890s, see Wishart, *The Last Days of the Rainbelt*, xvi.

[13] Conducted under the New Deal Civil Works Administration in eight eastern Colorado counties, these interviews were found by Wishart in Denver's Colorado Historical Society; see *The Last Days of the Rainbelt*, xviii. See ibid., 52, for the midwestern source area (some 20 percent were also foreign-born in Germany and Russia); 46 and 52, for young and poor; 58 and 64, for tree claims; 51, for purchase wagons and supplies; 85, for sod houses; 68, for "bountiful" first crop. For background on tree claims, see McIntosh, "Use and Abuse of the Timber Culture Act," 347–62.

[14] Wishart, *The Last Days of the Rainbelt*, 2, 66, for two or more people per square mile; xiv and 38, for rainfall following the plow; 8 and 75, for dry cycles of 1887, 1890, 1893–1895; xv, 39, for J. E. Payne; 158, for the fact that the rhythm of settlement and abandonment continues.

[15] Hudson, *Plains Country Towns*, 88–90, for town morphologies. Womack, *Norman: An Early History, 1820–1900*, esp. 26, 28, 31, 32–33, for the Santa Fe Railroad layout of Norman. For Norman's 320-acre townsite and station grounds, see Womack, *Cleveland County Oklahoma Place Names*, 28.

[16] Quote is from Templer, "The Legal Context for Groundwater Use," 65.

[17] Geographers have invited Frank Popper (Rutgers University) and Deborah Popper (City University of New York–Staten Island) twice (November 1990 and February 6, 2008) to present at the University of Oklahoma. On both occasions they delivered persuasive lectures on the need for a "Buffalo Commons."

20 The Dry West

In "The West," we identify additional reasons, besides dryness, the area's defining attribute, why the West is different from the East. We then look at three important geographical topics that apply to all the United States and that will broaden our context for understanding its regions: "State Capital Centrality," "The Century of Mass Migration," and "Cities and Suburbs." Developing these topics beyond 1900 seems appropriate for greater relevance.

The West

Our van is loaded with students and we are driving the five-hundred-mile (eight-hundred-kilometer) stretch of Interstate 40 west from Norman, Oklahoma, to El Cerrito, New Mexico. As good geographers, we are observing what is changing outside of our van windows. I point out one not-so-obvious transition we are experiencing early in our trip: that *elevation* is rising gradually from some 1,200 feet (365 meters) above sea level at Norman to 5,700 feet (1,735 meters) above sea level at El Cerrito. But I want everyone to be aware of five additional and, for the most part, tangible transitions. The first is fundamental: As we distance ourselves from the Gulf of Mexico as a source of moisture, average annual *precipitation* amounts decrease—from thirty-three inches at Norman to about fourteen inches at El Cerrito. Second, and reflecting the precipitation decreases, the post oak–woodland *natural vegetation* gives way to grassland, then to large areas of desert shrub, followed by juniper and piñon on our ascent to El Cerrito. Third, *land use* evolves from dry-farmed wheat to cattle ranching that is intermixed with small patches of irrigated agriculture. Fourth, rural *population* densities decrease—from more than six to less than two people per square mile. Finally, the increasing elevation translates to lower *temperatures*—at the rule-of-thumb rate of 3.5 degrees Fahrenheit per one thousand feet—a total change of about sixteen degrees Fahrenheit (nine degrees Celsius).

In the context of climatic zones, our van is crossing from subhumid to semi-arid conditions. If we continued west on I-40 to Arizona and California, we would reach an arid zone where average annual precipitation falls below ten inches—true desert conditions. In their collection of essays published as *The Mountainous West*, geographers William Wyckoff (see spotlight) and Lary Dilsaver made the point that the West is actually two Wests: Mountainous and Arid. Fragmented and discontinuous as it is, the Mountainous West, argue Wyckoff and Dilsaver, is a region unto itself. Substantiating their argument is the fact that the Mountainous West is where moisture collects before flowing by gravity to a lower Arid West. Imagining the West as two fragmented and discontinuous subregions makes good sense. Water is the critical issue in both subregions. But without water from the higher and wetter Mountainous West, life in the lower and drier Arid West is precarious at best. People in the West fight over water. They have come to blows over the use of watering holes for livestock, and legal battles pit community against community and state against state over surface-water allocations. As Wyckoff put it, "Every drop of water is money in the West."[1]

But the West is different from the East for reasons beyond water, and the area's more recent settlement by European Americans underlies at least three of these differences. First, the West is where a majority of Native Americans live, and it is where most

Spotlight

William Wyckoff (1955–)

William Wyckoff, photographed in 2012 in Scotland by his daughter, Katie. Photograph courtesy of Bill Wyckoff.

It is a safe bet that no geographer has excited more people about the geography of the American West than Bill Wyckoff. Born in Burbank, California, in 1955, Bill was turned on to cultural geography by James P. Allen at California State University–Northridge, where Bill earned his BA in geography in 1977. Allen's excellent advice landed Bill at Syracuse University (Allen's alma mater) where, under D. W. Meinig, Bill earned his MA (1979) and PhD (1982) degrees. (Recall that in Chapter 9, we drew upon Wyckoff's dissertation about upstate New York, published as *The Developer's Frontier* in 1988.) Bill taught briefly at the University of Georgia (1984–1986), and in 1986 he and his family sunk a deep taproot in Bozeman, Montana, where Bill teaches at Montana State University. His *The Mountainous West*, coedited with Lary M. Dilsaver, provides an excellent overview to the geography of the West. And his *How to Read the American West: A Field Guide* (2014) is a monumental effort at educating travelers about how to read western landscapes. To photograph and describe his one hundred landscape examples, Bill spent five years visiting all 414 counties in the 11 western states. Bill's engaging scholarship and his youthful enthusiasm have persuaded many people of the value that geography brings to an understanding of places.

---

of their reservations are located. The policy of the federal government after 1803 was to push Native Americans west. Indian Territory (see Chapter 14), created in the 1830s along the far western border with Mexico, became the early destination for Indian "removal." As the United States subsequently expanded to the Pacific, it became federal policy to create reservations, the first in 1867, in this newly acquired territory. Authorities forcibly relocated Native Americans to what would become approximately three hundred reservations. By 1890 total Native American numbers had reached an all-time low of 250,000. Soon after 1900 the policy of allotments resulted in the extinguishment of all but one (Osage) reservation title in Oklahoma, yet Native Americans continued to live there. Generally speaking, after 1900 two trends have characterized Native Americans: Numbers have increased, rather dramatically after 1960, with new census self-identification procedures; and in about 1950, Native Americans began to leave their reservations for urban areas. By 2000, of two million Native Americans enumerated, only about one-quarter still lived on reservations, and the majority had become urban residents. Because urban destinations were mostly eastern cities, close to half of the population now lives in the East, but the four leading states in Native American numbers continue to be Oklahoma, California, Arizona, and New Mexico.[2]

The second reason the West is different from the East is because so much of the West is owned and controlled by the federal government. The federal government owns 54 percent of all land in the eleven western contiguous states; percentages of federal ownership in these states range from 80 in Nevada to 33 in Montana (see map 20.1). Because the West is dry and mountainous, most of it was unsuited to the kind of homesteading by farmers that took place in the East. Thus in the 1870s

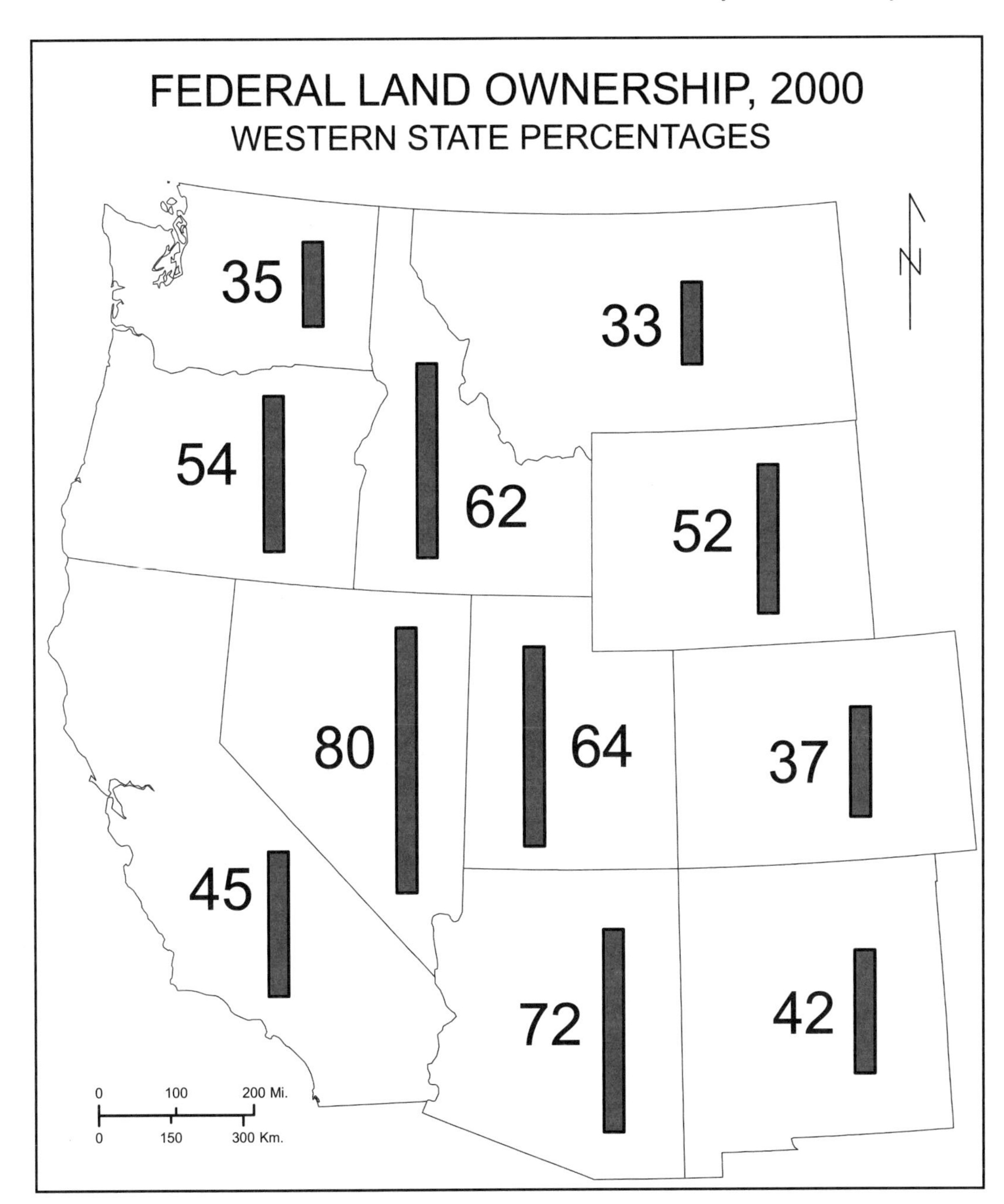

Map 20.1 Federal Land Ownership, 2000: Western State Percentages. Heights of bars represent percentages. The Alaska percentage (89) was highest; all other state percentages ranged from 1 (six states) to 18 (South Dakota).

Based on Jackson, "Federal Lands in the Mountainous West," 254. Data include Indian lands.

the federal government shifted its western land policy from one of disposing of land as private property to homesteaders, to one of reserving land. Geographer Richard Jackson reviews how, beginning with the setting aside of Yellowstone National Park in 1872, federal retention of western public lands evolved incrementally over the decades through legislation that targeted mining, grazing, and logging interests. The outcome, by 2000, found the following agencies in control of that 54 percent of the West: the Forest Service, the Bureau of Land Management (largely grazing lands and still public domain), the Department of Defense, the Fish and Wildlife Service, the National Park Service, and the Water and Power Resources Service. Ranchers, timber producers, miners, and others in these eleven states survive, Jackson explains, through the free or inexpensive use of these federal lands, but pressures are mounting from environmentally and recreationally oriented users to manage these lands for the benefit of a larger segment of Americans.[3]

Third and finally, images held about the West make it different from the East. As western historian Richard White put it in his *It's Your Misfortune and None of My Own*, "for more than a century the American West has been the most strongly imagined section of the United States." Battles between whites and Indians, cowboys driving longhorn cattle, and sheriffs gunning down outlaws constitute some of the imagery, as do the area's spectacular scenery and vast open spaces. Creating these images were artists, like Frederick Remington, photographers, like William Henry Jackson, performers, like Buffalo Bill, novelists, like Zane Grey, and actors in Westerns, like John Wayne. The West as imagined by Laura Ingalls Wilder (see Chapter 13) in her Little House books, notes White, is probably the "most popular" West created by a woman author. "The actual West and the imagined West," argues White, "are engaged in a constant conversation; each influences the other." Separating myth from reality about the West is difficult at best, but what seems to be clear is that Americans share powerful images about the West—and not about the East.[4]

State Capital Centrality

Most schoolchildren in the United States at some point must learn the fifty American state capitals. Students with a predilection for geography likely enjoy the assignment, and those who do learn the capitals will likely be rewarded in some way as life goes on. I learned the capitals—but not quite well enough to be rewarded. Years ago my young daughter, Suzanne, and I were on a long domestic flight that ended with a flight attendant announcing that he had two bottles of wine to give to the first person who could name the five state capitals that begin with the letter *A*. I immediately thought of four, which I told Suzanne: Austin, Atlanta, Albany, and Annapolis. "Come on, Dad," Suzanne said, elbowing me in the side. "You're a geographer; you can get it." But I simply could not think of Augusta, and the passenger who did won the bottles of wine. Had Christian Montès been on that flight, I am reasonably certain that he would quickly have won the prize.

Montès is a French geographer whose *American Capitals* (2014) will be the definitive study on the subject. Montès drew mainly upon a wealth of secondary literature to analyze how these capitals were selected and how they evolved. The story is far more complicated than just the selection of fifty capital cities. Each state had an average of 3.84 capitals, a fact that found Montès analyzing some 180 temporary and permanent capital choices. And he had to sort through 255 "stated causes" considered by legislators as they deliberated capital selections.[5]

Here is what Montès learned: If we limit our discussion to just the fifty permanent capitals, not the 130 communities that did not make it, we would find seven reasons behind their selection, beginning in colonial times. The three leading reasons, with their percentages, were centrality and/or accessibility (34.8), politics (30.3), and the economy (23.6); trailing those are defense (4.5), entry point (or establishment of a first presence in the New World [2.25]), religion (2.25), and anti–large-city sentiment (2.25). Competition to be selected the capital was usually fierce. Communities that lost out were often given the university, the penitentiary, or some other state institution (agricultural college, reform school, insane asylum) as a consolation prize. Only a handful of states, like Wisconsin (Madison) and Texas (Austin), ended up with both the capital and the state university in one city.[6]

Montès asked what the selection of America's fifty state capitals meant geographically. To answer this, he sorted the migration patterns or "wanderings" of the fifty capitals into five spatial categories, which he mapped. With minor modifications, this is what he found:

1. *Stability*
   Eight state capitals never moved: Boston, Cheyenne, Honolulu, Olympia, Reno, Salt Lake City, Santa Fe, St. Paul.
2. *Westward/Centrality*
   Twenty capitals relocated for accessibility to follow the trends of settlement: Albany, Augusta, Bismarck, Boise, Columbus, Des Moines, Harrisburg, Hartford, Indianapolis, Jackson, Jefferson City, Lansing, Lincoln, Little Rock, Madison, Pierre, Richmond, Springfield, Tallahassee, Trenton.
3. *Rotating*
   Five capitals evolved out of a complex system of wandering: Concord, Dover, Montpelier, Providence, Raleigh.
4. *Readjustment*
   Seven capitals readjusted short distances: Annapolis, Denver, Frankfort, Helena, Juneau, Oklahoma City, Salem.
5. *Apparently erratic*
   Ten capitals resulted from spatial patterns that, according to Montès, made no sense: Atlanta, Austin, Baton Rouge, Charleston, Columbia, Lincoln, Montgomery, Nashville, Phoenix, and Sacramento.[7]

Two additional findings are most interesting. Montès reported that thirty-nine of the fifty state capitals were selected during the nineteenth century, thirty-five of them before 1861. This means that 70 percent of our fifty capitals were made permanent before the Civil War. Indeed, the only capital to move after 1900 was Oklahoma's—from Guthrie to Oklahoma City in 1910. And from his analysis of names of given capitals, Montès found that nineteen of fifty, or nearly two-fifths, came from non-English languages. These capitals were French (6)—Baton Rouge, Boise, Des Moines, Montpelier, Pierre, Juneau; Native American (5)—Cheyenne, Honolulu, Oklahoma City, Tallahassee, Topeka; Greek (4)—Annapolis, Atlanta, Indianapolis, Olympia; German (2)—Bismarck, Frankfort; and Spanish (2)—Sacramento, Santa Fe. All our capitals, then, have been in place for more than a century, and their names reflect America's multicultural history.[8]

Thus, Montès found centrality to be the leading reason for selecting state capitals. Centrality also explained why Washington, DC, was chosen for the nation's capital, as noted in Chapter 8. But does the geographical concept of centrality, which can be applied with success to political units, also inform us about some of our culture regions? Ignoring for this discussion Texas and California, which we declared to be regions as well as states, centrality does inform us about the Spanish American Region and the Mormon Region. These two regions evolved as "organic societies," to use Meinig's term, and quite remarkably each of the regions had a strong central focus or "capital"—Santa Fe and Salt Lake City. When the state borders of New Mexico and Utah were superimposed in cookie-cutter fashion over each of these organic societies, Santa Fe and Salt Lake City were indeed preserved as central interactive nodes. Both, notes Montès, were capitals that never moved. But New Mexico's new borders excluded Spanish Americans especially in southern Colorado (the new borders also excluded the El Paso District, which historically had belonged with New Mexico), and Utah's new borders excluded Mormons, especially in Idaho, the "Arizona Strip" (located between the Grand Canyon and Utah's southern border), and elsewhere (see map 20.2). Both culture regions have remained intact despite their new political compartments. But each is an example of what many would argue was an unfortunate disregard by politicians for the existence of deeply rooted societies on the ground.[9]

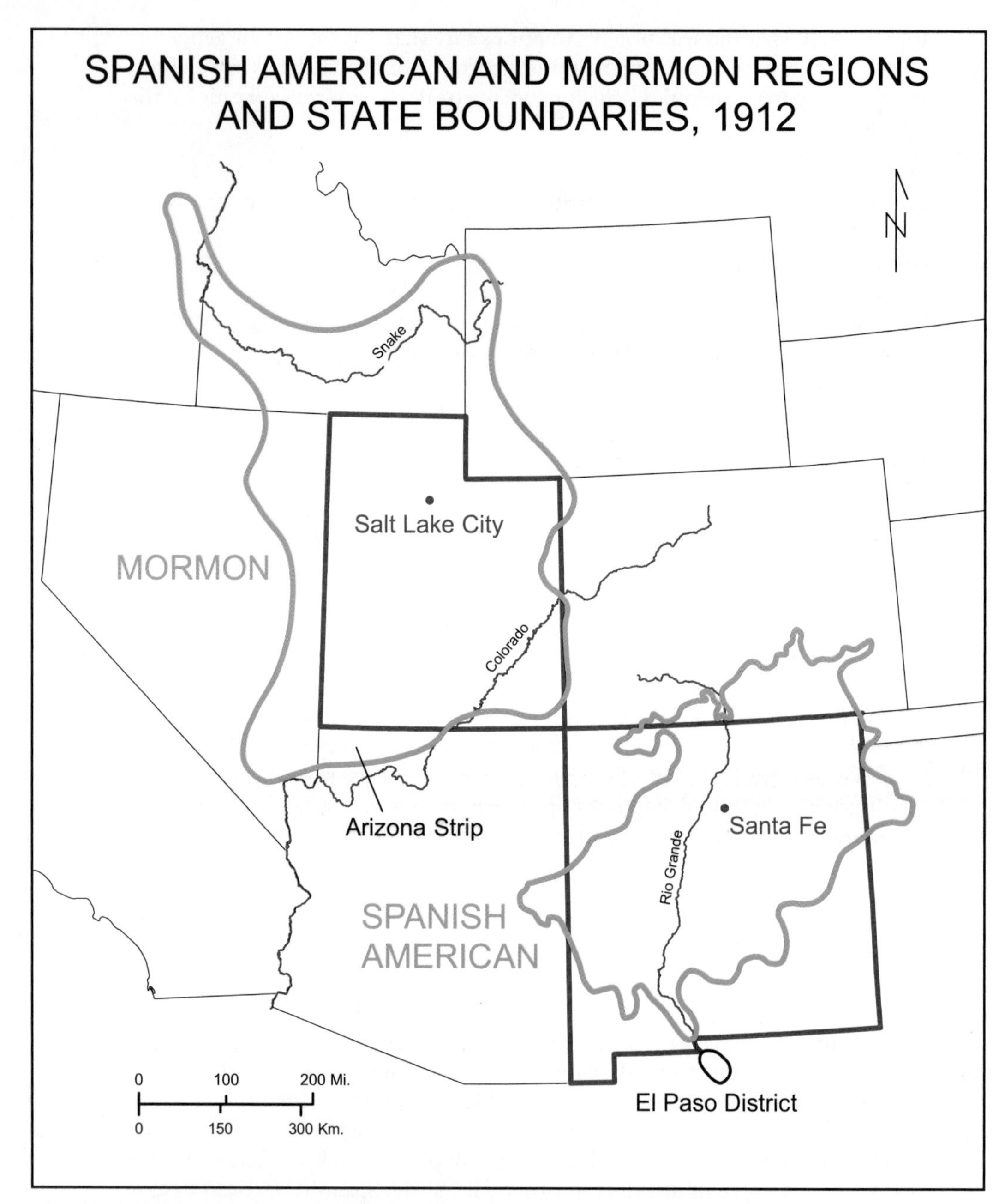

Map 20.2 Spanish American and Mormon Regions and State Boundaries, 1912. (New Mexico became a state in 1912; Utah in 1896). The Spanish American boundary is precise, and the Mormon boundary is generalized.

Sources: Maps 15.8 and 17.4.

The Century of Mass Migration

In the year 2006, the population of the United States officially reached three hundred million people, placing it third in world population ranking, behind China and India. Unlike China and India, however, America's population growth happened quite recently, and immigration was in large measure responsible. The so-called century of mass migration from 1820 to 1920 by itself found some fifty-five million people leaving Europe, of which thirty-three million went to the United States. America represented the land of opportunity, and until 1920 there was an "open door" with no restrictions on who could come and in what numbers. Reviewing briefly the historical geography of immigration—the changing European source areas, the changing American destinations, and the periods of arrival—gives us a needed context for understanding our regions.[10]

From about 1600 to 1820, Europe's emigrant source area was its west coast between Sweden and Spain. England was the principal contributor. Africa's western Guinea coast was the source area for slaves, whose numbers grew importantly after 1700. The destination for emigrants in this period was the Atlantic coast, except for the inland destinations for Spaniards and French, as noted in chapters 2 and 3. "Push" reasons why emigrants left Europe included political repression and, certainly for the English, religious persecution, and the "pull" factors included potential wealth from gold and furs and the desire to own land. The volume of flow across the Atlantic in any given year before 1820 might be described as a trickle. And Europeans along the Atlantic Seaboard who arrived before 1776 are generally referred to as "colonists," while "immigrants" applies to arrivals after 1776.

About 1820 Europe's emigrant source area broadened to especially include Germans and Irish—a larger northwestern Europe. The forced migration of slaves from Africa via Caribbean way stations essentially ended with the Civil War. And after 1820 the emigrant destination expanded to include the entire northeastern quadrant of the United States. Immigrants, many of whom were unskilled, avoided going to the Upland South and the Lowland South, regions where blacks already filled unskilled jobs. New York City was always the principal port of entry, and a large share of each immigrant group lived there. The Irish went in heavy numbers to almost entirely urban places in New England, New England Extended, and across the Old Northwest to Chicago. Avoiding New England, the Germans went in heavy numbers to the Ohio Valley and cities like Cincinnati, St. Louis, Chicago, and Milwaukee; they also took up farms in the Old Northwest and the New Northwest. Swedes and Norwegians generally headed to rural destinations: Swedes to the New Northwest and Norwegians to a wider area between Wisconsin and North Dakota. The availability of essentially free land with the passage of the Homestead Act in 1862 became a major pull factor, and the impact of this legislation, as noted in Chapter 13, was greatest in the New Northwest and the northern Great Plains. Large numbers arrived each year between 1820 and 1890; altogether some fifteen million in the seventy-year period. In the literature, northwestern Europeans arriving in this first phase of mass migration are referred to as the "old" immigrants.[11]

The 1880s saw a shift in source areas from northwestern Europe to eastern and southern Europe. Between 1890 and 1920, heavy numbers of Poles, Ukrainians, Italians, Greeks, and others passed through Ellis Island, the immigrant processing point after 1892. Urban areas of all sizes in the northeastern quadrant were the major destinations, and urban jobs of all kinds were the pull factor. Emigrants from eastern and southern Europe are known in the literature as the "new" immigrants, and between 1890 and 1920, some eighteen million arrived in this second phase of the century of mass migration. The peak year, 1907, saw 1.25 million arrivals. As Roman Catholics, the Italians and Poles heavily reinforced the earlier surge of Catholics represented by the Irish and Germans. Significantly, Catholics became the largest religious body in the United States and the dominant religion nearly everywhere between New England and Minnesota and Iowa.[12]

Laws passed in 1921 and 1924 almost slammed the door shut on those coming from Europe. Behind the legislation was a general public alarm about the growing number of new immigrants, notably the Italians, whom many considered "inferior." In 1921 Congress decreed that 3 percent of those foreign-born enumerated in 1910 would be allowed in annually. Considered too generous, legislation in 1924 reduced the quota to 2 percent of those foreign-born enumerated in 1890—before the new immigrant stream had gained momentum. The quotas in these bills did not apply to emigrants from other parts of the Western Hemisphere, and as we shall see in Chapter 21, to fill the need for unskilled labor, significant numbers of Mexicans began to arrive in the 1920s. Indeed, new legislation in the 1960s once again opened doors to Europeans and migrants elsewhere, and in the 1980s the United States entered a second period of mass migration of even greater magnitude than the century of mass migration.

Cities and Suburbs

We need to recognize that urbanization transformed the United States and each of its regions in fundamental ways. Generalizations about this phenomenon begin with thoughts on the rural to urban transition. Since the first census was taken in 1790, Americans have been moving from the farm to the city. In 1790 an estimated 95 percent of the population was rural and only 5 percent was urban. That small urban percentage, geographer James Vance Jr. reminds us, was present in the earliest port cities (see Chapter 7), performing the essential long-distance shipping services of exporting staple products and importing manufactured goods. Several years before the 1920 census was taken, the milestone that found 50 percent of Americans living in urban places was reached. In 1920 the most heavily urbanized part of the United States was its northeastern quadrant. Cities in a vibrant "Manufacturing Belt," whose corners were essentially Boston, Baltimore, St. Louis, and Milwaukee, had become the destinations for both rural Americans and foreign-born immigrants. And by the year 2000, about 80 percent of Americans were classified as urban.[13]

A second generalization concerns what is meant by "urban." In the United States, the Bureau of the Census defines urban using two basic criteria. To be urban, a place must have 2,500 or more inhabitants, a size definition used in the nineteenth century and officially recognized in the 1910 census. Thus, in the hierarchy of settlements that people live in—farm/ranch stead, hamlet, village, town, city, metropolis, even "megalopolis" (a term coined by French geographer Jean Gottmann in 1961)—2,500 people separates villages (rural) and towns (urban). To differentiate urban from rural around the peripheries of ever-growing urban places, a second criterion, a minimum of one thousand people per square mile, was adopted in the 1960 census. A long list of urban place categories created by the Census Bureau since about 1960 complicates the job of separating rural from urban today.[14]

A third generalization concerns the shapes of cities and the rise of suburbs. As American cities grew, their footprints on the ground evolved—from small circles about 1800 to large stars about 1900 to large circles about 1950. Social philosopher and historian Lewis Mumford (1895–1990) provides a framework based on energy sources and transportation to help understand these changes. About 1800, argued Mumford, animals (horses), people, and water were the power sources, and transportation was by horse and wagon or simply people on foot. Consequently, cities were basically small compact circles (semicircles if port cities), and population densities were high. A century later, coal-burning steam engines and electricity had been added to the power sources, and steam railroads, electric trolley cars, and horsecars did the transporting. City residents pushed out along the tracks of these new conveyances to create starlike urban patterns, and urban densities dropped to medium levels. Mumford shows that by 1950 petroleum had been added as a power source and automobiles and trucks now did the transporting. The flexibility of the automobile allowed for the spaces between the tracks to fill in, and cities became large circles of low densities (see figure 20.1).[15]

Geographer David Ward explains how developments in local transportation made possible the starlike expansions of cities and the creation of suburbs. In the 1840s, wrote Ward, high-income people began to push out along the tracks of steam railroads. Railway stations located half an hour from the central city became the nuclei around which small and exclusive suburbs formed. By the 1850s horsecars (conveyances on tracks pulled by horses) that offered more frequent stops and greater access in the urban core had much success in competing for passengers. High- and middle-income people in Boston, for example, now pushed out some four miles along the tracks of horsecars in a half-hour commute to work. In the late 1880s streetcars powered by overhead electricity (the first example was demonstrated in Richmond, Virginia, in 1887) then spread very quickly as an innovation. An eight- to ten-mile commute for middle-income people further enlarged suburbs by 1900 in, for example, Chicago. To alleviate increasing congestion in the core areas of New York, Chicago, and Boston, tracks elevated above downtown streets were built to accommodate larger amounts of traffic.[16]

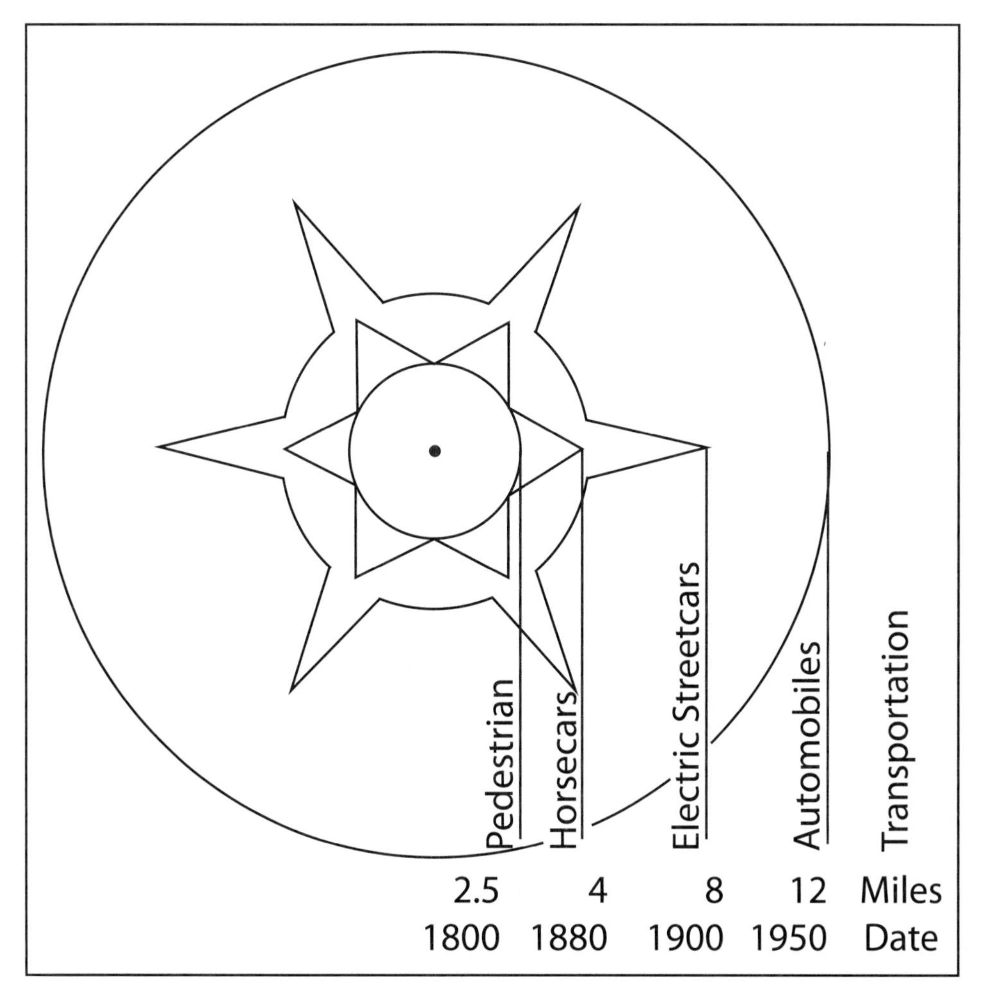

Figure 20.1 Generalized City Growth, 1800–1950. Miles represent distances from the central point of the community.

Source: Ward, *Cities and Immigrants*, 132.

After 1900 the automobile transformed cities from starlike shapes to large circles. When first sold in 1895, autos were largely confined to cities where virtually all the paved roads existed. Automobile sales as measured by vehicle registrations then soared—from fourteen thousand in 1900 to nine million in 1920 to forty-four million in 1950—and with autos came paved roads. Middle-income people who could afford new single-family houses on separate lots filled in the spaces between the tracks of electric streetcars, which rapidly declined with the rise of automobiles. The new suburban subdivisions merely added greater sprawl to the universally used gridiron pattern, except when curvilinear layouts were employed in high-income subdivisions. Meanwhile, the core areas of major cities built upward. Modern skyscrapers made possible by steel-frame construction and electric elevators identified city centers from miles away. New York City's famous Manhattan Island skyline grew from a twenty-nine-story building height in 1900 to one of fifty-seven stories in 1913, when chain-store magnate Frank Woolworth dazzled everyone with his new headquarters. But geographer John R. Borchert reminds us that the decentralization of cities that was made possible by the automobile also saw many a central city decline to the point where it was "no longer 'central.'"[17]

Notes

[1] See Wyckoff and Dilsaver, "Defining the Mountainous West," esp. 1–10. For the quote, "Every drop . . .," see Wyckoff, *How to Read the American West*, 6.

[2] Shumway and Jackson, "Native American Population Patterns," 186, for factual information, including the 1890 Indian population low point and the 1960 self-identification surge; 193, for the 1950 urban migration beginnings; 199, for the four state leaders in numbers.

[3] See Jackson, "Federal Lands in the Mountainous West," 253 and 255, for land percentages that are federally owned; 259–65, for the 1870s shift in federal policy; 267, for federal agencies in control; 256, 268–69, 274–75, for conflict between locals and all Americans.

[4] White, *It's Your Misfortune and None of My Own*, 613 and 615, for the two quotes; 628, for Laura Ingalls Wilder. Geographer Kevin Blake analyzes the writings of Zane Grey (1872–1939) in "Zane Grey and Images of the American West."

[5] Montès, *American Capitals*, 63, for 3.84 capitals; 8, for some 180 capitals; 85, for 255 "stated causes."

[6] Ibid., 156, for the seven reasons.

[7] Ibid., Chapter 3. Minor discrepancies exist in the numbers given by Montès in his figure (63), map (64), and text (64–79). See page 83, for "wanderings." A comment on Texas—Meinig, *Imperial Texas*, 42—argued that centrality for future development was behind Austin's choice as capital. But Montès, *American Capitals*, 75, put Austin within the "apparently erratic" category.

[8] Montès, *American Capitals*, 7, for thirty-nine of fifty capitals selected in the nineteenth century and thirty-five before 1861; 79, for only Oklahoma after 1900; 15–16, for the names of capitals.

[9] Meinig, *The Shaping of America*, 3:301, for Hispanos (Spanish Americans) and Mormons as "organic societies" set incongruously within "arbitrary compartments."

[10] See Hansen, *The Immigrant in American History*, 4, 21, 150, for the century of mass migration. Ward, *Cities and Immigrants*, 54, maps European source areas of "old" and "new" immigrants, 1820 to 1919.

[11] In *The Shaping of America*, 3: esp. 273–80, Meinig notes that New York City was always the principal immigrant port, and he discusses in some detail these immigrant groups, as well as lesser groups, their religions, and the dominant industries of employment.

[12] See Zelinsky, "An Approach to the Religious Geography of the United States," 155–56, for the spatial impact of Catholics in the northeastern quadrant.

[13] Vance, "Cities in the Shaping of the American Nation," 41.

[14] See U.S. Bureau of the Census, *Geographic Areas Reference Manual*, 12-2, for 2,500 in 1910; 12-3, for 1,000 in 1960. In *Megalopolis*, Gottmann was describing the "giant city" between Boston and Washington, DC. Five additional, solidly urbanized areas have since been designated in the world.

[15] In *Technics and Civilization*, Lewis Mumford develops the evolution of technics (machines) under three phases: ecotechnic (dawn), paleotechnic (old), and neotechnic (new).

[16] See Ward, *Cities and Immigrants*, 125–45, including 134, for Richmond.

[17] For the Woolworth building, see Meinig, *The Shaping of America*, 3:301. For central cities no longer "central," see Borchert, "American Metropolitan Evolution," 321.

In Perspective 21

As we argue in "The United States in 1900," the turn of the century marked the end of an era for Americans. A composite map that shows our "Culture Regions in 1900" follows, and then, under "Hispanicization," we focus on a major force that has been reshaping America's cultural geography mainly since about 1920. Finally we offer several "Conclusions."

The United States in 1900

The time around the turn of the century marked the end of an era for the United States. A rural society engaged in agriculture with little good land left to expand to was quickly giving way to an urban society focused on manufacturing and made mobile by the invention of the automobile. In 1889 government authorities announced that manufacturing had replaced agriculture as the dominant industry. In 1890 census officials reported the closing of the frontier, which meant an end to the availability of good farmland. The steady shift of Americans from the farm, where fewer people were needed, to the city, where workers were in heavy demand, was reflected in a rural population that declined from 72 percent in 1880 to 49 percent in 1920. And in a nation that now stretched three thousand miles (4,800 kilometers) from the Atlantic to the Pacific, automobiles and trucks would rapidly replace the horse and carriage and the electric trolley as principal modes of transportation.

Imagining what the United States was like in 1900 is facilitated by comparing Americans in 1900 and 1950, which is exactly what author-editor Frederick Lewis Allen did in *The Big Change*. In 1900 America's population stood at seventy-six million; the arrival of new immigrants helped it reach one hundred million by 1915, and high birth rates particularly in the 1940s pushed it to 150 million in 1950, which means that it almost doubled in a half-century. By 1900 the center of population for westward-moving Americans had reached central Indiana, and by 1950 it would move west one more state to central Illinois. This center, shown on map 21.1, indicates how weighted the country's population was to the northeastern quadrant where, by 1900, a vibrant "Manufacturing Belt" now produced approximately one-third of the world's industrial output. In 1900, American society was highly stratified: A small number of the very wealthy lived in palatial town houses or country villas that had indoor bathrooms, electric lights, and telephones, while the great majority of urban workers and farmers had no telephones and used gas or oil lamps and outside privies. By 1950 these class differences were in retreat, and society was less formal. Life expectancy in 1900 was forty-nine years of age; improved medicine and health care delivery saw it reach sixty-eight by 1950. Between 1900 and 1950, the number of students in institutions of higher education nationwide increased eight times. And the foreign languages spoken by immigrants on city streets in 1900 were all but gone in 1950.[1]

Of all these changes, the advent of the automobile had the biggest impact on transforming the American landscape. Manufacturer Henry Ford led the way by making cars affordable for the masses: Ford sold fifteen million inexpensive Model Ts between 1910 and 1927. With cars Americans enlarged their cities, and because private enterprise drove the platting and development of suburban tracts with very little government land-use planning to guide the process, urban sprawl was often the outcome. With cars came paved roads to connect suburbs with urban cores, regions

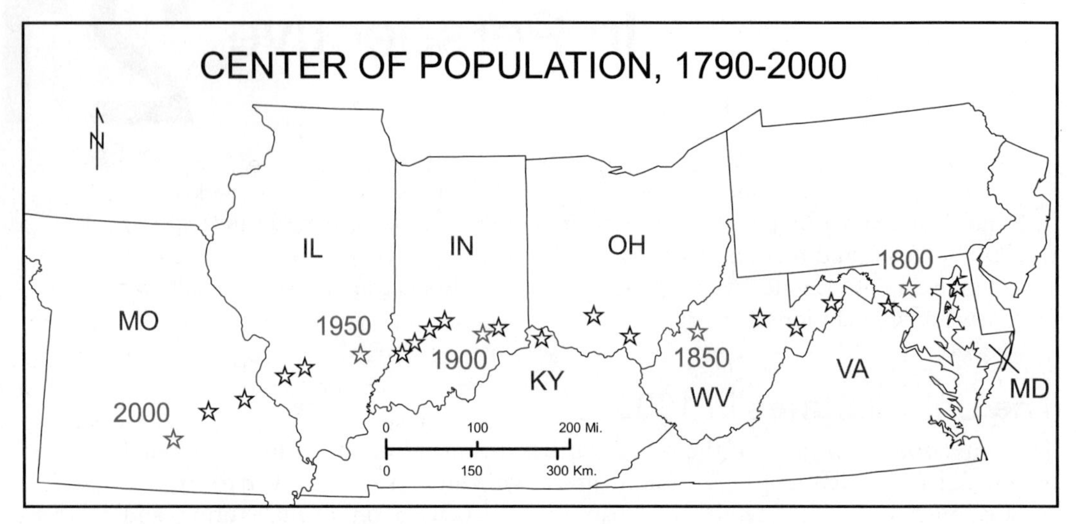

Map 21.1 Center of Population, 1790–2000. The Census Bureau calculates for each census the mean point of population (an equal number of people in every direction). These points show that Americans moved gradually west and south from the Eastern Shore of Maryland in 1790 to central Missouri in 2000. They also verify how heavily the country's population is weighted to the northeast.

Source: U.S. Census Bureau, *2000 Census of Population and Housing*, Population and Housing Unit Counts PHC-3-1, table A, 111–12.

with regions, and the East Coast with the West Coast. Along the roads came filling stations, garages, diners, auto courts (or motor lodges), and billboards. By about 1940 trucks were carrying much of the freight once transported by railroads. By 1950, only in the American West were there still areas of some size that remained inaccessible to motorists. In half a century, then, automobiles had gone from being a luxury to being a necessity, and for most of the fifty-nine million Americans who drove a vehicle in 1950, that vehicle symbolized prestige, authority, and freedom.[2]

Geographer Ralph Brown (see spotlight) was able to witness the many changes that were transforming the United States after 1900. As noted in Chapter 1, Brown was a master at reconstructing the geography of the past as that geography was viewed by its contemporaries. The scholarship and imagination that Brown poured into *Mirror for Americans*, in which he reconstructed the geography of the Eastern Seaboard in 1810, places that book on a short shelf of geography's classics. And according to Linda Jeanne Miles, Brown's biographer, Eunice Brown's pen-and-ink portraits and depictions together with her editing and typing of what Brown wrote, actually made *Mirror for Americans* a husband-and-wife effort. If only Brown had also reconstructed the geography of the United States as of 1900, it would certainly have enriched any study—including this one—of that time. Besides capturing America at a turning point, such a re-creation would have provided the setting for the time represented by our regions.[3]

Culture Regions in 1900

To represent our regions as of 1900 on one composite map presents several dilemmas, which we sort through in table 21.1. First, the Spanish Borderlands and New France, both huge and only thinly settled areas in their colonial periods, disappear as regions; in the United States the Spanish and French were overrun by other peoples, although their regional legacies continue. The three additional colonial-era regions, New England, the Middle Colonies, and the South persist, although the Middle Colonies

Spotlight

Ralph H. Brown (1898–1948)

Ralph and Eunice Brown at their home in St. Paul, Minnesota, August 1938. Photograph courtesy of Linda Miles Coppens, in "Ralph Hall Brown," 96. Coppens gave credit for the photograph to G. Burton Brown, son of Ralph and Eunice Brown.

Ralph Hall Brown's *Historical Geography of the United States* hooked me when I read it as an undergraduate. As a graduate student, I learned that in that book and in *Mirror for Americans*, Brown had pioneered in the use of primary sources and in conceptualizing environmental perception—the thought that in colonizing a new place, people reacted as much to what an environment was thought to be as to what it actually was (fancy versus fact). Imagine my thrill when, as a faculty member, my advisee, Linda Jeanne Miles, chose to write a dissertation about Brown.

Born in 1898 in Ayer, Massachusetts, located thirty miles northwest of Boston, Brown attended Massachusetts Agricultural College (now the University of Massachusetts) in Amherst from 1915 to 1917. From 1919 to 1921, he attended the University of Pennsylvania, earning a BS degree in economics. Brown immediately enrolled in the geography graduate program at the University of Wisconsin where, in 1925, he received a PhD in geography and economics. He and Eunice Rasmussen were married in Madison in 1924, and they had three children. Brown held two faculty positions: one in the University of Colorado (1925–1929) and the other at the University of Minnesota (1929–1948). In 1948 he died in St. Paul at the young age of fifty. Besides his two books, Brown authored fifty-two articles or parts of books, he served as secretary of the Association of American Geographers (1941–1945), and he edited the association's *Annals* (1947–1948). A noncontroversial scholar, Brown avoided theories like Turner's frontier thesis and research paradigms like Derwent Whittlesey's sequent occupance, and his research was not problem-oriented. He garnered respect and admiration through his meticulous use of maps and archival sources to reconstruct the geographies of past regions.

region comes to be known as the Middle Atlantic Region. Second, with the westward movement, the South contributes to an Upland South and to a Lowland South; New Englanders form a long New England Extended column west to North Dakota (as revealed in research by John Hudson); but a column of Midlanders moving west from essentially Pennsylvania is omitted here for insufficient documentation in the literature of historical geography. Because they are readily recognizable and commonly accepted as regions, the Old Northwest, the New Northwest, and the Great Plains need little explanation, and this is also true of Texas and California, if political units are accepted as regions. A final dilemma is what to call the three remaining western regions which, in the twentieth century, acquired new labels. We stick with the Spanish American Region, Oregon Country, and the Mormon Region rather than use their post-1900 labels of the Highland Hispano Region, the Pacific Northwest, and the Mormon Culture Region.[4]

Map 21.2, then, is a composite exhibit of our fourteen regions circa 1900. The boundaries of these regions are taken from foregoing maps, and because many of these regions overlap, regional boundaries alone are identified by the use of colors. The intended effect of map 21.2 is to reaffirm that cultural differences do indeed exist across the United States; and to understand these differences is to make sense

TABLE 21.1 Updated Culture Regions for 1900

Chapter Number	Name of Region	Comments and Updates
2	Spanish Borderlands	Colonial only; overrun and superseded
3	New France	Colonial only; overrun and superseded
4	New England	Recognized in colonial period; persists in general use
5	Middle Atlantic	Recognized in colonial period; known as Middle Colonies in colonial period; persists as Middle Atlantic Region
6	South	Recognized in colonial period; persists in general use
8	Upland South	Also known as Upper South and Border South
9	New England Extended	Term attributed to Peirce Lewis in 1967
10	Old Northwest	Textbook term; part of a larger Middle West
11	Lowland South	Also known as Lower South, Deep South, Dixie
12	Texas	Political unit and self-conscious region
13	New Northwest	Term is popular within the region
15	Spanish American Region	An academic label; another label, Highland Hispano Region, is attributed to Terry Jordan
16	Oregon Country	Becomes Pacific Northwest, a popular term
17	Mormon Region	D. W. Meinig's "Mormon Culture Region" has currency as an academic label
18	California	Political unit and self-conscious region
19	Great Plains	Term used since nineteenth century; also called High Plains

Sources: Zelinsky, "Classical Town Names," 466, attributes "New England Extended" to Peirce Lewis. "Highland Hispano" likely came from Terry Jordan in mid-1990s discussions that led to Nostrand and Estaville, eds., *Homelands*. Meinig's classic article, "The Mormon Culture Region," brought currency to that term.

of the United States. Furthermore, these fourteen regions continue to be the structure of America's cultural differences today. Wilbur Zelinsky suggests that in only "two or possibly three" of these regions is religion "immediately apparent to the casual observer and . . . generally apprehended by . . . inhabitants": The three regions Zelinsky had in mind are understood to be the Mormon, Lowland South (Baptist), and New Northwest (Lutheran) regions. Many would argue that besides these three religious regions, for a host of reasons, New England, the Great Plains, and the Spanish American Region would have to be added to a short list of America's most distinctive regions.[5]

Carving up the United States into sets of regions has been a major interest of American geographers. Following publication of John W. Powell's *Physiographic Regions of the United States* in 1895, geographers—whose young discipline had its roots in geology—emphasized identifying physiographic and physical regions. Notable examples published in the *Annals of the Association of American Geographers* in 1914 were Nevin M. Fenneman's set of physiographic "provinces" and Wolfgang L. G. Joerg's set of "natural" regions (using landforms, climate, and vegetation as his criteria). During the 1920s, however, at the time geographers were rejecting environmental determinism as a philosophy, they turned from mapping physical attributes to mapping cultural factors—regions showing, for example, agriculture, land use, and demographic patterns. A prominent example is J. Russell Smith's comprehensive regional geography, *North America*, published in 1925. In the decades that followed, geographers carved the United States into dozens of sets of cultural regions. The point is this: Whether physical or cultural, all regions are human inventions, and

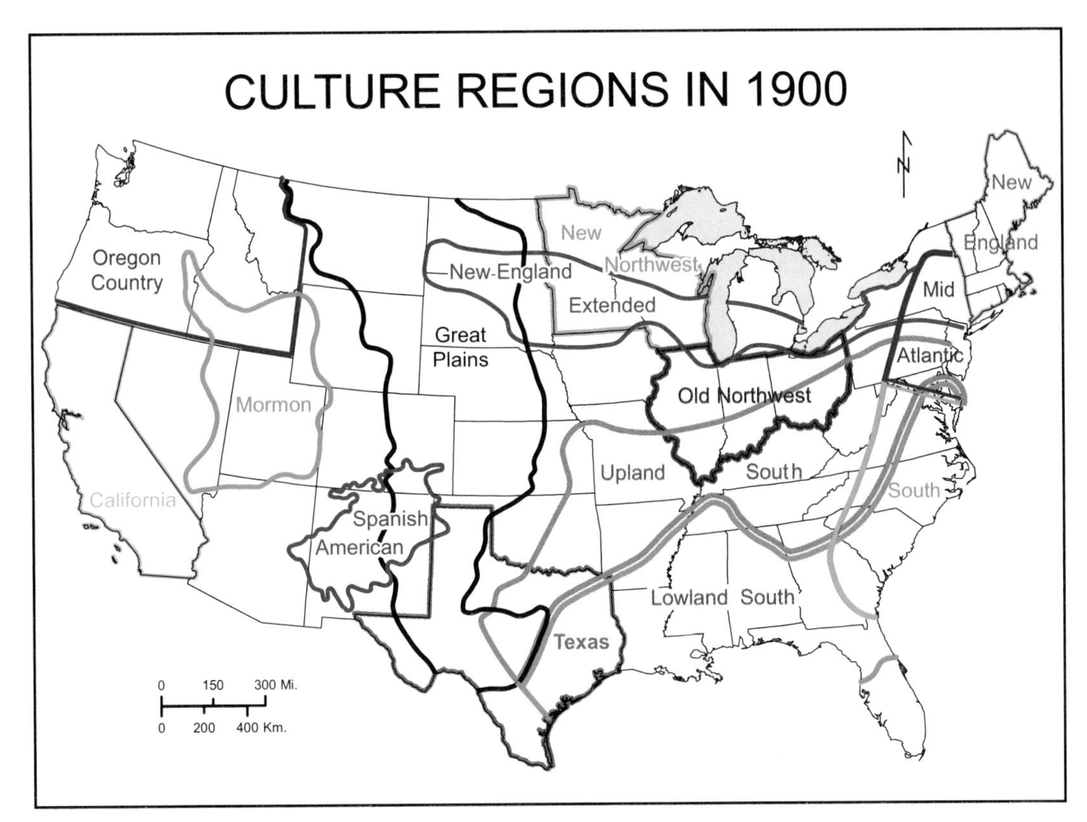

Map 21.2 Culture Regions in 1900.

Sources: See table 21.1.

any set of regions reflects the arbitrary judgments of its author. Culture regions are, quite simply, areas that exhibit cultural homogeneity; how they are conceived and the systems into which they are put reflect the perceptions and predilections of their framers.[6]

There is no simple answer to how America's culture regions developed, given that each region is unique. The three variables, common sense would suggest, are a *people* and the culture represented in the people's source area; a *place*, or the settlement destination of the people; and a *time*, or the era in which settlement occurred. In trying to come to grips with how America's culture regions developed, we have brought to our regions the three analytical themes of cultural diffusion, cultural ecology, and cultural landscape. The basic narrative seems to go as follows: First, migrants arrive with their cultural baggage. Second, drawing on their perceptions and the realities of the place, these migrants make adjustments to the new environment. And third, drawing on their values and practices, they build structures and lay out farms and communities and so leave their imprint. Recall the axiom expressed in the doctrine of first effective settlement (Chapter 2): The first people to effectively colonize an area will be of singular importance in shaping the cultural landscape of that area, no matter how small that group may have been or even if that group was overrun or displaced by others. In the making of America's culture regions, the regional character was set—granted, in degrees of relative importance—by the initial wave of colonists.

Finally, a comment must be made on contrived culture regions. All regions and the images they convey are constructs of people. But geographer William Wyckoff asks whether the regional images devised by some, although perhaps well grounded historically, may have been reinventions with an agenda. Visionaries or promoters may well have been motivated by reasons of boosterism or economic gain when they reimagined the attributes of a place and lobbied for the acceptance of their contrived imagery. We noted this to have happened in New England, where visionary Timothy Dwight and others, wishing to extol the spiritual and political virtues of their region, imagined New England's colonial villages to have been bustling nucleated centers, not the dispersed communities they were. Wyckoff cites examples beyond New England. In the early 1900s Edgar Hewett successfully reimagined for Santa Fe, New Mexico, a "Santa Fe style" of architecture modeled after local Pueblo Indian villages. In Santa Barbara, California, following an earthquake in the 1920s, town leaders reimagined with mercenary eyes a place to be rebuilt using a Spanish colonial mission style. Other examples of borderline-authentic place remaking exist in, for example, southeastern Pennsylvania; New Glarus, Wisconsin; Solvang, California; and Leavenworth, Washington. Thus, Wyckoff raises the legitimate question, informed by social theory, of the degree to which manipulation may have played in regional image making.[7]

Hispanicization

In 1919 historian Herbert Eugene Bolton proposed at the University of California–Berkeley that the history of the entire Western Hemisphere be taught to undergraduates as one yearlong course. The nation-states of Latin America and of Anglo-America, he argued, had all been colonies of European powers, they had all gained their independence in the space of about half a century, and their histories were too intertwined to be separated. For the quarter-century from 1919 to 1944, Bolton taught History 8 AB—History of the Americas—to sometimes as many as one thousand students a semester in Berkeley's large Wheeler Auditorium. Imagining the Western Hemisphere as a unit makes even more sense at the beginning of the twenty-first century. Since the 1980s the "Second Great Migration" to the United States has reintensified the Hispanic or Latino presence in the area where Latin America overlaps into Anglo America: Bolton's Spanish Borderlands. Indeed, the process of Hispanicization has spread a layer of Latin Americans beyond the Southwestern Borderlands to the entire United States. What follows is an outline of this transformation.[8]

Following the Mexican period, the 1850 census and supplemental estimates tell us that perhaps one hundred thousand Mexicans in the Southwestern Borderlands became Mexican Americans. Anglos soon overran the Mexican-descent populations in Texas and California at the opposite ends of the Borderlands, but in centrally positioned New Mexico, where fully two-thirds of the one hundred thousand people of Mexican descent lived in 1850, the old nuevomexicano subculture remained intact and evolved into the Spanish American Region discussed in Chapter 15. Meanwhile, Mexican immigrants in modest numbers crossed the northern international border each year. Texas, Arizona, and in the twentieth century, California, were the major destinations; relatively few were pulled to New Mexico and Colorado. Incomplete data on border crossings (no records whatever were kept between 1886 and 1893) and changing Census Bureau definitions for this population make it difficult to chart Mexican-descent population increases. What is clear, however, is that a surge of nearly five hundred thousand Mexicans arrived in the 1920s, following legislation that, while closing the door on emigrants from Europe, did not affect immigration from anywhere in the Western Hemisphere. A second surge arrived in the prosperous 1950s. And it is also clear that, save for a few distant cities like Chicago, Mexican-descent people were heavily concentrated in the five southwestern states. (Except for Cubans in greater Miami and Puerto Ricans in greater New York City, in 1960 the number of immigrants from Latin American countries other than Mexico was negligible.)[9]

The 1960 census provided the most comprehensive data yet gathered on persons of Mexican descent. Employing "white persons of Spanish surname" (WPSS) as its criterion, the Census Bureau enumerated a population of nearly 3.5 million in the five southwestern states. An additional 0.5 million people of Mexican descent were estimated to live in the non-southwestern states. For my dissertation in geography at UCLA, I used these data from 1960 to reconstruct patterns for "Hispanic Americans," an umbrella term that included Spanish Americans. The big picture was that, of four million Hispanic Americans in the United States in 1960, 3.5 million (87 percent) lived in the five southwestern states. I was particularly intent on defining the "northern border" of the Southwestern Borderlands: Where in the five states, I asked, did the significant presence of WPSS end? The wiggly line that I drew through the five states in map 21.3 omitted from the Hispanic American region those counties where WPSS were both less than five hundred in number and less than 5 percent of the total population. In the 1960s, then, it is fair to say that fully 85 percent of all four million people of Mexican descent in the United States lived south of that line. Interestingly, in 1960 four-fifths of all Hispanic Americans lived in California and Texas; in 1850, the preponderance had lived in New Mexico.[10]

Changes that have occurred since this 1960s reconstruction have been dramatic. A relaxing of restrictions on immigration that began with legislation in 1965 brought on a second era of mass migration. By 2000 some twenty million from around the world had entered the United States, and the influx continues. Those coming from Latin American countries constitute about half of the new arrivals; Mexico's share of that half is about 60 percent. In 2003 the Hispanic minority of thirty-nine million—including Mexicans, Puerto Ricans, Cubans, and others and amounting to 13 percent of all Americans—overtook African Americans to become America's largest

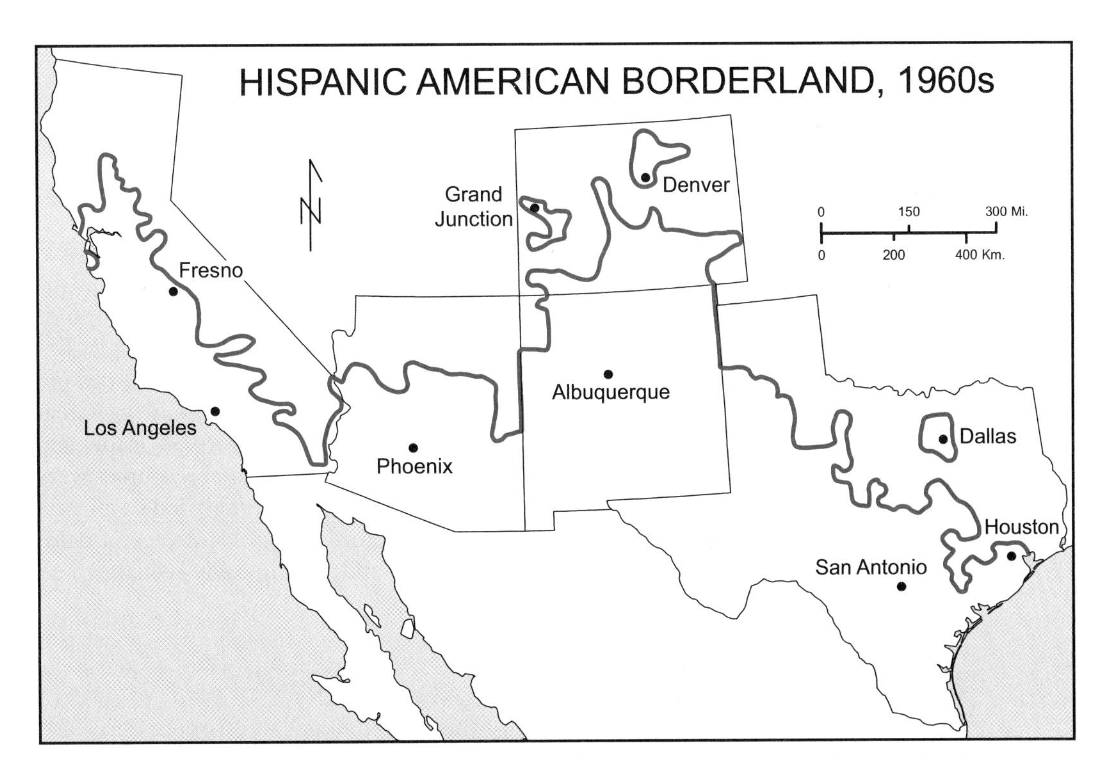

Map 21.3 Hispanic American Borderland, 1960s. Fully 85 percent of the four million people of Mexican descent in the United States in 1960 lived here. Shown are three regional outliers; four rather small nonregional inliers in Texas and New Mexico are omitted. Cities are given for locational purposes.

Source: Simplified from Nostrand, "The Hispanic-American Borderland," 657.

minority. Given the group's continued rapid growth rate, some project that by 2050 one-quarter of all Americans will be Hispanic. Since the 1960s the relative importance of the Mexican-descent concentration near the international border has weakened; three states in particular, Illinois, Florida, and Washington State, have become important Mexican immigrant destinations. Nonetheless, for Mexicans in Mexico who have not forgotten that the North American "Giant" took half their territory, all this outflow of compatriots to America must seem like poetic justice. For the resurgence of Mexicans in the five southwestern states can easily be construed as the group's redeeming lost territory.[11]

Conclusions

Our three analytical themes—cultural ecology, cultural diffusion, and cultural landscapes—suggest six generalizations which are offered by way of conclusion.

1. Physical Barriers

As Americans moved west from the Eastern Seaboard, wrote both Ellen Churchill Semple and Walter Prescott Webb, they encountered two longitudinal barriers. First, argued Semple, came the Appalachian landform barrier; then, contended Webb, came the Great Plains, which coincided with a climatic semiarid-zone barrier. Comparing the two, Appalachia was viewed as a barrier to transportation but not to settlement, while the Great Plains was considered to be a barrier to settlement but not to transportation. It makes sense that the rugged terrain of Appalachia did slow movement to the west, and we can also accept that scant precipitation in the Great Plains did prompt settler adaptations such as dry-farming techniques and tapping groundwater. But research conducted since Semple's and Webb's shows that neither Appalachia nor the Great Plains should be regarded as barriers. Having the environment determine human behavior, as it did for both Semple and Webb, had gone too far.

2. Resource Mismanagement

In the nineteenth century, most Americans seem to have had the attitude that natural resources were inexhaustible. Renewable resources like timber and water would always be there, and even nonrenewable resources like minerals and soils would be plentiful in such a vast land. Thus, lumbermen swept through the white-pine stands of the Northeast, the yellow-pine forests of the Lowland South, and the Douglas fir–covered slopes of western Oregon Country. Deforested areas, notably in the Lowland South, suffered severe soil erosion, and plowing under the grasslands in the Great Plains produced an eventual Dust Bowl. However, about 1900, under the leadership of Theodore Roosevelt and others, there was a realization that natural resources were indeed exhaustible and would have to be managed. This new mentality led to conservation practices such as sustained yield production in forestry, use of diversion dams and plowing on the contour to lessen soil erosion, and setting aside wetlands and other areas to protect wildlife.

3. Uneven Advance

The prevailing direction in the settlement of the United States was from east to west. But in this westward movement people did not surge across the land in one large wave. Rather, the westward movement was a highly selective and uneven settlement process. We saw many examples of this. In Appalachia, the limestone soils of the

Bluegrass and Nashville basins attracted colonists well before the intervening areas did. In the Lowland South, interfluves were chosen before the less well-drained bottomlands. In the dry West, exotic rivers and the presence of minerals drew people to settlements that became islands in a lightly populated inland sea. And on a grander scale, much of the Great Plains and the interior West were skipped over by those attracted to the humid Pacific-facing slopes. Americans did move west largely within bands of latitude, but as they did so their advance for various reasons was not orderly.

4. Pluralistic Society

The United States is a nation of immigrants. Leading the way by a large margin were Native Americans. Then came the Spanish and French, whose frontiers of inclusion, as we saw, arguably did less to displace and decimate Native Americans than did the frontiers of exclusion of the largely British and African American peoples. Subsequent emigrant waves brought to America Germans, Irish, Poles, and Italians, and, after about 1920, also Mexicans and other Latin Americans as well as people from Asia. The result was a pluralistic society—what Theodore Roosevelt called "hyphenated Americans." Some immigrant groups, notably those from Europe, assimilated more readily, and for them the term "melting pot" may have applied. Yet for others, notably groups from Africa, Latin America, and Asia, assimilation has been much slower, and the term "salad bowl" better represents their experience. The great American experiment, then, has been to successfully accommodate diverse people. Many would argue that cultural diversity is a source of strength, and there is general consensus that progress in accommodating diverse peoples in the United States versus other parts of the world has been admirable.

5. Geographical Persistence

One outcome of a diverse population is America's great variety of cultural landscapes. The examples are numerous: We have New England towns and their green-centered villages, Middle Colony courthouse squares that spread west with their various designs, Lowland South plantation headquarters and crossroad hamlets for taking care of business, New Mexican linear villages strung out along irrigation ditches, and Great Basin communities with their hierarchy of Mormon churches. The lead bullets fired by Coronado's soldiers are now being used to reconstruct Coronado's route and constitute an interesting if obscure way of adding new evidence to our understanding of the Spanish Borderlands landscape. The point is this: The first European American settlers in an area were of singular importance in creating present-day cultural landscapes. What those people built and the patterns they laid out have persisted geographically. And the geographical persistence of features and layouts underlie and are an important basis for the geographical exercise of delineating culture regions.

6. Regions versus Homelands

Regions have been our focus in this study. In selected cases, it can be argued that these regions are also homelands. A homeland is a place where people have adjusted to the natural environment, have stamped that environment with their cultural imprint, and have *bonded* with both the environment and their cultural overlay. What sets homelands apart from culture regions is the bonding with place—the emotional feelings of attachment and belonging that with time develop for people in a homeland. Although homelands can be viewed as kinds of culture regions, the issue of bonding with place suggests that they belong more squarely under the category of cultural ecology. In this present study, our Spanish American and Mormon regions qualify as homelands. Had they been treated as regional entities, French Louisiana and Mexican South Texas would also have qualified as homelands. And a number of Native American reservations, most conspicuously the Arizona-centered reservation of the Navajo, very much qualify as homelands as well.[12]

Notes

[1] See Allen, *The Big Change*, 14, for population totals; 27 and 219, for the gap between rich and poor; 27–48, for society in 1900; 202, for life expectancy; 222, for education. For the United States producing "one-third of the world's industrial output" in 1900, see Meinig, *The Shaping of America*, 3:240. Meinig calculates that "only a little more than 4 million of the 76 million enumerated in the 1900 census" lived in the American West; see ibid., 3:177.

[2] For fifteen million Model Ts sold from 1910 to 1927, see Meinig, *The Shaping of America*, 4:36–37. For the shift from luxury to necessity and fifty-nine million drivers in 1950, see Allen *The Big Change*, 121 and 130, respectively.

[3] For Brown's writing of *Mirror for Americans*, see Miles, "Ralph Hall Brown," esp. 85–93, for portraits and editing by Eunice, and John K. Wright's use of "gentlescholar" to characterize Brown's courteous, nonabrasive ways; and 55 and 147, for Eunice drawing maps and typing for her husband. Also see Coppens, "Ralph Hall Brown 1898–1948."

[4] For the New England column, Hudson, "Yankeeland in the Middle West," 196. The Midlander column or belt is discussed somewhat unconvincingly by Meinig in *The Shaping of America*, 2:279–84 and 3:267–68. Hudson, "North American Origins of Middlewestern Frontier Populations," 403–4, 400 (map), 411 (map) maps a Midlander stream between latitude 39° and 41° north, but showing how Midlanders fit within the Old Northwest or the Middle West is insufficiently developed.

[5] The quote is in Zelinsky, "An Approach to the Religious Geography of the United States," 165. Zelinsky leaves it to the reader to determine what the two or three regions are that he has in mind, but his discussion leaves few choices.

[6] Historian Vernon Carstensen, while tracing the evolution of regions, cites the Fenneman and Joerg examples in "The Development and Application of Regional-Sectional Concepts, 1900–1950," 105–7. Also see J. Russell Smith, *North America*.

[7] Wyckoff, "Creating Regional Landscapes and Identities," 335–57 (336, New England; 337, Santa Barbara; 351, Santa Fe).

[8] For the History of the Americas course, see Bannon, *Herbert Eugene Bolton*, esp. 142–45 and 182–89. Bolton retired in 1940, but in 1942–1944, he returned to teach History 8 (ibid., 216, 222).

[9] From the 1850 census and estimates for census gaps, I counted 80,302 people of Mexican descent in the Southwest circa 1850. Assuming that this number was an undercount, the figure 100,000 is perhaps more reasonable. See Nostrand, "Mexican Americans Circa 1850," 383. I review census data concerning Mexican descent people between 1850 and 1960 in Nostrand, "The Hispanic-American Borderland: A Regional, Historical Geography," 175–214, including 178, for the lack of records between 1886 and 1893.

[10] See Nostrand, "The Hispanic-American Borderland," esp. 638–40, 649, 655.

[11] See Meinig, *The Shaping of America*, 4:233–34, for the 1965 Immigration and Naturalization Act; 4:234, for at least twenty million by 2000; 4:237, for Mexico as the largest Latin American source. See Arreola, "Hispanic American Legacy, Latino American Diaspora," 18, for the Hispanic minority officially becoming the largest in 2003.

[12] See Nostrand and Estaville, *Homelands*, xiii–xxiii.

Bibliography

Books

Allen, Frederick Lewis. *The Big Change: America Transforms Itself 1900–1950.* New York: Harper and Brothers, 1952.

Allen, John Logan. *Passage through the Garden: Lewis and Clark and the Image of the American Northwest.* Urbana: University of Illinois Press, 1975.

Ambrose, Stephen E. *Nothing Like It in the World: The Men Who Built the Transcontinental Railroad 1863–1869.* New York: Simon and Schuster, 2000.

Andrews, Edward Deming. *The People Called Shakers: A Search for the Perfect Society.* New York: Dover Publications, 1963.

Bannon, John Francis. *Herbert Eugene Bolton: The Historian and the Man, 1870–1953.* Tucson: University of Arizona Press, 1978.

———. *The Spanish Borderlands Frontier 1513–1821.* New York: Holt, Rinehart and Winston, 1970.

Beck, Warren A., and Ynez D. Haase. *Historical Atlas of California.* Norman: University of Oklahoma Press, 1974.

Bolton, Herbert E. *The Spanish Borderlands: A Chronicle of Old Florida and the Southwest.* New Haven, CT: Yale University Press, 1921.

Bowen, William A. *The Willamette Valley: Migration and Settlement on the Oregon Frontier.* Seattle: University of Washington Press, 1978.

Brown, Ralph H. *Historical Geography of the United States.* New York: Harcourt, Brace & World, 1948.

———. *Mirror for Americans: Likeness of the Eastern Seaboard, 1810.* New York: American Geographical Society, 1943.

Carney, Judith A. *Black Rice: The African Origins of Rice Cultivation in the Americas.* Cambridge, MA: Harvard University Press, 2001.

Carter, George F. *Earlier Than You Think: A Personal View of Man in America.* College Station: Texas A&M University Press, 1980.

Clark, Andrew Hill. *Acadia: The Geography of Early Nova Scotia to 1760.* Madison: University of Wisconsin Press, 1968.

Colten, Craig E., and Geoffrey L. Buckley, eds. *North American Odyssey: Historical Geographies for the Twenty-First Century.* Lanham, MD: Rowman & Littlefield, 2014.

Conzen, Michael P., ed. *The Making of the American Landscape.* Boston: Unwin Hyman, 1990.

———. *The Making of the American Landscape.* 2nd ed. New York: Routledge, 2010.

Cronon, William. *Changes in the Land: Indians, Colonists, and the Ecology of New England.* New York: Hill and Wang, 1983.

Dana, Richard Henry, Jr. *Two Years before the Mast.* New York: World Publishing, 1946. First published 1840 by Harper and Brothers.

Denevan, William M., ed. *The Native Population of the Americas in 1492.* Madison: University of Wisconsin Press, 1976.

DeVoto, Bernard, ed. *The Journals of Lewis and Clark.* Boston: Houghton Mifflin, 1953.

Dumke, Glenn S. *The Boom of the Eighties in Southern California.* San Marino, CA: Huntington Library, 1944.

Earle, Carville. *The American Way: A Geographical History of Crisis and Recovery.* Lanham, MD: Rowman & Littlefield, 2003.

———. *Geographical Inquiry and American Historical Problems.* Stanford, CA: Stanford University Press, 1992.

Ensminger, Robert F. *The Pennsylvania Barn: Its Origin, Evolution, and Distribution in North America.* 2nd ed. Baltimore: Johns Hopkins University Press, 2003.

Estaville, Lawrence E., and Richard A. Earl. *Texas Water Atlas.* College Station: Texas A&M University Press, 2008.

Fenneman, Nevin M. *Physiography of Western United States.* New York: McGraw-Hill, 1931.

Fischer, David Hackett. *Albion's Seed: Four British Folkways in America.* New York: Oxford University Press, 1989.

Flint, Richard, and Shirley Cushing Flint, eds. *The Latest Word from 1540: People, Places, and Portrayals of the Coronado Expedition.* Albuquerque: University of New Mexico Press, 2011.

Franzwa, Gregory M. *The Oregon Trail Revisited.* St. Louis, MO: Patrice Press, 1972.

Furman, Necah Stewart. *Walter Prescott Webb: His Life and Impact.* Albuquerque: University of New Mexico Press, 1976.

Fusonie, Alan, and Donna Jean Fusonie. *George Washington: Pioneer Farmer.* Mount Vernon, VA: Mount Vernon Ladies' Association, 1998.

Gerlach, Russel L. *Immigrants in the Ozarks: A Study in Ethnic Geography.* Columbia: University of Missouri Press, 1976.

Gibson, Arrell M. *The Oklahoma Story*. Norman: University of Oklahoma Press, 1978.

Glassie, Henry. *Pattern in the Material Folk Culture of the Eastern United States*. Philadelphia: University of Pennsylvania Press, 1968.

Gottmann, Jean. *Megalopolis: The Urbanized Northeastern Seaboard of the United States*. New York: The Twentieth Century Fund, 1961.

Gregg, Josiah. *Commerce of the Prairies*. Edited by Max L. Moorhead. Norman: University of Oklahoma Press, 1954. First published 1844 by H. G. Langley.

Hansen, Marcus Lee. *The Immigrant in American History*. Cambridge, MA: Harvard University Press, 1940.

Harper, Robert A., and Frank O. Ahnert. *Introduction to Metropolitan Washington*. Washington, DC: Association of American Geographers, 1968.

Harris, Richard Colebrook. *The Reluctant Land: Society, Space, and Environment in Canada before Confederation*. Vancouver: University of British Columbia Press, 2008.

———. *The Resettlement of British Columbia: Essays on Colonialism and Geographical Change*. Vancouver: University of British Columbia Press, 1997.

———. *The Seigneurial System in Early Canada: A Geographical Study*. Madison: University of Wisconsin Press, 1966.

Harris, Richard Colebrook, ed. *Historical Atlas of Canada*. Vol. 1 *From the Beginning to 1800*. Toronto: University of Toronto Press, 1987.

Harris, Richard Colebrook, and John Warkentin. *Canada before Confederation: A Study in Historical Geography*. New York: Oxford University Press, 1974.

Hatcher, Harlan. *The Western Reserve: The Story of New Connecticut in Ohio*. Kent, OH: Kent State University Press, 1991.

Hilliard, Sam Bowers. *Atlas of Antebellum Southern Agriculture*. Baton Rouge: Louisiana State University Press, 1984.

———. *Hog Meat and Hoecake: Food Supply in the Old South, 1840–1860*. Carbondale and Edwardsville: Southern Illinois University Press, 1972.

Holbrook, Stewart H. *The Yankee Exodus: An Account of Migration from New England*. Seattle: University of Washington Press, 1968. First published 1950 by Macmillan.

Hubka, Thomas C. *Big House, Little House, Back House, Barn: The Connected Farm Buildings of New England*. Hanover, NH: University Press of New England, 1984.

Hudson, John. *Making the Corn Belt: A Geographical History of Middle-Western Agriculture*. Bloomington: Indiana University Press, 1994.

———. *Plains Country Towns*. Minneapolis: University of Minnesota Press, 1985.

Hunt, Charles B. *Physiography of the United States*. San Francisco: W. H. Freeman, 1967.

James, Preston E. *All Possible Worlds: A History of Geographical Ideas*. Indianapolis: Bobbs-Merrill, 1972.

Johansen, Dorothy O., and Charles M. Gates. *Empire of the Columbia: A History of the Pacific Northwest*. New York: Harper and Brothers, 1957.

Jordan, Terry G. *German Seed in Texas Soil: Immigrant Farmers in Nineteenth-Century Texas*. Austin: University of Texas Press, 1966.

———. *North American Cattle-Ranching Frontiers: Origin, Diffusion, and Differentiation*. Albuquerque: University of New Mexico Press, 1993.

———. *Trails to Texas: Southern Roots of Western Cattle Ranching*. Lincoln: University of Nebraska Press, 1981.

Jordan-Bychkov, Terry G. *The Upland South: The Making of an American Folk Region and Landscape*. Santa Fe, NM: Center for American Places, 2003.

Jordan-Bychkov, Terry G., and Mona Domosh. *The Human Mosaic: A Thematic Introduction to Cultural Geography*. 9th ed. New York: W. H. Freeman, 2003.

Judd, Sylvester. *History of Hadley*. Springfield, MA: H. R. Huntting, 1905.

Kelso, William M. *Jamestown: The Buried Truth*. Charlottesville: University of Virginia Press, 2006.

Kincaid, Robert L. *The Wilderness Road*. Middlesboro, KY, 1973. First published 1947 by Bobbs-Merrill.

Klingle, Matthew. *Emerald City: An Environmental History of Seattle*. New Haven, CT: Yale University Press, 2007.

Kniffen, Fred B. *Louisiana: Its Land and People*. Baton Rouge: Louisiana State University Press, 1968.

Krause, David J. *The Making of a Mining District: Keweenaw Native Copper, 1500–1870*. Detroit: Wayne State University Press, 1992.

Lemon, James T. *The Best Poor Man's Country: A Geographical Study of Early Southeastern Pennsylvania*. Baltimore: Johns Hopkins University Press, 1972.

Lewis, Peirce F. *New Orleans: The Making of an Urban Landscape*. 2nd ed. Santa Fe, NM: Center for American Places, 2003.

McIlwraith, Thomas F., and Edward K. Muller, eds. *North America: The Historical Geography of a Changing Continent*. 2nd ed. Lanham, MD: Rowman & Littlefield, 2001.

McKnight, Tom L. *Regional Geography of the United States and Canada*. 4th ed. Upper Saddle River, NJ: Pearson Prentice Hall, 2004.

McManis, Douglas R. *Colonial New England: A Historical Geography*. New York: Oxford University Press, 1975.

McWilliams, Carey. *Southern California Country: An Island on the Land*. New York: Duell, Sloan and Pearce, 1946.

Meinig, D. W. *The Great Columbia Plain: A Historical Geography, 1805–1910*. Seattle: University of Washington Press, 1968.

———. *Imperial Texas: An Interpretive Essay in Cultural Geography*. Austin: University of Texas Press, 1969.

———. *The Shaping of America: A Geographical Perspective on 500 Years of History*. Vol. 1, *Atlantic America, 1492–1800*. New Haven, CT: Yale University Press, 1986.

———. *The Shaping of America: A Geographical Perspective on 500 Years of History*. Vol. 2, *Continental America, 1800–1867*. New Haven, CT: Yale University Press, 1993.

———. *The Shaping of America: A Geographical Perspective on 500 Years of History*. Vol. 3, *Transcontinental America, 1850–1915*. New Haven, CT: Yale University Press, 1998.

———. *The Shaping of America: A Geographical Perspective on 500 Years of History*. Vol. 4, *Global America, 1915–2000*. New Haven, CT: Yale University Press, 2004.

———. *Southwest: Three Peoples in Geographical Change, 1600–1970*. New York: Oxford University Press, 1971.

Merrens, Harry Roy. *Colonial North Carolina in the Eighteenth Century: A Study in Historical Geography*. Chapel Hill: University of North Carolina Press, 1964.

Mitchell, Robert D. *Commercialism and Frontier: Perspectives on the Early Shenandoah Valley*. Charlottesville: University Press of Virginia, 1977.

Mitchell, Robert D., and Paul A. Groves, eds. *North America: The Historical Geography of a Changing Continent*. Totowa, NJ: Rowman & Littlefield, 1987.

Montès, Christian. *American Capitals: A Historical Geography*. Chicago: University of Chicago Press, 2014.

Moorhead, Max L. *New Mexico's Royal Road: Trade and Travel on the Chihuahua Trail*. Norman: University of Oklahoma Press, 1958.

———. *The Presidio: Bastion of the Spanish Borderlands*. Norman: University of Oklahoma Press, 1975.

Morgan, Murray. *Skid Road: An Informal Portrait of Seattle*. Seattle: University of Washington Press, 1951.

Morganstein, Martin, Joan H. Cregg, and Erie Canal Museum. *Erie Canal*. Erie Canal Museum. Charleston, SC: Arcadia Publishing, 2001.

Morris, John W., Charles R. Goins, and Edwin G. McReynolds. *Historical Atlas of Oklahoma*. 2nd ed. Norman: University of Oklahoma Press, 1976.

Mumford, Lewis. *Technics and Civilization*. New York: Harcourt, Brace & World, 1934.

Nesbit, Robert C., and William F. Thompson. *Wisconsin: A History*. 2nd ed. Madison: University of Wisconsin Press, 1989.

Nostrand, Richard L. *El Cerrito, New Mexico: Eight Generations in a Spanish Village*. Norman: University of Oklahoma Press, 2003.

———. *The Hispano Homeland*. Norman: University of Oklahoma Press, 1992.

Nostrand, Richard L., and Lawrence E. Estaville, eds. *Homelands: A Geography of Culture and Place across America*. Baltimore: Johns Hopkins University Press, 2001.

Ostergren, Robert C. *A Community Transplanted: The Trans-Atlantic Experience of a Swedish Immigrant Settlement in the Upper Middle West, 1835–1915*. Madison: University of Wisconsin Press, 1988.

Paullin, Charles O., and John K. Wright. *Atlas of the Historical Geography of the United States*. Washington and New York: Carnegie Institution of Washington and the American Geographical Society, 1932.

Peterson, Gary B., and Lowell C. Bennion. *Sanpete Scenes: A Guide to Utah's Heart*. Eureka, UT: Basin/Plateau Press, 1987.

Pike, Zebulon Montgomery. *The Journals of Zebulon Montgomery Pike, with Letters and Related Documents*. Edited and annotated by Donald Jackson. 2 vols. Norman: University of Oklahoma Press, 1966.

Pirkle, E. C., and W. H. Yoho. *Natural Landscapes of the United States*. 3rd ed. Dubuque, IA: Kendall/Hunt, 1982.

Pitt, Leonard. *The Decline of the Californios: A Social History of the Spanish-Speaking Californians, 1846–1890*. Berkeley: University of California Press, 1966.

Raitz, Karl, ed. *A Guide to the National Road*. Baltimore: Johns Hopkins University Press, 1996.

Rees, James C. *Mount Vernon: Official Guidebook*. Mount Vernon, VA: Mount Vernon Ladies' Association, 2001.

Reps, John W. *The Making of Urban America: A History of City Planning in the United States*. Princeton, NJ: Princeton University Press, 1965.

———. *Monumental Washington: The Planning and Development of the Capital Center*. Princeton, NJ: Princeton University Press, 1967.

Sale, Roger. *Seattle: Past to Present.* Seattle: University of Washington Press, 1976.
Sauer, Carl O. *Seventeenth Century North America.* Berkeley, CA: Turtle Island Foundation, 1980.
Semple, Ellen Churchill. *American History and Its Geographic Conditions.* Boston: Houghton Mifflin, 1903.
Shortridge, James R. *The Middle West: Its Meaning in American Culture.* Lawrence: University Press of Kansas, 1989.
Simmons, Marc. *Albuquerque: A Narrative History.* Albuquerque: University of New Mexico Press, 1982.
———. *The Last Conquistador: Juan de Oñate and the Settling of the Far Southwest.* Norman: University of Oklahoma Press, 1991.
Simmons, Marc, and Hal Jackson. *Following the Santa Fe Trail: A Guide for Modern Travelers.* 3rd ed. Santa Fe: Ancient City Press, 2001.
Singletary, Otis A. *The Mexican War.* Chicago: University of Chicago Press, 1960.
Smith, J. Russell. *North America: Its People and the Resources, Development, and Prospects of the Continent as the Home of Man.* New York: Harcourt, Brace and Company, 1925.
Spicer, Edward H. *Cycles of Conquest: The Impact of Spain, Mexico, and the United States on the Indians of the Southwest, 1533–1960.* Tucson: University of Arizona Press, 1962.
Stephens, A. Ray, and William M. Holmes. *Historical Atlas of Texas.* Norman: University of Oklahoma Press, 1989.
Stewart, George R. *Names on the Land: A Historical Account of Place-Naming in the United States.* New York: Random House, 1945.
Thrower, Norman J. W. *Original Survey and Land Subdivision: A Comparative Study of the Form and Effect of Contrasting Cadastral Surveys.* Fourth in the Monograph Series of the Association of American Geographers. Chicago: Rand McNally, 1966.
Tompkins, Walker A. *Santa Barbara's Royal Rancho: The Fabulous History of Los Dos Pueblos.* Berkeley, CA: Howell-North, 1960.
Turner, Frederick Jackson. *The Frontier in American History.* New York: Henry Holt, 1920.
Wacker, Peter O. *Land and People: A Cultural Geography of Preindustrial New Jersey: Origins and Settlement Patterns.* New Brunswick: Rutgers University Press, 1975.
Wade, Richard C. *The Urban Frontier: The Rise of Western Cities, 1790–1830.* Cambridge, MA: Harvard University Press, 1959.
Ward, David. *Cities and Immigrants: A Geography of Change in Nineteenth-Century America.* New York: Oxford University Press, 1971.
Webb, Walter Prescott. *The Great Plains.* Boston: Ginn and Company, 1931.
Weber, David J. *The Mexican Frontier, 1821–1846: The American Southwest under Mexico.* Albuquerque: University of New Mexico Press, 1982.
———. *The Spanish Frontier in North America.* New Haven, CT: Yale University Press, 1992.
White, E. B. *Charlotte's Web.* New York: HarperCollins, 1952.
White, Richard. *It's Your Misfortune and None of My Own: A History of the American West.* Norman: University of Oklahoma Press, 1991.
Williams, Michael. *Americans and Their Forests: A Historical Geography.* Cambridge: Cambridge University Press, 1989.
Wishart, David J. *The Last Days of the Rainbelt.* Lincoln: University of Nebraska Press, 2013.
———. *An Unspeakable Sadness: The Dispossession of the Nebraska Indians.* Lincoln: University of Nebraska Press, 1994.
Wishart, David J., ed. *Encyclopedia of the Great Plains.* Lincoln: University of Nebraska Press, 2004.
Womack, John. *Cleveland County Oklahoma Place Names.* Norman, OK: copyright by author, 1977.
———. *Norman: An Early History, 1820–1900.* Norman, OK: published by the author, 1976.
Wood, Joseph S. *The New England Village.* Baltimore: Johns Hopkins University Press, 1997.
Wyckoff, William. *The Developer's Frontier: The Making of the Western New York Landscape.* New Haven, CT: Yale University Press, 1988.
———. *How to Read the American West: A Field Guide.* Seattle: University of Washington Press, 2014.
Zelinsky, Wilbur. *The Cultural Geography of the United States.* Englewood Cliffs, NJ: Prentice-Hall, 1973.
Zochert, Donald. *Laura: The Life of Laura Ingalls Wilder.* New York: Avon Books (HarperCollins), 1976.

Articles

Aiken, Charles S. "The Evolution of Cotton Ginning in the Southeastern United States." *Geographical Review* 63, no. 2 (1973): 196–224.
———. "An Examination of the Role of the Eli Whitney Cotton Gin in the Origin of the United States Cotton Regions." *Proceedings of the Association of the American Geographers* 3 (1971): 5–9.

Alley, John. "City Beginnings in Oklahoma Territory." *Oklahoma Municipal Review* 9, nos. 3, 4, 6 (1935): 36–39, 51–54, 63, 84–86, and 96.
Barnes, C. P. "Economies of the Long-Lot Farm." *Geographical Review* 25, no. 2 (1935): 298–301.
Blake, Kevin S. "Zane Grey and Images of the American West." *Geographical Review* 85, no. 2 (1995): 202–16.
Borchert, John R. "American Metropolitan Evolution." *Geographical Review* 57, no. 3 (1967): 301–32.
———. "The Dust Bowl in the 1970s." *Annals of the Association of American Geographers* 61, no. 1 (1971): 1–22.
Bowden, Martyn J. "Culture and Place: English Sub-Cultural Regions in New England in the Seventeenth Century." *Connecticut History* 35, no. 1 (1994): 68–146.
———. "The Invention of American Tradition." *Journal of Historical Geography* 18, no. 1 (1992): 3–26.
———. "The Perception of the Western Interior of the United States, 1800–1870: A Problem in Historical Geosophy." *Proceedings of the Association of American Geographers* 1 (1969): 109–13.
Brightman, George F. "The Boundaries of Utah." *Economic Geography* 16, no. 1 (1940): 87–95.
Butzer, Karl. "French Wetland Agriculture in Atlantic Canada and Its European Roots: Different Avenues to Historical Diffusion." *Annals of the Association of American Geographers* 92, no. 3 (2002): 451–70.
Carlson, Alvar W. "Long-Lots in the Rio Arriba." *Annals of the Association of American Geographers* 65, no. 1 (1975): 48–57.
Carney, Judith A. "From Hands to Tutors: African Expertise in the South Carolina Rice Economy." *Agricultural History* 67, no. 3 (1993): 1–30.
Chavez, Fray Angelico. "Santa Fe Church and Convent Sites in the Seventeenth and Eighteenth Centuries." *New Mexico Historical Review* 24, no. 2 (1949): 85–93.
Comeaux, Malcolm L. "The Cajun Barn." *Geographical Review* 79, no. 1 (1989): 47–62.
Conzen, Michael P. "Spatial Data from Nineteenth Century Manuscript Censuses: A Technique for Rural Settlement and Land Use Analysis." *Professional Geographer* 21, no. 5 (1969): 337–343.
———. "A Transport Interpretation of the Growth of Urban Regions: An American Example." *Journal of Historical Geography* 1, no. 4 (1975): 361–382.
Crowley, William K. "Old Order Amish Settlement: Diffusion and Growth." *Annals of the Association of American Geographers* 68, no. 2 (1978): 249–264.
Cummings, Abbott Lowell. "Connecticut and Its Building Traditions." *Connecticut History* 35, no. 1 (1994): 192–233.
Dilsaver, Lary M. "After the Gold Rush." *Geographical Review* 75, no. 1 (1985): 1–18.
Doran, Michael F. "Negro Slaves of the Five Civilized Tribes." *Annals of the Association of American Geographers* 68, no. 3 (1978): 335–350.
Dunbar, Gary S. "Colonial Carolina Cowpens." *Agricultural History* 35, no. 3 (1961): 125–131.
Earle, Carville V. "Environment, Disease and Mortality in Early Virginia." *Journal of Historical Geography* 5, no. 4 (1979): 365–390.
———. "The First English Towns of North America." *Geographical Review* 67, no. 1 (1977): 34–50.
———. "A Staple Interpretation of Slavery and Free Labor." *Geographical Review* 68, no. 1 (1978): 51–65.
Estaville, Lawrence E., Jr. "Mapping the Cajuns." *Southern Studies: An Interdisciplinary Journal of the South* 25, no. 2 (1986): 163–171.
———. "Mapping the Louisiana French." *Southeastern Geographer* 26, no. 2 (1986): 90–113.
Francaviglia, Richard V. "The Mormon Landscape: Definition of an Image in the American West." *Proceedings of the Association of American Geographers* 2 (1970): 59–61.
Friis, Herman R. "A Series of Population Maps of the Colonies and the United States, 1625–1790." *Geographical Review* 30, no. 3 (1940): 463–470.
Gentilcore, R. Louis. "Missions and Mission Lands of Alta California." *Annals of the Association of American Geographers* 51, no. 1 (1961): 46–72.
———. "Vincennes and French Settlement in the Old Northwest." *Annals of the Association of American Geographers* 47, no. 3 (1957): 285–297.
Goldthwait, James Walter. "A Town That Has Gone Downhill." *Geographical Review* 17, no. 4 (1927): 527–552.
Gottschalk, L. C. "Effects of Soil Erosion on Navigation in Upper Chesapeake Bay." *Geographical Review* 35, no. 2 (1945): 219–38.
Gritzner, Charles. "Construction Materials in a Folk Housing Tradition: Considerations Governing Their Selection in New Mexico." *Pioneer America* 6, no. 1 (1974): 25–39.
Harris, Richard Colebrook. "Archival Fieldwork." *Geographical Review* 91, nos. 1 and 2 (2001): 328–34.
Hart, John Fraser. "Change in the Corn Belt." *Geographical Review* 76, no. 1 (1986): 51–72.

———. "The Demise of King Cotton." *Annals of the Association of American Geographers* 67, no. 3 (1977): 307–22.

Hewes, Leslie. "Making a Pioneer Landscape in the Oklahoma Territory." *Geographical Review* 86, no. 4 (1996): 588–603.

———. "The Oklahoma Ozarks as the Land of the Cherokees." *Geographical Review* 32, no. 2 (1942): 269–81.

Hornbeck, David. "Land Tenure and Rancho Expansion in Alta California, 1784–1846." *Journal of Historical Geography* 4, no. 4 (1978): 371–90.

———. "The Patenting of California's Private Land Claims, 1851–1885." *Geographical Review* 69, no. 4 (1979): 434–48.

Hoyt, Joseph B. "The Cold Summer of 1816." *Annals of the Association of American Geographers* 48, no. 2 (1958): 118–31.

Hudson, John C. "North American Origins of Middlewestern Frontier Populations." *Annals of the Association of American Geographers* 78, no. 3 (1988): 395–413.

———. "Yankeeland in the Middle West." *Journal of Geography* 85, no. 5 (1986): 195–200.

Jackson, Richard H. "Mormon Perception and Settlement." *Annals of the Association of American Geographers* 68, no. 3 (1978): 317–34.

———. "The Use of Adobe in the Mormon Cultural Region." *Journal of Cultural Geography* 1, no. 1 (1980): 82–95.

Jackson, Robert H. "The Impact of Liberal Policy on Mexico's Northern Frontier: Mission Secularization and the Development of Alta California 1812–1846." *Colonial Latin American Historical Review* 2, no. 2 (1993): 195–225.

James, L. Allan. "Sustained Storage and Transport of Hydraulic Gold Mining Sediment in the Bear River, California." *Annals of the Association of American Geographers* 79, no. 4 (1989): 570–92.

Jordan, Terry G. "Annals Map Supplement Number Thirteen: Population Origin Groups in Rural Texas." *Annals of the Association of American Geographers* 60, no. 2 (1970): 404–5 and map insert.

———. "Antecedents of the Long-Lot in Texas." *Annals of the Association of American Geographers* 64, no. 1 (1974): 70–86.

———. "Between the Forest and the Prairie." *Agricultural History* 38, no. 4 (1964): 205–16.

———. "German Houses in Texas." *Landscape* 14, no. 1 (1964): 24–26.

———. "The Imprint of the Upper and Lower South on Mid-Nineteenth-Century Texas." *Annals of the Association of American Geographers* 57, no. 4 (1967): 667–90.

———. "Population Origins in Texas, 1850." *Geographical Review* 59, no. 1 (1969): 83–103.

———. "A Reappraisal of Fenno-Scandian Antecedents for Midland American Log Construction." *Geographical Review* 73, no. 1 (1983): 58–94.

———. "The Texan Appalachia." *Annals of the Association of American Geographers* 60, no. 3 (1970): 409–27.

Kaatz, Martin R. "The Black Swamp: A Study in Historical Geography." *Annals of the Association of American Geographers* 45, no. 1 (1955): 1–35.

Kay, Jeanne, and Craig J. Brown. "Mormon Beliefs about Land and Natural Resources, 1847–1877." *Journal of Historical Geography* 11, no. 3 (1985): 253–67.

Kemble, John Haskell. "The West through Salt Spray: Rebirth of a Classic." *American West* 1, no. 4 (1964): 65–75.

Kenny, William Robert. "Mexican-American Conflict on the Mining Frontier, 1848–1852." *Journal of the West* 6, no. 4 (1967): 582–92.

Kniffen, Fred. "Folk Housing: Key to Diffusion." *Annals of the Association of American Geographers* 55, no. 4 (1965): 549–57.

———. "Louisiana House Types." *Annals of the Association of American Geographers* 26, no. 4 (1936): 179–93.

Kromm, David E., and Stephen E. White. "Interstate Groundwater Management Preference Differences: The Ogallala Region." *Journal of Geography* 86, no. 1 (1987): 5–11.

Lemon, James T. "The Agricultural Practices of National Groups in Eighteenth-Century Southeastern Pennsylvania." *Geographical Review* 56, no. 4 (1966): 467–96.

Lewis, G. Malcom. "William Gilpin and the Concept of the Great Plains Region." *Annals of the Association of American Geographers* 56, no. 1 (1966): 33–51.

Lewis, Peirce F. "Small Town in Pennsylvania." *Annals of the Association of American Geographers* 62, no. 2 (1972): 323–51.

———. "When America Was English." *Geographical Magazine* 52, no. 5 (1979–1980): 342–48.

Lindsey, David. "Place Names in Ohio's Western Reserve." *Names* 2, no. 1 (1954): 40–45.

Lineback, Neal G. "Pilgrims' Progress: The Environmental Challenges Faced by European Settlers in New England." *Focus* 46, no. 1 (2000): 14.

MacKinnon, Richard M. "The Sonoran Miners: A Case of Historical Accident in the California Gold Rush." *California Geographer* 11 (1970): 21–28.

Marcus, W. Andrew, and Michael S. Kearney. "Upland and Coastal Sediment Sources in a Chesapeake Bay Estuary." *Annals of the Association of American Geographers* 81, no. 3 (1991): 408–24.

Marschner, F. J. "Carleton P. Barnes, 1903–1962." *Annals of the Association of American Geographers* 53, no. 2 (1963): 233–34.

Mather, E. Cotton. "The American Great Plains." *Annals of the Association of American Geographers* 62, no. 2 (1972): 237–57.

McIntosh, C. Barron. "Use and Abuse of the Timber Culture Act." *Annals of the Association of American Geographers* 65, no. 3 (1975): 347–62.

Meinig, D. W. "The Mormon Culture Region: Strategies and Patterns in the Geography of the American West, 1847–1964." *Annals of the Association of American Geographers* 55, no. 2 (1965): 191–220.

Merrens, Harry Roy. "Historical Geography and Early American History." *William and Mary Quarterly* 22, no. 4 (1965): 529–48.

Mikesell, Marvin W. "Comparative Studies in Frontier History." *Annals of the Association of American Geographers* 50, no. 1 (1960): 62–74.

Mitchell, Robert D. "American Origins and Regional Institutions: The Seventeenth-Century Chesapeake." *Annals of the Association of American Geographers* 73, no. 3 (1983): 404–20.

———. "The Shenandoah Valley Frontier." *Annals of the Association of American Geographers* 62, no. 3 (1972): 461–86.

Moorhead, Max L. "Rebuilding the Presidio of Santa Fe, 1789–1791." *New Mexico Historical Review* 49, no. 2 (1974): 1023–42.

Muller, Edward K. "Distinctive Downtown." *Geographical Magazine* 52, no. 11 (1979–1980): 747–55.

———. "Early Urbanization in the Ohio Valley: A Review Essay." *Historical Geography Newsletter* 3, no. 2 (1973): 19–30.

———. "Regional Urbanization and the Selective Growth of Towns in North American Regions." *Journal of Historical Geography* 3, no. 1 (1977): 21–39.

———. "Selective Urban Growth in the Middle Ohio Valley, 1800–1860." *Geographical Review* 66, no. 2 (1976): 178–99.

Muller, Edward K., and Paul A. Groves. "The Emergence of Industrial Districts in Mid-Nineteenth Century Baltimore." *Geographical Review* 69, no. 2 (1979): 159–78.

Nelson, Howard J. "A Service Classification of American Cities." *Economic Geography* 31, no. 3 (1955): 189–210.

———. "The Spread of An Artificial Landscape over Southern California." *Annals of the Association of American Geographers* 49, no. 3, part 2 (1959): 80–100.

———. "Town Founding and the American Frontier." *Yearbook of the Association of Pacific Coast Geographers* 36 (1974): 7–23.

———. "Townscapes of Mexico: An Example of the Regional Variation of Townscapes." *Economic Geography* 39, no. 1 (1963): 74–83.

———. "Walled Cities of the United States." *Annals of the Association of American Geographers* 51, no. 1 (1961): 1–22.

Nelson, Howard, Cornelius Loesser, Eugene McMillan, Richard Reeves, Frank Scott, and Paul Zierer. "Remnants of the Ranchos in the Urban Pattern of the Los Angeles Area." *California Geographer* 5 (1964): 1–9.

Newcomb, Robert M. "Twelve Working Approaches to Historical Geography." *Yearbook of the Association of Pacific Coast Geographers* 31 (1969): 27–50.

Nostrand, Richard L. "The Colonial New England Town." *Journal of Geography* 72, no. 7 (1973): 45–53.

———. "The Hispanic-American Borderland: Delimitation of an American Culture Region." *Annals of the Association of American Geographers* 60, no. 4 (1970): 638–61.

———. "Mexican Americans Circa 1850." *Annals of the Association of American Geographers* 65, no. 3 (1975): 378–90.

———. "A Model for Geography." *Journal of Geography* 67, no. 1 (1968): 13–17.

Nuttall, Zelia. "Royal Ordinances Concerning the Laying Out of New Towns." *Hispanic American Historical Review* 5, no. 2 (1922): 249–54.

Ogden, Adele. "Hides and Tallow: McCulloch, Hartnell and Company, 1822–1828." *California Historical Society Quarterly* 6, no. 3 (1927): 254–64.

Olsen, Michael L., and Harry C. Myers. "The Diary of Pedro Ignacio Gallego Wherein 400 Soldiers Following the Trail of Comanches Met William Becknell on His First Trip to Santa Fe." *Wagon Tracks Santa Fe Trail Association Quarterly* 7, no. 1 (1992): 1, 15–20.

Pattison, William D. "The Pacific Railroad Rediscovered." *Geographical Review* 52, no. 1 (1962): 25–36.

———. "Westward by Rail with Professor Sedgwick: A Lantern Journey of 1873." *Historical Society of Southern California Quarterly* 42, no. 4 (1960): 1–15.

Peattie, Roderick. "The Isolation of the Lower St. Lawrence Valley." *Geographical Review* 5, no. 2 (1918): 102–18.

Peters, Bernard C. "Changing Ideas about the Use of Vegetation as an Indicator of Soil Quality: Example of New York and Michigan." *Journal of Geography* 72, no. 2 (1973): 18–28.

———. "Oak Openings or Barrens: Landscape Evolution on the Michigan Frontier." *Proceedings of the Association of American Geographers* 4 (1972): 84–86.

Pillsbury, Richard. "The Urban Street Pattern as a Culture Indicator: Pennsylvania, 1682–1815." *Annals of the Association of American Geographers* 60, no. 3 (1970): 428–46.

Price, Edward T. "The Central Courthouse Square in the American County Seat." *Geographical Review* 58, no. 1 (1968): 29–60.

Prunty, Merle, Jr. "The Renaissance of the Southern Plantation." *Geographical Review* 45, no. 4 (1955): 459–91.

Raup, H. F. "Transformation of Southern California to a Cultivated Land." *Annals of the Association of American Geographers* 49, no. 3, part 2 (1959): 58–78.

Rice, John G. "The Effect of Land Alienation on Settlement." *Annals of the Association of American Geographers* 68, no. 1 (1978): 61–72.

Rostlund, Erhard. "The Geographic Range of the Historic Bison in the Southeast." *Annals of the Association of American Geographers* 50, no. 4 (1960): 395–407.

———. "The Myth of a Natural Prairie Belt in Alabama: An Interpretation of Historical Records." *Annals of the Association of American Geographers* 47, no. 4 (1957): 392–411.

Schein, Richard H. "Unofficial Proprietors in Post-Revolutionary Central New York." *Journal of Historical Geography* 17, no. 2 (1991): 146–64.

———. "Urban Origin and Form in Central New York." *Geographical Review* 81, no. 1 (1991): 52–69.

Scofield, Edna. "The Origin of Settlement Patterns in Rural New England." *Geographical Review* 28, no. 4 (1938): 652–63.

Shortridge, James R. "The Emergence of 'Middle West' as an American Regional Label." *Annals of the Association of American Geographers* 74, no. 2 (1984): 209–20.

Shumway, J. Matthew, and Richard H. Jackson. "Native American Population Patterns." *Geographical Review* 85, no. 2 (1995): 185–201.

Spencer, J. E., and Ronald J. Horvath. "How Does an Agricultural Region Originate?" *Annals of the Association of American Geographers* 53, no. 1 (1963): 74–92.

Stanislawski, Dan. "Early Spanish Town Planning in the New World." *Geographical Review* 37, no. 1 (1947): 94–105.

———. "The Origin and Spread of the Grid-Pattern Town." *Geographical Review* 36, no. 1 (1946): 105–20.

Starrs, Paul F. "Meetinghouses in the Mormon Mind: Ideology, Architecture, and Turbulent Streams of an Expanding Church." *Geographical Review* 99, no. 3 (2009): 323–55.

Stump, Roger. "The Dutch Colonial House and the Colonial Revival." *Journal of Cultural Geography* 1, no. 2 (1981): 44–55.

Trewartha, Glenn T. "Types of Rural Settlement in Colonial America." *Geographical Review* 36, no. 4 (1946): 568–96.

Turner, Mark S. "Buttonwillow, Heart of Cotton Country: California's Center of Population Monument." *California Transportation Journal* 1, no. 2 (2005): 2–6.

Urban, Michael A. "An Uninhabited Waste: Transforming the Grand Prairie in Nineteenth Century Illinois, USA." *Journal of Historical Geography* 31, no. 4 (2005): 647–65.

Vance, James E., Jr. "Cities in the Shaping of the American Nation." *Journal of Geography* 75, no. 1 (1976): 41–52.

———. "The Oregon Trail and Union Pacific Railroad: A Contrast in Purpose." *Annals of the Association of American Geographers* 51, no. 4 (1961): 357–79.

Walsh, Margaret. "The Spatial Evolution of the Mid-Western Pork Industry, 1835–75." *Journal of Historical Geography* 4, no. 1 (1978): 1–22.

Webb, Walter Prescott. "Geographical-Historical Concepts in American History." *Annals of the Association of American Geographers* 50, no. 2 (1960): 85–93.

Winberry, John J. "Indigo in South Carolina: A Historical Geography." *Southeastern Geographer* 19, no. 2 (1979): 91–102.

Winsor, Roger A. "Environmental Imagery of the Wet Prairie of East Central Illinois, 1820–1920." *Journal of Historical Geography* 13, no. 4 (1987): 375–97.

Wishart, David J. "The Changing Position of the Frontier of Settlement on the Eastern Margins of the Central and Northern Great Plains, 1854–1890." *Professional Geographer* 21, no. 3 (1969): 153–57.

Young, Karl. "Brief Sanctuary: The Mormon Colonies of Northern Mexico." *American West* 4, no. 2 (1967): 4–11, 66–67.

Zelinsky, Wilbur. "An Approach to the Religious Geography of the United States: Patterns of Church Membership in 1952." *Annals of the Association of American Geographers* 51, no. 2 (1961): 139–93.

———. "Classical Town Names in the United States: The Historical Geography of an American Idea." *Geographical Review* 57, no. 4 (1967): 463–95.

———. "The New England Connecting Barn." *Geographical Review* 48, no. 4 (1958): 540–53.

———. "The Pennsylvania Town: An Overdue Geographical Account." *Geographical Review* 67, no. 2 (1977): 127–47.

Book Chapters

Allen, John L. "Geographical Knowledge and American Images of the Louisiana Territory." In *Geographic Perspectives on America's Past: Readings on the Historical Geography of the United States*, edited by David Ward, 33–39. New York: Oxford University Press, 1979.

Arreola, Daniel D. "Hispanic American Legacy, Latino American Diaspora." In *Hispanic Spaces, Latino Places: Community and Cultural Diversity in Contemporary America*, edited by Daniel D. Arreola, 13–35. Austin: University of Texas Press, 2004.

Bennion, Lowell C. "Ben." "Mormondom's Deseret Homeland." In *Homelands: A Geography of Culture and Place across America*, edited by Richard L. Nostrand and Lawrence E. Estaville, 184–209. Baltimore: Johns Hopkins University Press, 2001.

———. "Plural Marriage, 1841–1904." In *Mapping Mormonism: An Atlas of Latter-day Saint History*, edited by Brandon S. Plewe, 122–25. Provo, UT: Brigham Young University Press, 2012.

Bowden, Martyn J. "Desert Wheat Belt, Plains Corn Belt: Environmental Cognition and Behavior of Settlers in the Plains Margin, 1850–99." In *Images of the Plains: The Role of Human Nature in Settlement*, edited by Brian W. Blouet and Merlin P. Lawson, 189–201. Lincoln: University of Nebraska Press, 1975.

———. "The Great American Desert and the American Frontier, 1800–1882: Popular Images of the Plains." In *Anonymous Americans: Explorations in Nineteenth-Century Social History*, edited by Tamara K. Hareven, 48–79. Englewood Cliffs, NJ: Prentice-Hall, 1971.

———. "The New England Yankee Homeland." In *Homelands: A Geography of Culture and Place across America*, edited by Richard L. Nostrand and Lawrence E. Estaville, 1–23. Baltimore: Johns Hopkins University Press, 2001.

Butzer, Karl W. "Retrieving American Indian Landscapes." In *The Making of the American Landscape*, 2nd ed., edited by Michael P. Conzen, 32–57. New York: Routledge, 2010.

Carstensen, Vernon. "The Development and Application of Regional-Sectional Concepts, 1900–1950." In *Regionalism in America*, edited by Merrill Jensen, 99–118. Madison: University of Wisconsin Press, 1951.

Conzen, Michael P. "Ethnicity on the Land." In *The Making of the American Landscape*, edited by Michael P. Conzen, 221–48. Boston: Unwin Hyman, 1990.

Earle, Carville. "Beyond the Appalachians, 1815–1860." In *North America: The Historical Geography of a Changing Continent*, 2nd ed., edited by Thomas F. McIlwraith and Edward K. Muller, 165–87. Lanham, MD: Rowman & Littlefield, 2001.

Harris, Richard Colebrook. "The Extension of France into Rural Canada." In *European Settlement and Development in North America: Essays on Geographical Change in Honour and Memory of Andrew Hill Clark*, edited by James R. Gibson, 27–45. Toronto: University of Toronto Press, 1978.

———. "France in North America." In *North America: The Historical Geography of a Changing Continent*, 2nd ed., edited by Thomas F. McIlwraith and Edward K. Muller, 65–88. Lanham, MD: Rowman & Littlefield, 2001.

———. "Retracing French Landscapes in North America." In *The Making of the American Landscape*, 2nd ed., edited by Michael P. Conzen, 73–90. New York: Routledge, 2010.

Hilliard, Sam B. "Plantations and the Moulding of the Southern Landscape." In *The Making of the American Landscape*, edited by Michael P. Conzen, 104–26. Boston: Unwin Hyman, 1990.

———. "A Robust New Nation, 1783–1820." In *North America: The Historical Geography of a Changing Continent*, edited by Robert D. Mitchell and Paul A. Groves, 149–71. Totowa, NJ: Rowman & Littlefield, 1987.

Holdsworth, Deryck. "Revaluing the House." In *Place/Culture/Representation*, edited by James Duncan and David Ley, 95–109. New York: Routledge, 1993.

Hornbeck, David. "Spanish Legacy in the Borderlands." In *The Making of the American Landscape*, edited by Michael P. Conzen, 51–62. Boston: Unwin Hyman, 1990.

Jackson, Richard H. "Federal Lands in the Mountainous West." In *The Mountainous West: Explorations in Historical Geography*, edited by William Wyckoff and Lary M. Dilsaver, 253–77. Lincoln: University of Nebraska Press, 1995.

Johnson, Hildegard Binder. "Gridding a National Landscape." In *The Making of the American Landscape*, 2nd ed., edited by Michael P. Conzen, 142–61. New York: Routledge, 2010.

Jordan, Terry G. "Vegetational Perception and Choice of Settlement Site in Frontier Texas." In *Pattern and Process: Research in Historical Geography*, edited by Ralph E. Ehrenberg, 244–57. Washington, DC: Howard University Press, 1975.

Lewis, Peirce F. "Americanizing English Landscape Habits." In *The Making of the American Landscape*, 2nd ed., edited by Michael P. Conzen, 91–114. New York: Routledge, 2010.

Meyer, David R. "The National Integration of Regional Economics, 1860–1920." In *North America: The Historical Geography of a Changing Continent*, 2nd ed., edited by Thomas F. McIlwraith and Edward K. Muller, 307–31. Lanham, MD: Rowman & Littlefield, 2001.

Mood, Fulmer. "The Origin, Evolution, and Application of the Sectional Concept, 1750–1900." In *Regionalism in America*, edited by Merrill Jensen, 5–98. Madison: University of Wisconsin Press, 1951.

Muller, Edward K. "From Waterfront to Metropolitan Region: The Geographical Development of American Cities." In *American Urbanism: A Historigraphical Review*, edited by Howard Gillette Jr. and Zane L. Miller, 105–33. New York: Greenwood Press, 1987.

Myers, Michael D., and William E. Doolittle. "The New Narrative on Native Landscape Transformations." In *North American Odyssey: Historical Geographies for the Twenty-first Century*, edited by Craig E. Colten and Geoffrey L. Buckley, 9–26. Lanham, MD: Rowman & Littlefield, 2014.

Nostrand, Richard L. "The Spanish Borderlands." In *North America: The Historical Geography of a Changing Continent*, edited by Thomas F. McIlwraith and Edward K. Muller, 47–63. Lanham, MD: Rowman & Littlefield, 2001.

Ogden, Adele. "New England Traders in Spanish and Mexican California." In *Greater America: Essays in Honor of Herbert Eugene Bolton*, edited by Adele Ogden, 395–413. Berkeley: University of California Press, 1945.

Sauer, Carl O. "The Barrens of Kentucky." In *Land and Life*, edited by John Leighly, 23–31. Berkeley: University of California Press, 1965.

———. "European Backgrounds of American Agricultural Settlement." In *Selected Essays 1963–1975 Carl O. Sauer*, edited by Bob Callahan, 16–44. Berkeley, CA: Turtle Island Foundation, 1981.

———. "The Settlement of the Humid East." In *Climate and Man: Yearbook of Agriculture*, 157–66. Washington, DC: US Department of Agriculture, 1941.

Steinitz, Michael P. "The Architectural Landscape." In *The New England Village*, by Joseph S. Wood, 71–87. Baltimore: Johns Hopkins University Press, 1997.

Templer, Otis W. "The Legal Context for Groundwater Use." In *Groundwater Exploitation in the High Plains*, edited by David E. Kromm and Stephen E. White, 64–87. Lawrence: University Press of Kansas, 1992.

Trewartha, Glenn T. "Climate and Settlement of the Subhumid Lands." In *Climate and Man: Yearbook of Agriculture*, 167–76. Washington, DC: United States Department of Agriculture, 1941.

Williams, Michael. "Clearing the Forests." In *The Making of the American Landscape*, 2nd ed., edited by Michael P. Conzen, 162–87. New York: Routledge, 2010.

Wishart, David J. "Settling the Great Plains, 1850–1930: Prospects and Problems." In *North America: The Historical Geography of a Changing Continent*, 2nd ed., edited by Thomas F. McIlwraith and Edward K. Muller, 237–60. Lanham, MD: Rowman & Littlefield, 2001.

Wyckoff, William. "Creating Regional Landscapes and Identities." In *North American Odyssey: Historical Geographies for the Twenty-first Century*, edited by Craig E. Colten and Geoffrey L. Buckley, 335–57. Lanham, MD: Rowman & Littlefield, 2014.

Wyckoff, William, and Lary M. Dilsaver. "Defining the Mountainous West." In *The Mountainous West: Explorations in Historical Geography*, edited by William Wyckoff and Lary M. Dilsaver, 1–59. Lincoln: University of Nebraska Press, 1995.

Parts of Series

Adkins, Howard G. "The Geographic Base of Urban Retardation in Mississippi, 1800–1840." In Geographic Perspectives on Southern Development, *West Georgia College Studies in the Social Sciences* 7 (1973): 35–49. Carrollton, GA.

Barrows, Harlan H. *Lectures on the Historical Geography of the United States as Given in 1933.* Edited by William A. Koelsch. Department of Geography Research Paper No. 77. Chicago: University of Chicago, 1962.

Bieber, Ralph P., ed. *Southern Trails to California in 1849.* Southwest Historical Series, vol. V. Glendale, CA: Arthur H. Clark, 1937.

Bushong, Allen D. "Ellen Churchill Semple 1863–1932." *Geographers Biobibliographical Studies* 8 (1984): 87–94.

Coppens, Linda Miles. "Ralph Hall Brown 1898–1948." *Geographers Biobibliographical Studies* 9 (1985): 15–20.

Dodge, Stanley D. "The Frontier of New England in the Seventeenth and Eighteenth Centuries and Its Significance in American History." *Papers of the Michigan Academy of Science Arts and Letters* 28 (1942): 435–39.

Dunbar, Gary S. "Fall Line." In *Encyclopedia of Southern History*, edited by David C. Roller and Robert W. Twyman, 419–20. Baton Rouge: Louisiana State University Press, 1979.

Hafen, LeRoy R., and Ann W. Hafen. *Old Spanish Trail: Santa Fé to Los Angeles.* Volume 1 of *The Far West and the Rockies Historical Series, 1820–1875.* Glendale, CA: Arthur H. Clark, 1954.

Hilliard, Sam B. "The Tidewater Rice Plantation: An Ingenious Adaptation to Nature." In *Coastal Resources*, Geoscience and Man vol. 12, edited by H. J. Walker, 57–66. Baton Rouge: Louisiana State University, 1975.

Lavender, David. "Bent's Fort: Outpost of Manifest Destiny." In *The Santa Fe Trail: New Perspectives*, Essays in Colorado History, no. 6: 11–25. Denver: Colorado Historical Society, 1987.

Lecompte, Janet. "The Mountain Branch: Raton Pass and Sangre de Cristo Pass." In *The Santa Fe Trail: New Perspectives.* Essays in Colorado History, no. 6: 55–66. Denver: Colorado Historical Society, 1987.

Lineback, Neal. "(Geography in the News) Celebrating Lewis and Clark." *Perspective* 32, no. 2 (2003): 8 and 11.

———. "(Geography in the News) Deadly Tornadoes." *Perspective* 26, no. 5 (1998): 11.

McManis, Douglas R. *European Impressions of the New England Coast 1497–1620.* Department of Geography Research Paper No. 139. Chicago: University of Chicago, 1972.

Meinig, Donald W. "The American Colonial Era: A Geographic Commentary." *Proceedings of the Royal Geographical Society of Australia, South Australian Branch*, vol. 59 (1958): 1–22.

Nostrand, Richard L., and Sam B. Hilliard, eds. *The American South.* Geoscience and Man, vol. 25. Baton Rouge: Louisiana State University, 1988.

Pattison, William D. *Beginnings of the American Rectangular Land Survey System, 1784–1800.* Department of Geography Research Paper No. 50. Chicago: University of Chicago, 1957.

Petersen, James F. "Along the Edge of the Hill Country: The Texas Spring Line." In *A Geographic Glimpse of Central Texas and the Borderlands: Images and Encounters*, edited by James F. Petersen and Julie A. Tuason, 20–30. Pathways No. 12. Indiana, PA: National Council for Geographic Education, 1995.

Raitz, Karl B. *The Kentucky Bluegrass: A Regional Profile and Guide.* Studies in Geography, no. 14. Chapel Hill: Department of Geography, University of North Carolina, 1980.

Sandoval, David A. "Who is Riding the Burro Now? A Bibliographic Critique of Scholarship on the New Mexican Trader." In *The Santa Fe Trail: New Perspectives*, Essays in Colorado History, no. 6: 75–92. Denver: Colorado Historical Society, 1987.

Wacker, Peter O. "New Jersey's Cultural Landscape before 1800." *Proceedings of the Second Annual Symposium of the New Jersey Historical Commission*, Trenton, December 5, 1970, 35–62. Newark: New Jersey Historical Society, 1971.

Wolfskill, George. "Introduction." In *Essays on Walter Prescott Webb and the Teaching of History*, edited by Dennis Reinhartz and Stephen E. Maizlish, 3–10. *The Walter Prescott Webb Memorial Lectures* 19. College Station: Texas A&M University Press, 1985.

Interviews and Personal Correspondence

Jordan, Terry G. Professor of Geography, North Texas State University. Personal correspondence, April 27, 1978.

Muller, Edward K. Professor of History, University of Pittsburgh. Personal correspondence, March 31 and April 28, 2011.

Spaeth, Hans-Joachim. Professor of Geography, University of Oklahoma. Interview by the author, October 14, 1983.

Zelinsky, Wilbur. Professor of Geography, the Pennsylvania State University. Personal correspondence, March 30, 2000.

Maps

"American Indians." In *Hammond United States History Atlas*, map B, page U-4. Maplewood, NJ: Hammond, 1973.

"Indians of North America." Map supplement in *National Geographic* 142, no. 6 (December 1972): 739A.

Raisz, Erwin. *Landforms of the United States*. 6th ed. Cambridge, MA: E. Raisz, 1957.

Newspapers and Magazines

Bryce, Robert. "Critics Remember Alamo Differently." *Christian Science Monitor*, December 21, 1994, 1, 6.

Collins, James. "Domesticated Daring." *Time*, November 3, 1997, 106.

Cowen, Robert C. "Global Warming: How It Works." *Christian Science Monitor*, December 3, 1997, 10–11.

DeVoto, Bernard. "Sea to Shining Sea." *Fortune* 43, no. 2 (1951): 101–8.

Gibbons, Boyd. "The Itch to Move West: Life and Death on the Oregon Trail." *National Geographic* 170, no. 2 (1986): 146–77.

Gore, Rick. "Cascadia." *National Geographic* 193, no. 5 (1998): 6–37.

Hummel, Carol L. "I Tore Open the Envelope—'E. B. White!' I Blurted." *Christian Science Monitor*, November 6, 1985, 38.

Newman, Cathy. "The Shakers' Brief Eternity." *National Geographic* 176, no. 3 (1989): 302–25.

Wilford, John Noble. "Corn in the New World: A Relative Latecomer." *New York Times*, March 7, 1995, B5, B8.

Theses, Dissertations, and Fieldtrip Guides

Estaville, Lawrence Ernest, Jr. "The Louisiana French Culture Region: Geographic Morphologies in the Nineteenth Century." PhD diss., University of Oklahoma, 1984.

Glass, Joseph W. "The Pennsylvania Culture Region: A Geographic Interpretation of Barns and Farmhouses." PhD diss., Pennsylvania State University, 1971.

Grim, Ronald E. "The Origins and Early Development of the Virginia Fall-Line Towns." Master's thesis, University of Maryland, 1971.

Holder, Gerald Leon. "The Fall Zone Towns of Georgia: An Historical Geography." PhD diss., University of Georgia, 1973.

Miles, Linda Jeanne. "Ralph Hall Brown: Gentlescholar of Historical Geography." PhD diss., University of Oklahoma, 1982.

Mitchell, Robert D. "The Colonial Tidewater: Southern Maryland and Virginia." In *Association of American Geographers Field Trip Guide*, 1–7. Washington, DC: The Association, 1984.

Nostrand, Richard Lee. "The Hispanic-American Borderland: A Regional, Historical Geography." PhD diss., University of California–Los Angeles, 1968.

———. "A Settlement Geography of the Santa Ynez Valley, California." Master's thesis, University of California–Los Angeles, 1964.

Roth, Jeffery Edwin. "Long Lots in New Mexico and Texas: The French Connection, 1693–1731." PhD diss., University of Oklahoma, 2005.

Torbenson, Craig Laron. "College Fraternities and Sororities: A Historical Geography, 1776–1989." PhD diss., University of Oklahoma, 1992.

Comments, Lectures, and Book Reviews

Jordan, Terry G. "Reviews of *The Making of a History*, by Gregory M. Tobin, and *Essays on Walter Prescott Webb (W. P. Webb Memorial Lecture)*, edited by K. R. Philip and Elliot West." *Historical Geography* 8, no. 1 (1978): 29–32.

Popper, Frank, and Deborah Popper. "Buffalo Commons." Lectures presented at the University of Oklahoma, November 1990 and February 6, 2008.

Roberts, Jeffrey P. "Historical Geography of Philadelphia." Lecture presented at the Atwater Kent Museum, Philadelphia, April 23, 1979.

Ward, David. "Comment at Eastern Historical Geography Association," Boston University, November 3, 1972.

Government Publications

Johnson, Willard D. "The High Plains and Their Utilization." *Twenty-First Annual Report of the United States Geological Survey, 1899–1900*, part 4, Hydrology, 609–741. Washington DC: Government Printing Office, 1901. Continued in *Twenty-Second Annual Report of the United States Geological Survey, 1900–1901*, part 4, 637–69. Washington, DC: Government Printing Office, 1902.

U.S. Bureau of the Census. *Geographic Areas Reference Manual.* Washington, DC: U.S. Department of Commerce, Economics and Statistics Administration, Bureau of the Census, 1994.

———. *2000 Census of Population and Housing.* Population and Housing Unit Counts PHC-3-1, United States Summary. Washington, DC: Government Printing Office, 2004.

Index

Note: Letters following page numbers indicate illustrations: *f* for figures, *m* for maps, and *t* for tables.

About the Author

Richard L. Nostrand grew up in a faculty family in Seattle, Washington. He discovered geography as an academic discipline at the University of Washington, where he earned his BA degree in 1961. Graduate work in geography took him to UCLA for his MA (1964) and PhD (1968) degrees; he also spent a valuable year of study (1964–1965) at Syracuse University. Nostrand began his teaching career at UMass–Amherst (1967–1973), but proximity to the southwestern borderland pulled him to a second and final position at the University of Oklahoma (1973–2004). Throughout his career his publications have focused on the changing geography of Hispanic or Latino peoples, especially in New Mexico. And his major teaching interest has always been the historical geography of the United States—out of which this book developed.